# A2 LEVEL
# BIOLOGY

Phil Bradfield, John Dodds, Judy Dodds and Norma Taylor

Longman

**Pearson Education**
Edinburgh Gate
Harlow
Essex

Sixth impression 2007
ISBN 978-0-582-42945-1


Concept designed by Hardlines Ltd, Charlbury, Oxford.

Edited, designed and produced by Gecko Ltd, Cambridge.

Illustrations by Lizzie Harper, Helen Humphreys, Raith Overhill, Peter Simmonett, Raymond Turvey and Pete Welford.

Printed in China
GCC/06

**Acknowledgments**

The authors would like to thank many people for their help, support and encouragement in writing this book. In particular: Alan Clamp, Liz Jones, Steve Potter, Jean Smith, Alison Waldron and Judy Webster.

We are grateful to the following examination boards for permission to reproduce their copyright material: AQA; Edexcel and OCR.

Figure acknowledgements are on page 461.

Dr Phil Bradfield is Head of Biology at Davenant Foundation School, Loughton, Essex. He is also an examiner for A-level Biology.

John Dodds is Assistant Principal and Head of Biology at The Albany College, Hendon.

Judy Dodds is Coordinator of Sixth Form Biology at Cardinal Wiseman School, Ealing and also teaches A-level Biology at The Albany College, Hendon.

Norma Taylor has retired after teaching A-level Biology for 26 years, most recently at The Albany College, Hendon.

# Contents

# Introduction

This book is for any student studying Biology at A2 level. Together with the AS book it covers the core content of all the major specifications. We have used a number of different features to help you with your studies.

The summary at the end of each chapter condenses the content of the chapter into a series of bullet points, to help you with revision.

End of chapter questions test your knowledge of the whole chapter. Most are exam-style questions, to give you plenty of practice. Some of these questions also give you an opportunity to

demonstrate Key Skills. These questions are marked with a symbol.

🔑 **C2.2** This symbol means you can fully demonstrate Communications skill 2.2 with the question.

🔑 **N3.2** This example means that you can partially demonstrate Application of Number skill 3.2 with the question.

At the end of this book you will find some Synoptic questions. This style of question tests your knowledge of a variety of topics and draws together AS and A2 Biology.

---

The introduction to the chapter gives you an idea of the content of the chapter, and reminds you about any earlier work from AS Biology.

## (10) Excretion and water balance

The importance of maintaining a constant internal optimum state was explained in Chapter 9. Two further mechanisms which are involved in achieving this are the control of the level of nitrogenous waste products and the control of water balance (**osmoregulation**). In mammals, nitrogenous waste products are processed in the liver and passed out of the body (**excreted**) via the kidneys. The correct water potential of the blood plasma and body fluids is maintained homeostatically by the kidneys under the influence of antidiuretic hormone (ADH).

### 10.1 What is excretion?

The chemical reactions which take place in the body of an organism (**body metabolism**) produce some substances which are of no use to the body and which would poison the body if they were allowed to build up. In other words they are **toxic**. For example:

- **carbon dioxide** is produced during respiration
- various **nitrogen containing compounds** are produced during the breakdown of excess amino acids and nucleic acids
- bile pigments **bilirubin** and **biliverdin** are produced during the breakdown of haemoglobin.

Such products must be removed from the body so that their levels do not rise to the point where they become dangerous. The process of eliminating such substances from the body is called **excretion**.

In mammals, carbon dioxide is excreted from the **lungs** during breathing. The bile pigments are produced from haemoglobin in the **liver** and passed into the **bile**. They pass into the **duodenum** with the bile and are passed out of the body in the **faeces**. In vertebrates, the nitrogenous waste products are produced in the liver and excreted from the kidneys. The excretion of nitrogenous waste is discussed in detail in this chapter.

**?**

**1** Why is it true to say that passing faeces out of the body involves excretion as well as egestion? *(2 marks)*

### 10.2 Nitrogenous metabolic waste

Proteins in the diet are digested, producing amino acids which are absorbed and transported in the blood. Also, any body proteins which are no longer required,

**This chapter includes:**
- what is excretion?
- nitrogenous metabolic waste
- the structure and function of mammalian kidneys
- the role of the kidneys and ADH in water balance
- control of the water budget in small desert mammals.

This list gives you the main ideas in the chapter. It is not a full contents list - that is given on pages iii - v.

These 'pies' tell you whether or not the section or box applies to your specification.

**Definition**

**Excretion** is removal of the waste products of metabolic processes from the body. These products would be toxic if they were allowed to accumulate.

**Remember**

It is important to distinguish between excretion and **egestion** (sometimes called **defaecation**), which is the removal from the body of waste material, such as undigested food, which has not been part of body metabolism.

**Remember** boxes include exam hints, definitions of new words or phrases, or reminders of facts that some students may find difficult to remember.

# Key for labels

 AQA A (A2)   OCR (A2)

 Edexcel (A2)   AQA B (A2)

Note on pies in Extension boxes: In some A2 Biology specifications there are option blocks. If the pie is coloured for your specification, it may be that the material covered is in one of the option blocks. Please check your specification carefully

Extension boxes contain material that is usually only in one specification but should be of interest to all students. The 'pie' tells you which specification the material applies to.

Cross references are provided in the main text and in extension boxes, to make it easy for you to look up related ideas within this book and the AS book.

Each chapter is divided into sections. These numbers tell you which sections appear on the pages you are looking at.

---

④ **Evolution**

**Extension**

**Remember**

The mean is the average value and the mode is the most commonly occurring value. Modal class refers to the most commonly occurring group of values.

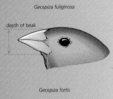

*Geospiza fuliginosa*

depth of beak

*Geospiza fortis*

depth of beak

*(a)*

**Figure 4.35** *(a) The heads of Geospiza fuliginosa and Geospiza fortis, (b) graphs showing range of beak size on three islands*

**Did You Know**

Biologists have recently discovered another variety of finch on the Galapagos. A vampire finch has been observed that feeds on blood from open wounds on cattle and other large animals, again exploiting another available niche.

**Box 4.6  A case study of finches on the Galapagos**

A study of finches on three islands in the Galapagos h[...] some interesting findings regarding two species, *Geos[...] fuliginosa* and *Geospiza fortis*, and variations in their [...] (Figure 4.35). Both species feed on plant seeds; the ty[...] depending on the size of beak. Only *G. fortis* is found [...] Island. It has a mean beak depth and modal class of 10[...] mm, with a range on either side. In this example the m[...] the modal class are the same.

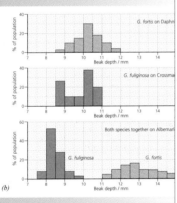

On Crossman Island, only *G. fuliginosa* is present, wit[...] majority of birds, i.e. the modal class, also having a be[...] of 10 to 10.5 mm. The mean value, however, is smalle[...] modal class.

When these two finch species are together on Albema[...] the modal classes are quite different. *G. fuliginosa* has [...] class of 8 to 8.5 mm and *G. fortis* a modal class of 12.[...] 13 mm. They can therefore exploit different food sour[...] occupy different niches, so reducing the competition [...] would arise with similar-sized beaks and increasing th[...] of survival. Directional selection has taken place favo[...] smaller beak in *G. fuliginosa* and a larger beak in *G. f[...] Selection has favoured those birds whose beak size wi[...] competition and allow feeding to be more successful – [...] the 'fittest' here.

142

---

In bright light visual acuity is therefore greatest at the fovea due to the cones, and we move our eyes so as to keep the image we are interested in focused on the fovea. However, in dim light acuity is poorest at the fovea as the cones have low sensitivity. We see best if we look slightly to one side – out of the 'corners' of our eyes. Light from the object we are interested in then falls mainly on the rods which are away from the fovea. The rods can be stimulated by the dimmer light. Rods cannot detect colour so we lose colour vision in dim light.

To summarise, in bright light the cones at the fovea are stimulated, giving colour vision with high acuity, whereas in dim light the rods away from the fovea are sensitive enough to be stimulated, giving black and white vision with low acuity. At the blind spot there are no rods or cones (see page 236) and therefore light cannot be detected at all.

**?**

4  Give one difference and one similarity between the fovea and the blind spot.
*(2 marks)*

**Interpretation by the brain**

Information from the eyes is carried as nerve impulses by the thousands of neurones of the optic nerve to a part of the brain called the **visual cortex**, each part of which represents a part of the retina. Impulses from the different parts of the retina stimulate the corresponding part of the visual cortex. Because of convergence, rods have fewer neurones going to the brain and therefore a smaller area of cortex is devoted to impulses from rods (Figure 7.17). The 'pattern' of stimulation at the visual cortex is interpreted and we can 'see'. The significance and meaning of what we see depends upon reference to other parts of the brain where previous visual information is stored (see page 251 for further information on brain function and Box 7.9 on page 255 for more detail on the brain and vision).

**7.7  Colour vision – the trichromatic theory**

There are three types of cone cell, each containing a different form of iodopsin. Each is sensitive to different wavelengths of light corresponding to the colours blue, green and red (Figure 7.18). In effect the three types of cone are blue cones, green cones and red cones. The names refer to the colour of light absorbed and *not* to the colour of their appearance.

Pure red light will only break down the red iodopsin and only the red cones will fire impulses to the brain. This is interpreted by the brain as red. However, yellow light will break down some of the red iodopsin and some of the green iodopsin and so both red and green cones will fire impulses to the brain. This is interpreted by the brain as yellow. In this way the full range of colours can be seen depending on the relative proportions of the different cones which are stimulated. White light stimulates all three types of cone equally, all three types of cone fire and the brain interprets this as the colour white. Some of the colour combinations are given in Table 7.4 on page 244 and Figure 7.18.

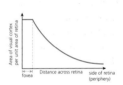

**Note:** a much bigger area of the visual cortex is allocated to the impulses arriving from the fovea. Each cone supplies a separate neurone to the visual cortex. Rods have convergence so fewer neurones supplied to visual cortex.

**Figure 7.17** *Allocation of area of visual cortex per unit area of retina (only half of the retina is shown)*

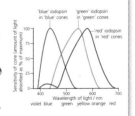

**Figure 7.18** *Cones and the wavelength of light*

243

---

**Did You Know** boxes include facts related to the chapter that we think you will find interesting!

Self-assessment questions are provided at various places within each chapter. They should help you to check your understanding of the work you have done so far. Answers to these questions are provided at the end of the chapter.

# 1 Respiration

If you ask someone with a limited knowledge of biology the meaning of the word 'respiration', the answer they will usually give is 'breathing'. Although these two processes are linked, the real meaning of respiration is much more involved. It is a series of biochemical reactions resulting in the release of energy. Breathing is the 'end' of the process, the way that some animals exchange gases with their environment. The oxygen obtained through breathing is used in respiration, and the carbon dioxide produced by respiration is removed by breathing. In fact all organisms, bacteria, protoctists, fungi and plants, as well as animals, need a source of energy to 'drive' their biological processes, and all carry out respiration to provide this energy.

> **This chapter includes:**
> - what is energy?
> - the role of ATP
> - enzyme cofactors
> - glycolysis
> - the Krebs cycle
> - the electron transport chain
> - anaerobic respiration
> - respiratory quotients
> - redox indicators.

## 1.1 Energy

What is energy? We use this word a great deal in everyday language. For instance we might talk about a person having 'a lot of energy'. We would mean that they are always on the move, always getting things done. This is very similar to the scientific meaning of the word, where an object has a lot of energy if it can 'do' things. Take the example of a tennis ball. If it is just sitting on the ground, it is not doing much. To make it 'do' things, we could throw it, which might make it bounce off a wall or break a window. Alternatively we could roll it down a hill, or drop it from a height. In these circumstances the ball has energy. The energy has to be given to the ball, passed from the person to the object.

In scientific terms, energy is defined as the 'ability to do work' and calculated from the product of the force and the distance moved by the force (i.e. energy = force × distance). If we move a kilogram weight through a distance of 5 metres, we need to use a certain amount of energy. If we double the distance to 10 metres, this will need twice as much energy, as will doubling the weight to 2 kilograms. The standard unit of energy is the joule (J).

Going back to the example of our tennis ball, there are other ways we could move it. For instance, we could use a machine powered by an electric motor, or even a petrol engine, to 'throw' the ball. These are using different forms of energy (electrical and chemical) to produce the energy of movement (kinetic) of the ball. Different types of energy (with some biological examples) include:

- **kinetic** – when an animal uses contraction of its muscles to run
- **chemical** – the energy stored in a fuel or food
- **heat** – where energy is lost to the surroundings from a warm body
- **sound** – the air vibrations produced by mammalian vocal cords
- **light** – the energy 'trapped' in photosynthesis (Chapter 2) or produced by glow-worms
- **electrical** – the impulses propagated along a nerve cell.

Energy can be converted from one form into another. This is a fundamental idea that we must understand if we are to follow the process of respiration. Take the example of a boy throwing a tennis ball, which then breaks a window. There are

> **Remember**
>
> The joule (J) is the amount of energy used when a force of one newton moves an object through a distance of one metre. This is a small amount of energy, and in practice energy is often measured in kilojoules (kJ) where 1 kJ = 1000 J.

several energy conversions which take place during this event. First of all the boy gains his energy from his food (chemical) which is converted into kinetic energy of the ball, through the contraction of the arm muscles. The ball hits the window, where its energy is converted into sound (breaking glass!) and heat (the glass, air and ball all warm up slightly). We could trace the energy in the boy's food further back, as chemical energy of plant material, originally gained from the light energy of the sun during photosynthesis.

**?**

1 List the main energy conversions which take place when:
   (a) a car is driven
   (b) a glow-worm glows
   (c) a leaf photosynthesises.                                    *(3 marks)*

The point is that life processes involve numerous conversions of energy from one form into another. However, energy is **never created nor destroyed**, only made available through these conversions. This is a fundamental law of physics, called the **law of conservation of energy**.

The idea of **chemical energy** is central to an understanding of respiration. Any substance contains a certain amount of energy in this form. When the substance undergoes a chemical reaction, the products also contain energy. If the amount of energy in the products is less than that in the reactants, the law of conservation of energy predicts that energy will be given out, as heat, light or some other form. For example, imagine burning some ethanol. The reaction is represented by the equation:

$$C_2H_5OH + 3O_2 \rightarrow 2CO_2 + 3H_2O$$
(ethanol)

The products, carbon dioxide and water, contain less energy than the reactants (ethanol and oxygen) and so energy is released, mainly as heat (about 1400 kJ per mole of ethanol). This is called an **exergonic** reaction (meaning 'gives out energy'). Since the energy is mainly lost as heat, it is also known as an **exothermic** reaction (meaning 'gives out heat').

On the other hand, some reactions form products which contain more energy than the reactants. These reactions will be **endergonic** (take in energy) needing an external source of energy to 'drive' them. One example of an endergonic reaction is the synthesis of proteins from amino acids (see *AS Biology*, Section 5.7, page 132). As we will see, the reactions of respiration are exergonic, releasing energy for endergonic processes which need it.

There are a large number of endergonic and exergonic reactions taking place within a cell. It is very important that the energy released from exergonic reactions can be made available for endergonic ones. Cells have achieved this as a result of the evolution of a molecule which is used as an energy 'intermediate'. This is the compound adenosine triphosphate, or ATP.

**Remember**
The sum of all the reactions taking place within cells is called **metabolism**.

## 1.2  Adenosine triphosphate

In the combustion of ethanol reaction described above, energy is released as heat. In the reactions taking place within cells, although some heat is produced as a by-product, heat would be useless as a source of energy, and too much heat could damage cells. Instead, most energy is passed from exergonic to endergonic reactions through the formation of an intermediate source of chemical energy, **adenosine triphosphate** (**ATP**). ATP can be formed from a related compound, adenosine diphosphate (ADP). The formation of ATP needs an input of energy, and the reverse process, the breakdown of ATP, gives out energy.

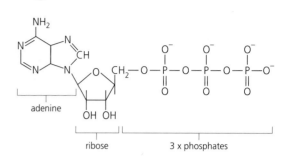

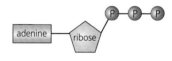

*(a)*

*(b)*

ATP is a type of compound called a phosphorylated nucleotide (Figure 1.1a). It consists of the base adenine attached to the pentose sugar ribose, which is in turn attached to three inorganic phosphate groups (adenine + ribose = **adenosine**, three phosphates = **triphosphate**). Adenine is an organic base found in DNA and RNA, and ribose is the sugar found in RNA (see *AS Biology*, Section 5.2, page 120). You do not need to know the full structural formula of ATP, but an understanding of the simplified structural diagram (Figure 1.1b) is useful.

**Figure 1.1** *(a) Structural formula of ATP (note that some hydrogen atoms have been omitted for simplicity), (b) Simplified diagram of ATP molecule*

ATP can be broken down into ADP and inorganic phosphate ($P_i$) by the addition of water (a **hydrolysis** reaction) as shown in Figure 1.2. The reaction is catalysed by an enzyme called an ATPase.

A simpler way of writing the equation is:

$$ATP + H_2O \rightleftharpoons ADP + P_i$$

This reaction yields 30.6 kJ of energy per mole of ATP hydrolysed. The cells of *all* organisms so far investigated, from bacteria to elephants and oak trees, contain ATP. It has a **universal** role as an **immediate source of energy** in cells, acting as an intermediate energy 'currency'. However, ATP remains within cells and cannot be transported around the body of an animal or plant. It cannot be stored for longer than a few minutes, so that more ATP must be continuously produced wherever energy is needed.

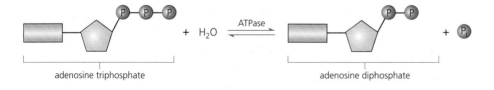

adenosine triphosphate                                adenosine diphosphate

( Note that the hydrolysis of ATP to form ADP, the reaction from left to right, *releases* energy. The reverse reaction, where ATP is synthesised from ADP and $P_i$ *takes in* the same amount of energy – this energy is derived from respiration.)

**Figure 1.2** *The interconversion of ATP and ADP*

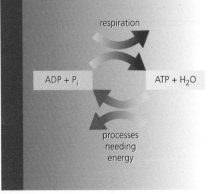

respiration

ADP + P$_i$          ATP + H$_2$O

processes needing energy

The function of respiration is to release the chemical energy in organic molecules, such as glucose, for the formation of ATP. The ATP is then available for use in other processes such as:

● contraction of muscle fibres, producing movement (see Section 8.6, page 276)
● active transport of ions against a concentration gradient (see *AS Biology*, Section 2.6, page 48)
● synthesis of large organic molecules, such as nucleic acids and proteins
● cell division (see *AS Biology*, Chapter 6).

**1.3  Metabolic pathways**

The reactions of respiration form a **metabolic pathway**. A metabolic pathway is a series of reactions which take place in a cell. The product of one reaction is the substrate for the next reaction, and each reaction is catalysed by a particular enzyme (see *AS Biology*, Section 4.1, page 101). We can show this diagrammatically:

$$\text{enzyme 1  enzyme 2  enzyme 3  enzyme 4}$$
$$A \longrightarrow B \longrightarrow C \longrightarrow D \longrightarrow E$$

A to E are substrates (and products) in the pathway. Most of the reactions taking place within a cell are part of metabolic pathways. The enzymes catalysing these reactions are often bound to membranes within the cell so that they are arranged close to each other, and in the sequence of the metabolic pathway. This arrangement is called a **multi-enzyme complex**. This makes the reactions of the pathway more likely to occur.

> **2** Explain, using your knowledge of the way that enzymes work, why the presence of a multi-enzyme complex will make reactions of the metabolic pathway more likely to take place. *(2 marks)*

There are several advantages to having reactions occur via metabolic pathways:

● The direct conversion of substrate A to product E may involve a large change of energy. If, for example, large amounts of energy were given out in 'one go', this could damage the cell. By releasing the energy gradually, in small steps, this can be more controlled, so that the delicate structure of the cell is not damaged.
● The intermediate products (B to D) of the metabolic pathway may themselves be useful, or may be substrates for other products, forming branching pathways:

$$A \rightarrow B \rightarrow C \rightarrow D \rightarrow E$$
$$\downarrow$$
$$F$$
$$\downarrow$$
$$G \rightarrow H$$

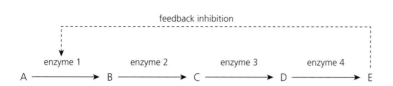

*Figure 1.3 Feedback inhibition: the end product (E) of the metabolic pathway inhibits an enzyme at an earlier stage in the pathway, acting to control or regulate the pathway, so that as E builds up the conversion from A→ B is inhibited. As E is used up, the inhibitor is removed*

● The final product of the metabolic pathway (E in our example above) may act as an inhibitor of an earlier enzyme in the pathway. This is called **feedback inhibition** or **end product inhibition**, and is an important way that metabolic pathways are regulated by the cell (Figure 1.3).

In feedback inhibition, the product acts as what is called an **allosteric inhibitor**, reversibly binding to the enzyme at a specific region away from its active site, called the **allosteric site** (Figure 1.4).

### Anabolism and catabolism

We have seen that a cell's reactions, or metabolism, largely consist of metabolic pathways. These are of two types: **anabolic** and **catabolic**. Anabolic reactions (anabolism) involve the build up of small molecules into larger ones, requiring an input of energy. Examples of this include protein synthesis and photosynthesis. Catabolic reactions are the converse: they involve the breakdown of large molecules into smaller ones, releasing energy. Respiration is the most obvious example of a catabolic pathway. Digestive reactions, such as the hydrolysis of starch or protein in the gut, are also catabolic.

**Key**
E = enzyme
S = substrate
AS = active site
A = allosteric site
I = inhibitor

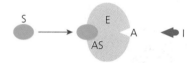

When inhibitor is not in allosteric site, substrate can enter the active site and enzyme-catalysed reaction can occur.

### 1.4  Cofactors and coenzymes

Many enzymes need particular extra non-protein substances in order to carry out their catalysis. These are called **cofactors**, and there are three main types of them: inorganic ions, prosthetic groups and coenzymes.

**Inorganic ions** work by combining with the enzyme or the substrate, making the enzyme–substrate complex form more easily. For example, the enzyme salivary amylase needs chloride ions to catalyse the breakdown of starch to maltose efficiently.

**Prosthetic groups** are non-protein organic cofactors which are permanently attached to a particular enzyme. For example, the enzyme catalase, which catalyses the decomposition of hydrogen peroxide to water and oxygen, contains a prosthetic group called a haem group. This contains an iron(II) ion (similar to the haem of haemoglobin: see *AS Biology*, Box 1.2, page 314) which is involved in the reaction. As we shall see, **cytochromes** of the **electron transport chain** of respiration also contain prosthetic haem groups. Here the haem acts as an electron 'carrier', accepting electrons which reduce the iron to $Fe^{2+}$, then passing them on to other carriers so that the iron is re-oxidised to $Fe^{3+}$. In other words, they carry out **redox** (**red**uction/**ox**idation) reactions.

When inhibitor binds with allosteric site, shape of active site changes and substrate can no longer enter – so enzyme-catalysed reaction stops.

*Figure 1.4 Allosteric inhibition*

**Cofactors and diet**

As well as the substances which make up the bulk of our food (carbohydrates, proteins, lipids, fibre and water) we need to eat much smaller amounts of vitamins and minerals. Some of these components of food are an essential part of our diet because they are used by the body for making enzyme cofactors. Iron is a part of the haem group, present in catalase and cytochromes. Smaller amounts of other trace elements, such as cobalt, copper, magnesium, manganese and zinc are needed in our diet as enzyme activators.

Several vitamins are also needed as precursors for coenzymes and prosthetic groups. NAD is derived from the vitamin nicotinic acid, and the prosthetic group FAD (flavin adenine dinucleotide), which we will meet in Section 1.8, from riboflavin (vitamin $B_2$).

**Coenzymes** are particularly significant in respiration. They are small, non-protein, organic molecules which temporarily bind with the enzyme molecule when it forms an enzyme–substrate complex. The coenzyme acts as a carrier molecule, transferring chemical groups or atoms from one substrate to another. An important coenzyme in respiration is **nicotinamide adenine dinucleotide (NAD)**.

### Nicotinamide adenine dinucleotide (NAD)

The structure of NAD is shown in Figure 1.5. Its role is to work with dehydrogenase enzymes, which catalyse the removal of hydrogen atoms from their substrates. In the metabolic pathways of respiration the NAD, attached to the dehydrogenase, accepts the hydrogen atoms and passes them to another hydrogen carrier.

The chemistry involved is rather complicated, but it is worth explaining, since it will be relevant later when we discuss respiration in more detail. The pair of hydrogen atoms removed by the dehydrogenase dissociate into hydrogen ions (protons) and electrons:

$$2H \rightleftharpoons 2H^+ + 2e^-$$

In the cell the NAD molecule exists as $NAD^+$ (it has lost an electron) and as a result carries the hydrogens as NADH and a proton:

$$NAD^+ + 2H^+ + 2e^- \rightarrow NADH + H^+$$

When the $(NADH + H^+)$ is reoxidised, the reactions go into reverse and $NAD^+$ is regenerated.

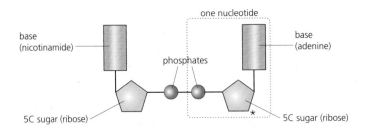

The (base – sugar– phosphate) groups constitute a nucleotide. Hence nicotinamide adenine *di*nucleotide.

\* Note that if an additional phosphate is attached to the ribose group at this point, this produces the very similar compound NADP (nicotinamide adenine dinucleotide phosphate) which is an important coenzyme in photosynthesis (Section 2.5, page 39)

***Figure 1.5*** *Structure of NAD*

## 1.5  An overview of respiration

### Remember

To understand respiration, it is important to remember the chemist's definitions of oxidation and reduction reactions. These occur through the transfer of either oxygen atoms, hydrogen atoms or electrons during a reaction. It's easiest to remember as a table:

Table 1.1 *The meanings of oxidation and reduction*

| Oxidation is: | Reduction is: |
| --- | --- |
| • addition of oxygen | • removal of oxygen |
| • removal of hydrogen | • addition of hydrogen |
| • removal of electrons | • addition of electrons |

In the metabolic pathway we call respiration, glucose is oxidised to produce energy in the form of ATP. This is summarised in the familiar equation:

$$C_6H_{12}O_6 + 6O_2 \rightarrow 6CO_2 + 6H_2O \text{ (plus energy released as ATP)}$$

In normal circumstances, if plenty of oxygen is available, this method of **aerobic** respiration (using oxygen) goes on in animal and plant cells to provide the energy needed for the processes of life. Before we look in detail at this, we should note that under certain circumstances cells can respire **anaerobically** (without using oxygen) producing reduced amounts of ATP. This happens in:

- very active **muscle** cells, when not enough oxygen can reach the cells, and they carry out a partial oxidation of glucose, forming a substance called **lactate**

- yeast cells, and some plant cells (e.g. roots deprived of oxygen) when glucose is oxidised to **ethanol** and **carbon dioxide**:

$$C_6H_{12}O_6 \rightarrow 2C_2H_5OH + 2CO_2$$
$$\text{ethanol}$$

We will look at these anaerobic processes in more detail in Section 1.11, page 18. For now we will be concerned with the complete (aerobic) oxidation of glucose. This takes place in four stages, making up a metabolic pathway of numerous reactions (Figure 1.6). The stages are as follows:

- **Glycolysis** (pronounced gly-colly-sis) where the 6C sugar glucose is split into two molecules of a 3C compound called **pyruvate**. Some energy is released to make a little ATP at this stage.

- The **link reaction**, where the 3C pyruvate loses a carbon (as $CO_2$) and the 2C 'fragment' is carried into the next stage.

- The **Krebs cycle**, where the 2C compound is completely broken down to $CO_2$. In this stage hydrogen atoms are also removed from the carbon compounds of the cycle, in a series of **dehydrogenation** reactions, and passed to the next stage. Some more ATP is made during the reactions of the Krebs cycle.

### Remember

In descriptions of respiration (and photosynthesis) biochemistry, it is usual to describe the molecules involved as 6C, 4C, 3C (etc.) compounds. This refers to the number of carbon atoms that the substance contains. For instance, glucose is a hexose sugar, with six carbon atoms in each molecule. We therefore call it a 6C compound. Another substance, oxaloacetate, has only four carbon atoms per molecule, so this is a 4C compound. Carbon dioxide is a 1C compound.

● **The electron transport chain** (sometimes abbreviated to e.t.c.) where initially hydrogen atoms, and subsequently just their electrons, are passed through a series of carrier molecules. This is coupled to the synthesis of ATP. Most of the ATP produced by respiration is formed in the electron transport chain.

Notice in Figure 1.6 the locations of these four stages in the cell. Only the first stage, glycolysis, takes place in the liquid part of the cytoplasm (the cytosol).

*Figure 1.6 Outline summary of respiration*

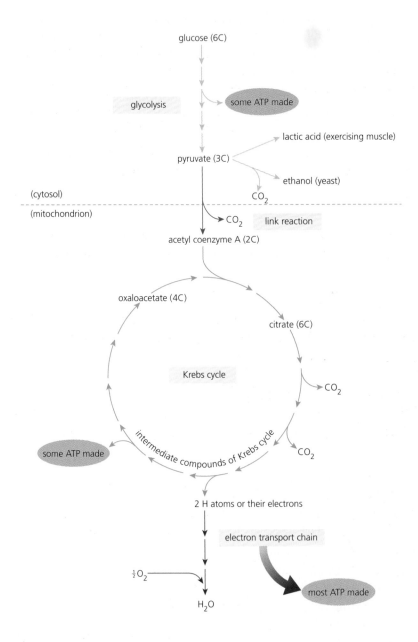

The remaining three stages, the link reaction, Krebs cycle and electron transport chain, take place in the mitochondrion. The link reaction and the Krebs cycle take place in the matrix of the mitochondrion, and the electron transport chain with its coupled production of ATP is situated on the inner mitochondrial membrane, or cristae (Figure 1.7).

## 1.6  Glycolysis

This pathway begins with the 6C sugar glucose and ends with the formation of two molecules of 3C pyruvate. Hence its name, glycolysis, which means 'sugar splitting'. It consists of several steps, each catalysed by a different enzyme. You do not need to know all the reactions and intermediate products, but there are four main events that you should notice happening during glycolysis:

1  At the start of glycolysis, hexose (6C) sugars have phosphates (and energy) added to them (they are **phosphorylated**). This *uses up* ATP.

2  The phosphorylated hexose sugar is then broken down into two molecules of phosphorylated triose (3C) sugars (**triose phosphate**).

3  Each triose phosphate loses its phosphate to ADP, regenerating a molecule of ATP.

4  Each triose phosphate is also **oxidised**, losing hydrogen atoms to the coenzyme NAD, which becomes reduced (see Section 1.4 above):

$$NAD^+ + 2H^+ + 2e^- \rightarrow NADH + H^+$$

The reducing power of the reduced NAD can, under aerobic conditions, be used in the mitochondria to generate more ATP (see Section 1.9 on page 14).

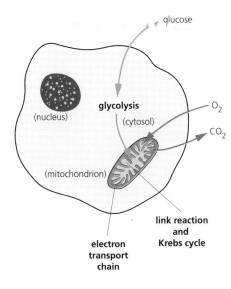

**Figure 1.7** *Location of stages of respiration in a cell*

> ### Remember
>
> #### Pyruvate or pyruvic acid?
>
> The substance we have so far called pyruvate can also be called pyruvic acid. This is because pyruvic acid, when it dissolves in water, forms pyruvate ions (as well as hydrogen ions):
>
> $$CH_3(CO)COOH \rightleftharpoons CH_3(CO)COO^- + H^+$$
>
> This is the same as the way that stronger acids such as sulphuric acid and hydrochloric acid form (respectively) sulphate and chloride ions:
>
> $$H_2SO_4 \rightleftharpoons SO_4^{2-} + 2H^+$$
> $$HCl \rightleftharpoons Cl^- + H^+$$
>
> In this chapter we will deal with other organic compounds, such as lactate, citrate and oxaloacetate. We will continue to use the names of these salts, rather than the acids they derive from (lactic acid, citric acid, oxaloacetic acid).

The complete reactions of glycolysis are shown in Figure 1.8, along with a simplified version showing the main stages. For exam purposes *the simplified stages and events are the only ones that you must remember.*

Glycolysis begins with the transfer of phosphate to glucose, forming glucose 6-phosphate (the 6 refers to the particular carbon atom in the glucose molecule to which the phosphate group is attached). This has two functions. Firstly, glucose 6-phosphate, unlike glucose, cannot pass through the plasma membrane of the cell, so the glucose 6-phosphate is trapped within the cell. Secondly, phosphorylation of the glucose makes it more reactive, so that it is easier for it to take part in subsequent reactions. This phosphorylation uses up ATP, donating

***Figure 1.8*** *Stages of glycolysis (a) the metabolic pathway of glycolysis, (b) simplified summary of the main events.*

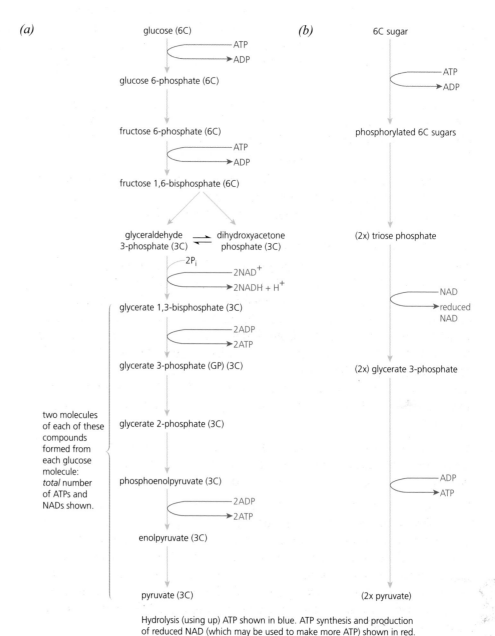

Hydrolysis (using up) ATP shown in blue. ATP synthesis and production of reduced NAD (which may be used to make more ATP) shown in red.

a phosphate and energy to the sugars. The glucose 6-phosphate is converted to an isomer, fructose 6-phosphate, which has another phosphate added, forming fructose 1,6-diphosphate and using up another molecule of ATP.

This phosphorylated 6C sugar is now broken down into two molecules of phosphorylated 3C sugars (triose phosphate), called glyceraldehyde 3-phosphate and dihydroxyacetone phosphate. These two compounds are isomers of each other, and can be readily interconverted. Only glyceraldehyde 3-phosphate continues the glycolysis pathway, but dihydroxyacetone phosphate forms a link with other metabolic pathways.

The glyceraldehyde 3-phosphate is next oxidised to glycerate 3-phosphate (GP) through an intermediate compound which is also further phosphorylated by inorganic phosphate (this does not use up ATP). Oxidation of the GP yields H atoms, which are used to reduce NAD. The intermediate compound (glycerate 1,3-bisphosphate) easily loses one of its phosphates to regenerate ATP from ADP.

Towards the end of the pathway, a second phosphate is lost from each glycerate 3-phosphate, forming another molecule of ATP. This production of ATP is called **substrate level phosphorylation**, because the ADP is phosphorylated using a phosphate from the compounds (substrates) of the glycolysis pathway.

Remember that each glucose molecule forms **two** molecules of triose phosphate, and two molecules of pyruvate at the end of the pathway. In Figure 1.8, the *total* ATPs used up or produced per glucose and the *total* NADs reduced, are shown.

### The energy budget of glycolysis

The final products of glycolysis are pyruvate, ATP and reduced NAD (NADH + $H^+$). The reduced NAD can be used to generate more ATP in the electron transport chains located in the mitochondria. However, NADH is unable to pass across the mitochondrial membranes. This problem is overcome by molecular 'shuttles' which carry electrons from NADH across the membrane, rather than NADH itself. These electrons supply the 'reducing power' to form more ATP in the electron transport chain. There are two of these shuttles. One provides enough reducing power to make three molecules of ATP, the other only two molecules. So each NADH formed during glycolysis is later 'worth' three (or two) ATPs.

From Figure 1.8, you can see that two ATPs are used up in phosphorylation of one glucose molecule during glycolysis. Four ATPs are produced (by substrate level phosphorylation), leaving a net gain of two.

However, in addition to this there are the three (or two) ATPs which can be formed from each reduced NAD produced during glycolysis. These ATPs will only be generated by the electron transport chain if the cell is respiring aerobically. If this is the case, an extra six (or four) ATPs will be made per glucose molecule as a result of the 'reducing power' of these reduced NADs. Hence, although strictly speaking glycolysis directly produces a net yield of two ATPs, it eventually leads to the formation of six to eight ATPs under aerobic conditions.

## 1.7 The link reaction

If oxygen is available, pyruvate produced by glycolysis in the cytosol enters the matrix of the mitochondrion, where the **link reaction** takes place. This gets its name from the fact that it links glycolysis with the Krebs cycle. Each 3C pyruvate molecule loses a carbon atom (it is **decarboxylated**), the carbon atom forming carbon dioxide. This leaves a 2C 'fragment' called an **acetyl** group, which is picked up by a coenzyme called **coenzyme A** (abbreviated to CoA). The CoA with its attached acetyl group is called **acetyl CoA**. At the same time, the pyruvate is oxidised, losing hydrogen to NAD, forming reduced NAD:

$$\text{pyruvate} + \text{CoA} + \text{NAD}^+ \rightarrow \text{acetyl CoA} + CO_2 + \text{NADH} + H^+$$

This reduced NAD can subsequently be oxidised in the electron transport chain to produce more ATP. However, the acetyl group still contains energy from the original glucose molecule which started the respiration pathway, and this energy is released in the Krebs cycle.

## 1.8 The Krebs cycle

This cyclical sequence of reactions was discovered by the famous biochemist Sir Hans Krebs (Figure 1.9) in 1937. It was a major advance in our understanding of cellular biochemistry, and it later became clear that not only was it responsible for the breakdown of pyruvate, but also acted as a metabolic 'hub' or central pathway, into which other food materials passed in order for them to be broken down (see Section 1.12, page 20).

Krebs recognised that the findings of his research team which led to the discovery of the cycle were built upon the results of many years of study by other biochemists. In addition to having the cycle named after him, he was awarded the Nobel Prize for medicine in 1953.

The Krebs cycle has a number of other names, including the **tricarboxylic acid (TCA) cycle**, and the **citric acid cycle**. Both of these names derive from compounds which are part of the cycle. However, most people use the easier term Krebs cycle.

The reactions of the Krebs cycle (Figure 1.10), along with the enzymes which catalyse each stage, occur in the matrix of the mitochondrion. Note that Figure 1.10 shows detailed information about the stages of the cycle. For exam purposes, you are not expected to remember all these details, just the main points (see Figure 1.6, page 8, and the summary below).

The cycle begins with the 2C acetyl group, attached to acetyl CoA, combining with the 4C compound **oxaloacetate**. The CoA is released by this reaction to collect more acetyl groups for the link reaction. Combining a 2C group with a 4C compound produces the 6C compound **citrate**.

The 6C citrate is then converted back to oxaloacetate by a series of steps which include two types of reactions:

*Figure 1.9 Sir Hans Krebs*

**Figure 1.10** *The Krebs cycle*

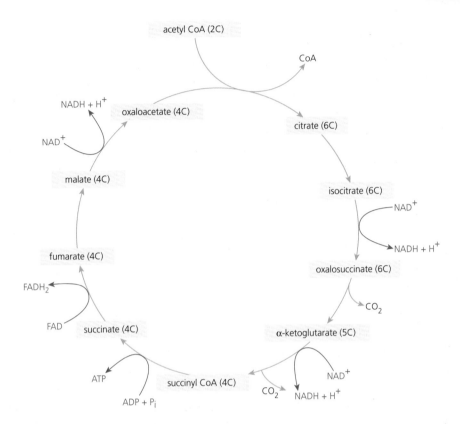

- **Decarboxylation** reactions, catalysed by **decarboxylase** enzymes, involve carbon atoms being removed from the Krebs cycle intermediate compounds, forming carbon dioxide.
- **Dehydrogenation** reactions, catalysed by **dehydrogenase** enzymes, involve the intermediates being oxidised by the removal of hydrogen atoms. The hydrogens are taken up by hydrogen acceptor molecules such as NAD, as well as another molecule called FAD (flavin adenine dinucleotide).

Reduced NAD and reduced FAD are oxidised in the electron transport chain, yielding energy to synthesise ATP from ADP. One molecule of ATP is also directly synthesised in the cycle.

It is probably a good idea at this stage to look back at Figure 1.6 (page 8) to remind yourself of the overall process of respiration. Remember that two molecules of pyruvate are produced from each glucose molecule. In other words, the Krebs cycle has to 'turn' *twice* for each glucose oxidised.

A summary of the events in the cycle is:

- a 2C acetyl group combines with a 4C compound to form a 6C compound
- the 6C compound undergoes a series of reactions, eventually losing two C atoms, regenerating the 4C compound

- the C atoms are lost as $CO_2$, a waste product of respiration
- the 6C compound is oxidised by removal of H atoms
- the H atoms pass to hydrogen acceptor molecules: three molecules of reduced NAD and one of reduced FAD ($FADH_2$) are formed
- a molecule of ATP is synthesised directly, by substrate level phosphorylation.

We must now look at the final stage of the process of respiration, to understand how reduced NAD or FAD is used to generate ATP in the presence of oxygen. This process, called **oxidative phosphorylation**, happens when electrons from NADH or $FADH_2$ are passed through the series of carrier molecules called the **electron transport chain**.

### 1.9 The electron transport chain

Remember the summary equation for respiration:

$$C_6H_{12}O_6 + 6O_2 \rightarrow 6CO_2 + 6H_2O$$

So far we have seen glucose used up in glycolysis, and $CO_2$ produced in the link reaction and Krebs cycle. We have not yet seen exactly where oxygen is needed for respiration, or where water is produced. Both these events happen in the electron transport chain. Electrons from NADH or $FADH_2$ are passed through a system, or chain, of carrier molecules. At the end of the chain, molecular oxygen is reduced to water, giving us our missing two components of the equation.

However, electron transport is linked, or **coupled** to another process: the formation of ATP from ADP and inorganic phosphate. This means that the two processes take place simultaneously.

The electron carriers are large protein complexes located in the inner membrane of the mitochondrion. They are of three types: flavoproteins, quinones and cytochromes. They are arranged in a particular sequence in the inner membrane (Figure 1.11), in order of their increasing **electron affinity**, or 'tightness' with which they bind with electrons.

At the start of the chain, NADH loses electrons, passing them to the first carrier, and returning to its oxidised state ($NAD^+$):

$$NADH + H^+ \rightarrow NAD^+ + 2H^+ + 2e^-$$

The electrons are then passed from carrier 1 to carrier 2 and so on, down the chain. At the end of the chain, molecular oxygen accepts the electrons, together with the protons produced from the oxidation of the NADH at the start of the chain:

$$\tfrac{1}{2}O_2 + 2H^+ + 2e^- \rightarrow H_2O$$

This takes place at the final electron carrier, an enzyme called **cytochrome oxidase**.

But how is this coupled to ATP production? As the electrons are passed along the chain they lose energy. This energy is used to pump protons ($H^+$) through the

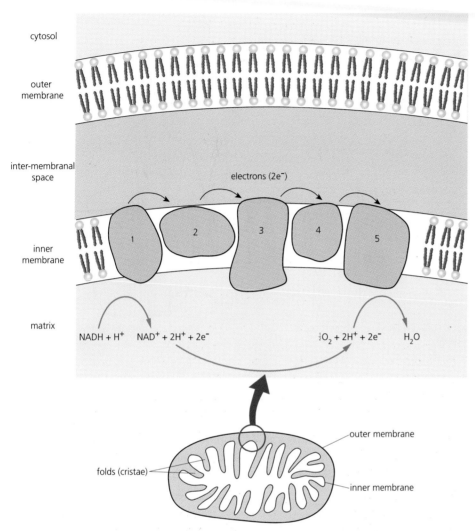

**Figure 1.11** *The carriers of the electron transport chain. The upper diagram shows an enlarged part of the outer wall of the mitochondrion. Structures 1–5 represent carrier molecules embedded in the inner membrane*

inner mitochondrial membrane into the space between the two membranes of the mitochondrion (the inter-membranal space). Some of the electron carrier molecules also act as ion pumps for this purpose (Figure 1.12).

**3**  Where would you expect the pH to be lower, in the matrix of the mitochondrion, or in the inter-membranal space?          *(1 mark)*

In this way a proton concentration gradient is set up across the inner mitochondrial membrane. This effectively acts as a store of energy, which can be released if the protons are allowed to pass back into the matrix from the inter-membranal space. There are other protein channels in the membrane which do allow this. They have an **ATP synthase** enzyme associated with them, and as the protons re-enter the matrix through the protein channel, their energy is used to synthesise ATP from ADP and inorganic phosphate, catalysed by the ATP synthase (Figure 1.12).

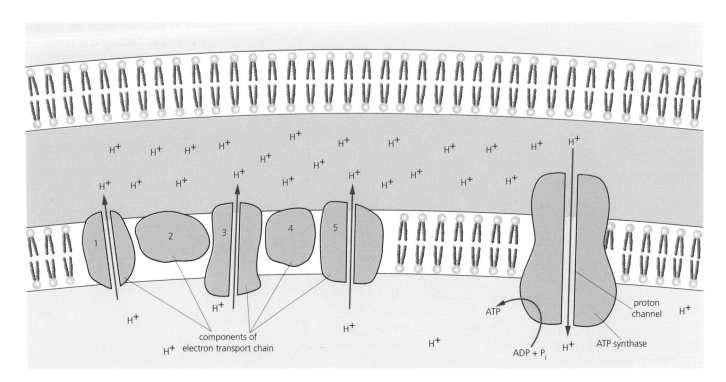

*Figure 1.12 Proton pumping by the carriers of the electron transport chain leads to a high concentration of hydrogen ions in the inter-membranal space. When these re-enter the matrix of the mitochondrion through proton channels, their energy is used to make ATP*

This way by which electron transport and ATP synthesis are coupled is called **Mitchell's chemiosmotic theory**. It was first proposed as a hypothesis by the British biochemist Peter Mitchell in 1961. Initially it was treated with scepticism by other scientists, but we now know that proton gradients supply the energy for a number of processes in biology, including the synthesis of ATP in chloroplasts and in bacteria. Mitchell was awarded a Nobel Prize for his work in 1978.

**?**

4 Which of the following statements apply to glycolysis, the Krebs cycle or the electron transport chain? (Some statements can apply to more than one process.)

(a) ATP is produced

(b) $CO_2$ is formed

(c) a 6C compound is converted into two molecules of a 3C compound

(d) it takes place in the mitochondrion

(e) $NAD^+$ is reduced to $NADH + H^+$. *(5 marks)*

**1.10** **Total ATPs produced from aerobic respiration of glucose**

Each molecule of reduced NAD (NADH) formed in the Krebs cycle can generate the production of three ATP molecules through oxidative phosphorylation. Reduced FAD ($FADH_2$) contains less energy, and can only produce two ATPs. The total ATPs which can be made by aerobic respiration are shown in Table 1.2.

| Source of ATP | ATP molecules made per glucose molecule |
|---|:---:|
| glycolysis (substrate-level phosphorylation) | 2 |
| $2 \times$ (NADH + H$^+$) from glycolysis | 6 (or 4) |
| $2 \times$ ATP in Krebs cycle (substrate-level phosphorylation) | 2 |
| $2 \times$ (NADH + H$^+$) from link reaction | 6 |
| $6 \times$ (NADH + H$^+$) from Krebs cycle | 18 |
| $2 \times$ FADH$_2$ from Krebs cycle | 4 |
| **Total** | **38 (or 36)** |

*Table 1.2 The total yield of ATP from the aerobic respiration of glucose*

Respiration is a relatively efficient way of producing energy from a fuel (glucose). A car engine at best only converts 20% of the energy in petrol into kinetic energy. We can compare this with respiration by doing a simple calculation:

The complete combustion of glucose in oxygen in the laboratory gives out 2870 kJ of energy. Thirty-eight molecules of ATP yield ($38 \times 30.6$) kJ or 1162.8 kJ.

Therefore the maximum percentage of energy of the glucose which is converted to ATP is:

$$\frac{1162.8}{2870} \times 100$$

which is about 40%. The rest of the energy is lost as heat, but the process is still about twice as efficient as the energy conversion of a car engine.

**Box 1.1 Does aerobic respiration really yield 38 ATPs?**

Since the 1940s, it has been generally accepted that the number of molecules of ATP made when reduced NAD is reoxidised in the mitochondrion is three. Similarly, oxidation of FADH$_2$ has been thought to produce two ATPs. These figures (used in Section 1.10 above) have been repeated in all textbooks of biochemistry, and have become a central dogma of biology. However, following Mitchell's chemiosmotic theory, and its subsequent acceptance by the scientific community, many biochemists began to look again at the yields of ATP, and by the mid-1990s the traditional integral (whole number) values of ATPs per reduced NAD or FADH$_2$ were being challenged. The consensus of opinion among many biochemists now is that the yields of ATP are non-integral.

Oxidation of one molecule of NADH + H$^+$ supplies enough energy to make about 2.5 molecules of ATP, and oxidation of one molecule of FADH$_2$ enough for about 1.5 molecules of ATP (or to put it more simply, oxidation of two molecules of reduced NAD is needed to make five molecules of ATP). This results in a total yield of ATP from the complete oxidation of glucose, of 31 molecules, rather than the 38 suggested above. If correct, it also means that the efficiency of the respiration process is rather less than previously estimated (see Section 1.10).

> **?**
>
> **5** Calculate the revised percentage efficiency of respiration based on a total of 31 ATPs. *(2 marks)*

## 1.11 Anaerobic respiration

We saw in Section 1.5, page 7, that cells can respire, deriving energy from organic molecules, without the use of oxygen. This is known as **anaerobic respiration**. Because anaerobic respiration does not need oxygen, it can be used by organisms which live in oxygen-deficient environments, such as bacteria living in stagnant water. But it is also of use to some cells to maintain supplies of ATP when they are temporarily deprived of oxygen.

In anaerobic respiration, a lack of oxygen means that any NADH which is formed cannot be oxidised in the electron transport chain. In turn, this means that the Krebs cycle will not function, and pyruvate will not enter the mitochondria to take part in the link reaction. The pyruvate remains in the cytosol, where it is converted into either ethanol and carbon dioxide (in yeast and plant cells) or lactate (in animal cells).

We will look at two situations where anaerobic respiration takes place: in yeast cells and in active muscle cells.

### Anaerobic respiration in yeast

Yeast is a single-celled fungus (Figure 1.13a). It normally lives in well-aerated places, such as the surface of fruits. For example, the 'bloom' or powder on the surface of plums and grapes consists of yeast cells (Figure 1.13b).

Yeast normally respires aerobically, using oxygen to break down the sugar from the fruit into carbon dioxide and water. However, if yeast is deprived of oxygen it can respire anaerobically, producing ethanol (ethyl alcohol) and carbon dioxide. Glycolysis takes place as normal, breaking glucose down into two molecules of pyruvate, and generating two reduced NADs. The pyruvate is now decarboxylated (forming $CO_2$) to produce ethanal. Ethanal is then reduced to ethanol by the $NADH + H^+$ generated from glycolysis. This re-oxidises the $NADH + H^+$ back to $NAD^+$. The regeneration of $NAD^+$ allows glycolysis to continue, but since the $NADH + H^+$ does not enter the mitochondrion for oxidative phosphorylation only two ATPs are produced, by substrate-level phosphorylation. These events are summarised in Figure 1.14.

*Figure 1.13 (a) The single-celled fungus yeast. Notice the yeast cells reproducing by 'budding' forming new cells,
(b) The powder or 'bloom' on the surface of these plums contains 'wild' yeast cells*

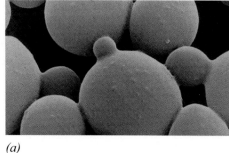

*(a)*

*(b)*

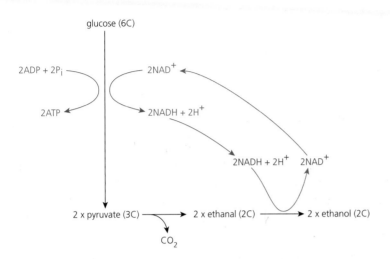

glucose (6C)

2ADP + 2P$_i$

2ATP

2NAD$^+$

2NADH + 2H$^+$

2NADH + 2H$^+$    2NAD$^+$

2 x pyruvate (3C) → 2 x ethanal (2C) → 2 x ethanol (2C)

CO$_2$

*Figure 1.14* *Anaerobic respiration of glucose by yeast. Pyruvate is reduced to ethanol by NADH and does not enter the Krebs cycle. This regenerates NAD$^+$ for use in glycolysis. Since the reducing power of NADH is not passed to the electron transport chain, there is a net gain of only two ATPs*

This metabolic pathway is also known as **alcoholic fermentation**, and has been used by humans for thousands of years in the production of bread and alcoholic drinks. In bread making, the yeast produces carbon dioxide, which makes the dough rise. In wine making, the sugars in grape juice are fermented to produce alcohol (Figure 1.15). The process only stops when the concentration of alcohol rises above about 15%, which kills the yeast cells. For beer making, barley grain is allowed to start germinating. The seeds produce sugars from stored starch, which are then fermented by the yeast.

### Anaerobic respiration in muscle

During vigorous exercise, such as sprinting, not enough oxygen can reach respiring muscles to allow aerobic respiration to take place. Muscles then begin to respire anaerobically, reducing pyruvate to **lactate** (sometimes called lactic acid). This produces a little ATP, allowing the muscles to go on working despite the shortage of oxygen. As with alcoholic fermentation, reduced NAD from glycolysis is used to reduce the pyruvate, so only two ATPs are produced from each glucose molecule respired. Notice that lactate is a 3C compound, and so decarboxylation does not happen, and CO$_2$ is not produced during lactate formation:

$$CH_3COCOO^- + NADH + H^+ \rightarrow CH_3CHOHCOO^- + NAD^+$$

pyruvate                                    lactate

During the period of exercise lactate builds up in the muscles, and if levels become too high it can produce muscle fatigue (see 'Did You Know' box). The lactate does not remain in the muscles. After the exercise it is carried away by the blood to the liver, where it is oxidised back to pyruvate and then aerobically respired to carbon dioxide and water. The oxygen needed to fully oxidise the lactate produced by anaerobic respiration is called the **oxygen debt**.

The total oxygen debt a person can tolerate before muscle fatigue begins to be experienced is relatively fixed. A sprinter in a 100 metre race might run 95% of it 'anaerobically', that is using ATP from lactate production (Figure 1.16).

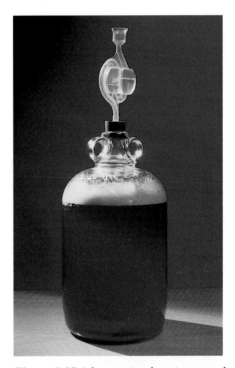

*Figure 1.15* *A home wine brewing vessel*

*Figure 1.16* *This sprinter is repaying her oxygen debt after the race*

*Figure 1.17* The oxygen debt is the difference between the oxygen demand during exercise and the available oxygen supply to the muscles. The debt is 'repaid' after the period of exercise

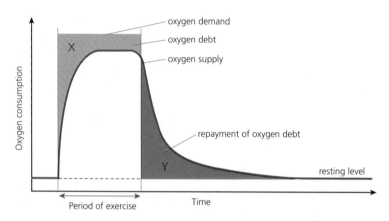

Note: area of X = area of Y

On the other hand, athletes in a 10 000 metre race will have to run relatively slowly, since only a few per cent of their energy will come from anaerobic respiration. If they sprint from the start, they will soon build up intolerable levels of lactate, leading to muscle fatigue. 'Middle distance' runners, competing in distances such as 800 or 1500 metres, have perhaps the most difficult task in judging their oxygen debt correctly. They must not run too fast at first, or the build up of lactate in their blood will restrict their ability to sprint in the final stages of the race. The build up of the oxygen debt and its repayment can be illustrated graphically (Figure 1.17).

It is not just sprinting which relies on lactate production to generate ATP. When the muscles carry out any sudden, or 'burst' activity they will overstretch the capacity of the blood to supply enough oxygen for aerobic respiration. Events such as weight-lifting, throwing the javelin or putting the shot also use anaerobic ATP production and result in an oxygen debt. One aspect of an athlete's training is aimed at improving his or her ability to sustain a large oxygen debt without experiencing muscle fatigue. Athletes also reduce their oxygen debt through training, which improves their circulation and heart development, allowing more oxygen to reach the muscles.

## 1.12 Other substrates for respiration

In addition to glucose, other carbohydrates can be used as sources of energy for ATP synthesis. Hydrolysis of polymers such as starch or glycogen yields glucose or glucose phosphate for glycolysis, and other sugars such as fructose and galactose can also be chemically modified to enter the glycolysis pathway at various points.

Lipids and proteins can also be oxidised by respiration to yield energy (Table 1.3). Oxidation of protein produces a little more energy per gram for ATP synthesis than does carbohydrate, while oxidation of lipid produces over twice as much energy per unit mass.

When demands of energy are great, or carbohydrate is in short supply, lipids (triglycerides stored in fatty tissue) are respired. They are first hydrolysed to glycerol and fatty acids. The glycerol (a 3C compound) is converted into the triose sugar dihydroxyacetone phosphate, which then forms glyceraldehyde 3-phosphate, an intermediate in the glycolysis pathway. Respiration of glycerol yields a total of 19 ATPs per molecule oxidised. Fatty acids are oxidised in a complex process which removes 2C (acetyl) fragments and feeds them into the Krebs cycle as acetyl coenzyme A. The energy yield from fatty acids depends on the length of their hydrocarbon chains, and since these can be very long, some fatty acids can produce over 150 ATPs per molecule oxidised. Fatty acids therefore provide an important source of metabolic energy.

Proteins are not normally respired, except in cases of severe starvation when carbohydrate and lipid stores are used up. Protein is first hydrolysed to amino acids, then these have their amino group removed in a process called deamination (see Section 10.2, page 321). The amino group is converted to ammonia, then (in mammals) to urea for excretion. The carbon 'backbone' of the amino acid molecule can then be fed into the respiration pathway at a number of points. Sometimes this 'backbone' is itself an intermediate in glycolysis or the Krebs cycle, such as pyruvate or oxaloacetate, or it may need further chemical modification.

In summary, aerobic respiration of glucose forms a **central pathway** into which other pathways lead, so that all major classes of nutrients can be metabolised. This is shown diagrammatically in Figure 1.18.

| Substrate | Approximate energy yield/kJ g$^{-1}$ |
|---|---|
| carbohydrate | 17 |
| lipid | 39 |
| protein | 23 |

**Table 1.3** *Energy available for ATP synthesis from complete oxidation of different substrates*

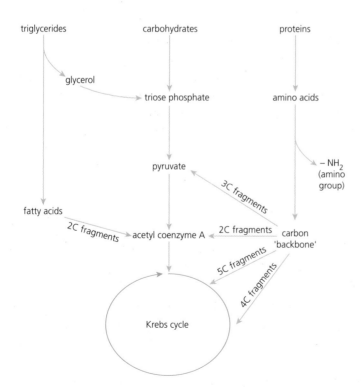

**Figure 1.18** *How other nutrients may be fed into the glucose respiration pathway*

### 1.13 Respiratory quotients

The concept of respiratory quotients has been dealt with in *AS Biology* (Box 9.6, page 250) but it is worth revising at this point. The respiratory quotient (RQ) is the ratio of the volume of carbon dioxide produced by respiration divided by the volume of oxygen used up. The RQ is a useful measure, because it can sometimes tell us the type of substrate (i.e. food) being respired. For example the aerobic respiration of glucose can be shown by the summary equation:

$$C_6H_{12}O_6 + 6O_2 \rightarrow 6CO_2 + 6H_2O \text{ (+ energy as ATP)}$$

The equation tells us that six molecules of oxygen are needed to oxidise a glucose molecule, producing six molecules of carbon dioxide. Since equal numbers of molecules of a gas occupy the same volume at the same temperature and pressure, the volume of $O_2$ used must equal the volume of $CO_2$ produced. So the respiratory quotient is found from:

$$RQ = \frac{\text{volume of } CO_2 \text{ produced}}{\text{volume of } O_2 \text{ used}} = \frac{6}{6} = 1$$

**?**

6 The fatty acid stearic acid has the formula $C_{18}H_{36}O_2$. The following equation shows the reaction which takes place when stearic acid is fully oxidised during respiration:

$$C_{18}H_{36}O_2 + 26O_2 \rightarrow 18CO_2 + 18H_2O \text{ (+ energy)}$$

Use the equation to calculate the RQ for the respiration of stearic acid.
*(1 mark)*

Respiration of other lipid molecules gives a RQ of about 0.7. Carbohydrates produce a RQ of 1, and proteins give a RQ of about 0.8–0.9. So if we find that an organism is exchanging $CO_2$ and $O_2$ with a RQ of 1, we can infer that it is mainly respiring carbohydrate. However, organisms rarely respire just one type of substrate, so their RQ value may not be a reliable indicator of the substrate used. For example, a RQ of 0.9 could mean that the organism is respiring a mixture of carbohydrate and lipid, or carbohydrate and protein, or all three!

**?**

7 What would be the RQ for an organism which is fermenting glucose to produce ethanol and $CO_2$?
*(1 mark)*

In fact complete anaerobic respiration rarely occurs. A yeast which is respiring glucose partly aerobically and partly anaerobically will produce a RQ greater than 1.

Respiratory quotients can be calculated for an organism by measuring the volume of $CO_2$ that the organism produces over a certain time, and the volume of oxygen that it uses in the same time interval. For small animals or plants, such as insects or germinating seeds, this is done using a piece of apparatus called a **respirometer**. Simple respirometers are covered in *AS Biology*, Box 9.7, page 250. A revision question about respirometers is included at the end of this chapter (page 26).

## 1.14  Redox indicators

We have seen that respiration consists of a series of reactions which oxidise substrates such as glucose to release energy. Often the oxidation reaction is accompanied by another reaction, where a second substance is reduced at the same time (a **redox** reaction). Examples of this are in the Krebs cycle, where dehydrogenase enzymes catalyse the removal of hydrogen atoms from the intermediate compounds of the cycle, passing them to coenzymes such as NAD. Redox reactions can be followed by the use of chemicals called **redox indicators**. Two commonly used redox indicators are TTC and methylene blue.

**TTC** (triphenyl tetrazolium chloride) is a redox indicator which can act as an artificial hydrogen acceptor. In its oxidised state it is colourless, but when reduced it becomes a pink colour, the colour deepening as more of the TTC becomes reduced. This substance can be used to show the activity of dehydrogenase enzymes in respiring cells. For example, if 1 cm$^3$ of 0.5% TTC solution is added to 10 cm$^3$ of a 10% suspension of respiring yeast cells, and the mixture incubated at room temperature, a pink colour will develop in the tube after some minutes. Dehydrogenase enzymes in the yeast result in hydrogen atoms being removed from respiratory substrates. Instead of the hydrogens passing to NAD, they will reduce the TTC.

The dye **methylene blue** (often used to stain cells for viewing under the light microscope) is also a redox indicator. In this case, the indicator is coloured blue in the oxidised state, and colourless when reduced. If a 1% solution of methylene blue is added to 10 cm$^3$ of milk and incubated at 30 °C, the mixture will soon turn colourless as a result of the activity of dehydrogenase enzymes.

**?**

8  What is the source of the dehydrogenase enzymes in the milk? *(1 mark)*

Redox indicators can be used in this way to study the activity of dehydrogenase enzymes, for example by altering the incubation temperature to find the effect on the rate of reaction.

## Summary – ① Respiration

- Respiration is the process by which energy contained within organic molecules is made available for living organisms, through the formation of the compound adenosine triphosphate (ATP).

- When ATP is hydrolysed to form adenosine diphosphate (ADP) and inorganic phosphate, energy is liberated, which can be used to drive biological processes which require it.

- Respiration is made up of a series of metabolic pathways, which each consist of a sequence of reactions catalysed by enzymes. Metabolic pathways can be anabolic (building up larger molecules) or catabolic (breaking down molecules to form smaller ones).

- Many enzymes need other substances called cofactors to enable them to work. An important group of cofactors in respiration is composed of coenzymes such as nicotinamide adenine dinucleotide, or NAD, which acts as a hydrogen carrier.

- The first metabolic pathway in respiration is glycolysis, where 6C monosaccharides such as glucose are converted into two molecules of a 3C compound called pyruvate. This process, which takes place in the cytoplasm of the cell, yields reduced coenzyme ($NADH + H^+$) and ATP by substrate-level phosphorylation.

- If oxygen is available, pyruvate is converted to the 2C compound called acetyl coenzyme A, and its acetyl group enters the Krebs cycle by combining with a 4C compound to form a 6C compound. The Krebs cycle takes place in the matrix of the mitochondrion.

- The intermediate compounds of the Krebs cycle are decarboxylated to give $CO_2$ and also dehydrogenated. The hydrogens are used to reduce more coenzyme ($NAD^+$ and FAD). Some ATP is generated by substrate-level phosphorylation.

- The reducing power of the reduced coenzymes ($NADH + H^+$ and $FADH_2$) is passed through the electron transport chain and used to generate ATP by oxidative phosphorylation. This occurs on the inner membrane of the mitochondrion.

- In the absence of oxygen, anaerobic respiration takes place. In yeast and some plant cells, pyruvate is converted to ethanol and carbon dioxide, re-oxidising some $NADH + H^+$. In muscle cells, lactate is formed. Both processes result in a net yield of just two ATPs per molecule of glucose respired.

- Carbohydrate, lipid and protein can all be respired under different circumstances, 'feeding in' to the respiratory pathway at different points. Of the three substrates, lipid yields the most energy per unit mass.

- The respiratory quotient is the ratio of the volumes of carbon dioxide produced to oxygen used by an organism or a tissue. The RQ can provide information as to the type of substrate being respired.

- Redox indicators are chemicals which can be used to follow the course of a redox reaction.

## Answers

1  (a)  Chemical (in petrol) $\rightarrow$ kinetic *(1)*.

   (b)  Chemical (in food) $\rightarrow$ light *(1)*.

   (c)  Light $\rightarrow$ chemical *(1)*.

2  A multi-enzyme complex will make the metabolic pathway more likely to take place because as each product is formed, it is close to the next enzyme in the pathway *(1)* so is likely to collide with its active site *(1)* and produce the next reaction (there is also less likelihood of interference from other metabolites in the cell).

3  The concentration of $H^+$ is greater in the inter-membranal space, so the pH is lower there *(1)*.

4  (a)  Glycolysis, Krebs cycle and electron transport chain *(1)*.

   (b)  Krebs cycle *(1)*.

   (c)  Glycolysis *(1)*.

   (d)  Krebs cycle and electron transport chain *(1)*.

   (e)  Glycolysis and Krebs cycle *(1)*.

5  The complete combustion of glucose in oxygen in the laboratory gives out 2870 kJ of energy. Thirty-one molecules of ATP yield ($31 \times 30.6$ kJ) or 948.6 kJ *(1)*. Therefore the maximum percentage of energy of the glucose which is converted to ATP is:

$$\frac{1162.8}{2870} \times 100 = 33.1\% \ (1)$$

6  From the equation we see that 26 molecules of oxygen are used in the reaction, and 18 molecules of carbon dioxide are made. Therefore the RQ is $18 \div 26$, which is 0.69 *(1)*.

7  Alcoholic fermentation produces carbon dioxide but uses no oxygen, so if respiration were completely anaerobic, the RQ would be equal to the volume of $CO_2$ produced divided by zero, i.e. infinity *(1)*.

8  The dehydrogenases must be present in organisms living in the milk. These are bacteria *(1)*. (If the milk is exposed to the air for longer periods, more bacteria will be found in it.)

## End of Chapter Questions

1   The diagram shows a simplified section through a mitochondrion of a muscle cell.

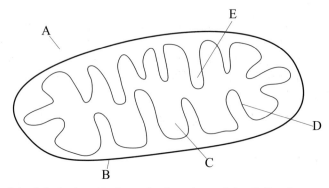

Which of the labels A to E show the location of the following:

**(a)**  the enzymes catalysing the Krebs cycle reactions    *(1 mark)*

**(b)**  phosphorylation of glucose by ATP    *(1 mark)*

**(c)**  most ATP production by the cell    *(1 mark)*

**(d)**  a high $H^+$ concentration as a result of proton pumping    *(1 mark)*

**(e)**  where lactate may be formed.    *(1 mark)*

*(Total 5 marks)*

2   The inner membrane of a mitochondrion forms many folds called cristae. Explain how this is an adaptation which helps the mitochondrion to carry out its function.    *(2 marks)*

3   Explain the difference between substrate level phosphorylation and oxidative phosphorylation.    *(4 marks)*

4   The diagram shows a simple respirometer being used to measure the oxygen uptake of some germinating seeds.

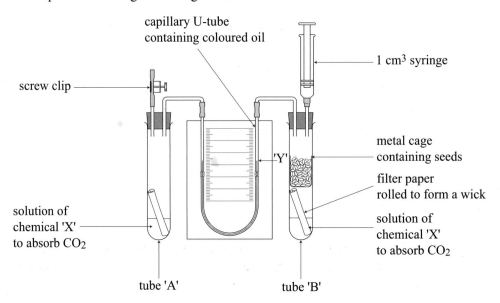

**(a)** Name a suitable chemical 'X' which will absorb carbon dioxide.

*(1 mark)*

**(b)** As the seeds respire, will the level of the oil in the U-tube at 'Y' move up or down? *(1 mark)*

**(c)** The control tube 'A' compensates for small changes in external air temperature and pressure which might alter the volume of gas in tube 'B'. However, this will not compensate for temperature changes which only occur inside tube 'B'. What may cause the temperature inside 'B' to change? *(1 mark)*

**(d)** Suggest a way of improving temperature control in tubes 'A' and 'B'.

*(1 mark)*

**(e)** Explain the purpose of the 1 cm$^3$ syringe. *(1 mark)*

*(Total 5 marks)*

**5** Explain the meaning of the following terms:

**(a)** a coenzyme *(2 marks)*

**(b)** alcoholic fermentation *(2 marks)*

**(c)** a redox indicator. *(2 marks)*

*(Total 6 marks)*

**6** The diagram represents an outline of cellular respiration.

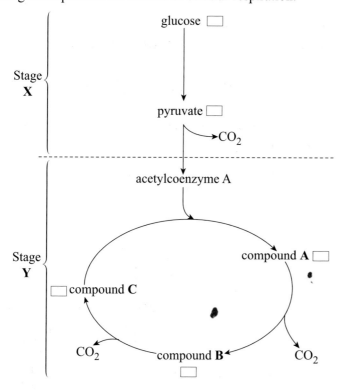

**(a)** (i) Complete the boxes on the diagram to show the number of carbon atoms contained in each molecule. *(2 marks)*

(ii) Complete the table to show the names of stages **X** and **Y**, and the part of the cell in which each stage occurs.

|  | Name of stage | Part of cell in which it occurs |
|---|---|---|
| Stage X |  |  |
| Stage Y |  |  |

*(2 marks)*

**(b)** Explain what happens to the hydrogen produced during stage **Y**.

*(3 marks)*

*NEAB 1998*

*(Total 7 marks)*

**7** The diagram below summarises the electron-transport chain, which is a stage of cell respiration. A, B and C represent electron carriers.

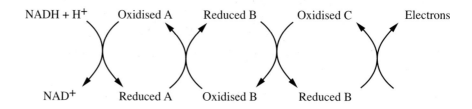

**(a)** (i) Describe what happens to the electrons and hydrogen ions at the end of this stage. *(2 marks)*

(ii) State where in a mitochondrion the electron transfer chain is situated. *(1 mark)*

**(b)** Name *one* process which produces NADH + H$^+$. *(1 mark)*

**(c)** ATP is synthesised as a result of this stage of respiration. State *two* uses of ATP in a cell. *(2 marks)*

*Edexcel 1999*

*(Total 6 marks)*

8  The diagram below shows apparatus used to investigate anaerobic respiration in yeast.

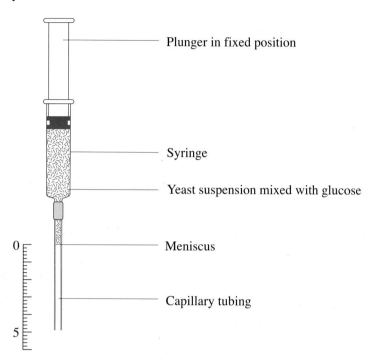

A yeast suspension containing glucose solution was drawn up into the syringe. The syringe was attached to a length of capillary tubing and supported in an upright position. Pressure was applied to the plunger until the meniscus was level with the top of the scale. The initial position of the meniscus was recorded.

**(a)**  Explain why the meniscus moved further down the scale during the investigation.                      *(2 marks)*

**(b)**  Describe how you would use this apparatus to compare the rate of anaerobic respiration using glucose with that using sucrose.      *(4 marks)*

*Edexcel 2000*                                                      *(Total 6 marks)*

# 2 Photosynthesis

Green plants make their own food by **photosynthesis**. This chemical reaction is 'driven' by light energy from the sun, and converts the simple inorganic compounds carbon dioxide and water into more complex organic compounds. It also releases oxygen, a requirement of aerobic respiration. The first product of this process is a carbohydrate which can be converted into other carbohydrates, lipids and proteins needed for respiration and growth. The light energy is converted into chemical energy by the green pigment **chlorophyll**. Photosynthesis therefore takes place in the green parts of plants, in particular the leaves.

**This chapter includes:**
- the significance of photosynthesis
- the adaptations of leaves
- the structure of the chloroplast
- photosynthetic pigments
- the light-dependent reactions
- the light-independent reactions
- plant mineral requirements
- limiting factors.

## 2.1 An overview of photosynthesis

The use of light energy in photosynthesis to synthesise complex organic chemicals is an example of a **transduction** process. This is where one sort of energy is converted into another form. We have met this principle before, with respiration, where the chemical energy in food molecules is converted into other forms, such as heat or kinetic energy. In photosynthesis, the energy of visible light, either from the sun or artificial sources, is converted into chemical energy by a plant. In doing this, plants make their own food: they are **autotrophs**. Since they use *light* to provide the energy for the process, they are **photo**autotrophs. Plants are not the only photoautotrophs. Algae, and many species of bacteria also contain photosynthetic pigments like chlorophyll. This chapter will mainly deal with photosynthesis in green plants.

Photosynthesis can be summarised by the familiar equation:

$$\text{carbon dioxide} + \text{water} \xrightarrow{\text{light}} \text{glucose} + \text{oxygen}$$
$$6CO_2 + 6H_2O \rightarrow C_6H_{12}O_6 + 6O_2$$

**1** What is the relationship between this equation and the summary equation for respiration? (See Section 1.5, page 7.) *(1 mark)*

From this simple summary equation we can see two important facts about photosynthesis. First of all, the carbon dioxide is chemically reduced during the reaction, by the addition of hydrogen: it is a **reduction reaction**. Secondly, the incorporation of the inorganic $CO_2$ into the glucose molecule is known as **fixation**. We say that the $CO_2$ has been **fixed** in photosynthesis.

However, the summary reaction is so over-simplified as to be misleading. If you look at the symbol equation above, you will notice that there are six oxygen molecules (i.e. 12 atoms) on the right, which the equation implies must have come wholly or partly from the six molecules of carbon dioxide on the left.

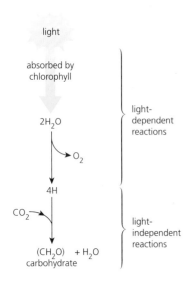

light

absorbed by chlorophyll

$2H_2O$

light-dependent reactions

$O_2$

$4H$

$CO_2$

light-independent reactions

$(CH_2O) + H_2O$
carbohydrate

*Figure 2.1 Summary of the reactions of photosynthesis*

They cannot all be from the water, since there are only six atoms of oxygen available in the water molecules. So does the oxygen produced by photosynthesis come from the $CO_2$ or the water molecules? This question can be answered by 'labelling' the $CO_2$ or water with an **isotope** of oxygen called heavy oxygen ($^{18}O$).

> **Remember**
>
> An element often exists naturally as a mixture of different **isotopes**. These have different masses, due to different numbers of neutrons in the nucleus of the atom. For example, there are three isotopes of carbon, $^{12}C$, $^{13}C$ and $^{14}C$. Each has six protons in its nucleus, but six, seven or eight neutrons respectively. Since only the electrons take part in chemical reactions, and not the nucleus, the different isotopes have identical chemical properties. However, they have different physical properties. Some, such as $^{14}C$, are radioactive (hence its use in 'carbon dating'). Others are not, but can be identified by their different masses.

Although an isotope used in labelling is often radioactive, with $^{18}O$ this is not the case. As with the more common isotope of oxygen, $^{16}O$, it is not radioactive, but it can be traced by its greater mass, using a machine called a mass spectrometer.

If heavy oxygen is used to label the carbon dioxide taken up by a plant during photosynthesis, none of the $^{18}O$ ends up in the oxygen produced by the plant. However, if the $^{18}O$ is instead used to label the water that the plant needs, the $^{18}O$ does become part of the oxygen gas produced. In other words, *all* of the oxygen produced by photosynthesis originates from the water molecules. There must be 12 water molecules on the left of the equation for this to be possible. So the equation would be better written as:

$$6CO_2 + 12H_2O* \rightarrow C_6H_{12}O_6 + 6O_2* + 6H_2O$$

where the * represents $^{18}O$ labelling.

The oxygen is therefore produced from the 'splitting' of water during photosynthesis. This takes place during the first stage of the pathway, which, because it uses light energy, is called the **light-dependent stage**. Hydrogen atoms from the water are ultimately used to reduce carbon dioxide in the second part of the pathway, called the **light-independent stage** (Figure 2.1).

As its name suggests, this stage does not use energy directly from sunlight. Instead it uses chemical energy in the form of ATP, generated by the light-dependent reactions. The hydrogen atoms derived from the water are also not used directly to reduce $CO_2$, but through an intermediate carrier molecule, the coenzyme **NADP** (nicotinamide adenine dinucleotide phosphate). NADP transfers H atoms in its reduced form, NADPH.

> **Remember**
>
> A summary of the metabolic pathway of photosynthesis is as follows. In the light-dependent reactions, light energy is absorbed by chlorophyll and used to generate ATP and NADPH, using H atoms from water, and releasing oxygen gas as a waste product. The NADPH is used in the light-independent reactions to reduce $CO_2$ to carbohydrate, driven by energy from hydrolysis of the ATP (Figure 2.2).

## 2.2  Leaves: the site of photosynthesis

Photosynthesis occurs in the green parts of plants, which contain chlorophyll. All the aerial (above ground) parts of a plant, such as stems, may contain chlorophyll, but in most species it is the leaves which have evolved the best adaptations for this function. If we were to design an 'ideal' organ for photosynthesis, it would need to include these features:

● a large flat surface area with chlorophyll located close to the top surface next to the incident light, for maximum light absorption

● a thin structure, since light would be absorbed in the first millimetre or so of tissue

● adaptations allowing gas exchange through the surface (pores) and a network of air spaces inside for gases to reach the photosynthetic tissue

● vessels to transport water to the cells, and others to transport products of photosynthesis away to non-photosynthetic tissues.

A typical leaf has all these adaptations (Figure 2.3). It provides a large surface area exposed to the sun, and specialised photosynthetic tissues containing large numbers of chloroplasts.

A continuous system of **xylem** tissue carries water and minerals from the roots and through the stem to the leaves, and **phloem** takes away the soluble products of photosynthesis to other parts of the plant (see *AS Biology*, Chapter 12). Pores in the epidermis (**stomata**) allow exchange of carbon dioxide and oxygen, and

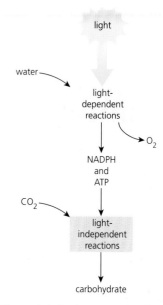

*Figure 2.2 Outline of light-dependent and light-independent reactions*

*Figure 2.3 Structure of a leaf from a typical dicotyledonous plant*
*(a) whole leaf,*
*(b) part of a longitudinal section through the leaf,*
*(c) a single palisade cell*

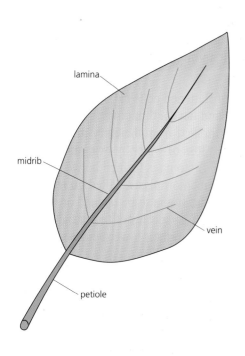

*(a)*

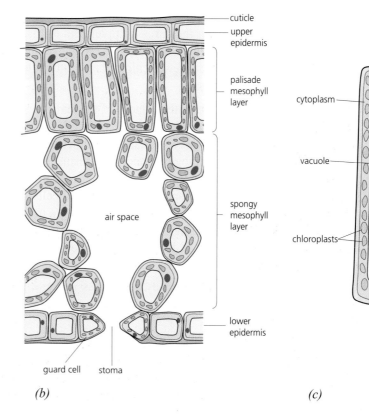

*(b)*

*(c)*

loss of water vapour through transpiration, which also cools the leaf, avoiding cellular damage through overheating. Inside the leaf, gases can diffuse through a system of sub-stomatal **air spaces**, giving access to the photosynthetic tissues.

**2** Why is overheating likely to be a particular problem for a leaf?

*(1 mark)*

The xylem and phloem are located in a system of veins branching from the central **midrib** of the leaf. Other tissues with thick or lignified cell walls are often located around the midrib and veins, supporting the leaf and maintaining its flat shape. The leaf stalk or **petiole** attaches the leaf blade or **lamina** to the stem and can cause the orientation of the lamina to be varied, so that the leaves 'follow' the direction of the sun throughout the day.

Most photosynthesis takes place in the **palisade mesophyll** tissue of the leaf. This usually consists of a single layer of columnar, vertically orientated cells with many chloroplasts, located just underneath the upper epidermis. Sometimes there are two layers of palisade cells. The upper epidermis (see below) is effectively transparent, so that the palisade layer is the first tissue to absorb the light. The palisade cells have thin cellulose walls, and their upright orientation allows the light to reach the chloroplasts with little obstruction (see *AS Biology*, Figure 1.25, page 25). Below the palisade layer is the **spongy mesophyll**, so called because of the air spaces which permeate this tissue. These intercellular air spaces extend to the lower parts of the palisade cells, allowing gases to diffuse to and from the cells. Since leaves are thin, there is only a short distance for the gases to travel, and diffusion is adequate for this. Spongy mesophyll cells tend to be less elongated and more equi-dimensional than palisade cells, with fewer chloroplasts, although they are still an important photosynthetic tissue.

On both surfaces of the leaf is a single layer of flattened cells called the **epidermis**. Epidermal cells lack chloroplasts, and have a thin covering layer of a waxy substance called **cuticle**. This protects the leaf from the entry of pathogenic microorganisms, and greatly reduces loss of water by evaporation. Water mainly exits the leaf through the stomata. A stoma is formed between two very specialised epidermal cells which do contain chloroplasts, called **guard cells**. These cells can change in shape, allowing the stoma to be opened or closed (see *AS Biology*, Section 12.3, page 348). Most species of terrestrial broad-leaved plants have larger numbers of stomata in their lower epidermis.

### 2.3 Chloroplasts

Chloroplasts are a type of cell organelle known as a **plastid**. These are membrane-bound bodies, generally containing pigments or food storage materials, or both. In the cells of petals, fruits and other brightly coloured parts of plants are plastids called **chromoplasts**, which attract insects or other animals. In cells of storage organs such as potatoes, as well as some seeds, are **amyloplasts** which store starch. Chloroplasts have both these functions, since they contain the photosynthetic pigments collectively called chlorophyll, and they also contain starch grains derived from photosynthesis.

Chloroplasts were briefly dealt with in *AS Biology* (Section 1.17, page 17) but it is now time to consider their structure and function in more detail. Most chloroplasts of plant cells are biconvex, disc-shaped structures (like a discus) about 5 μm in diameter (Figure 2.4).

However, their size and shape can vary a good deal. They range from 2 to 10 μm in diameter, and may be unusual shapes, such as the elongate or ribbon-like forms of some algal cells. The chloroplast is surrounded by a double membrane or envelope, which is not directly involved in the reactions of photosynthesis, and contains a complex system of internal membranes called **lamellae** or **thylakoids** (Figure 2.5).

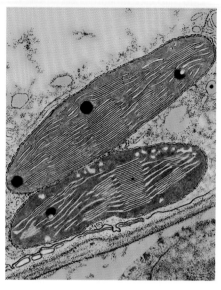

**Figure 2.4** *Electron micrograph showing two chloroplasts (× 15 000)*

**Figure 2.5** *Diagram of chloroplast structure*

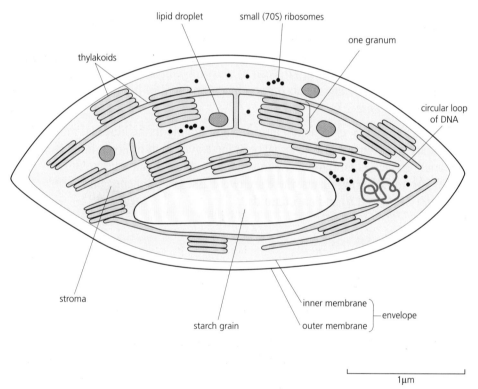

It is on these membranes, and in the fluid-filled matrix of the chloroplast, called the **stroma**, that the photosynthesis reactions take place. Notice that chloroplasts are similar to mitochondria in their structure, although mitochondria only consist of two membranes, whereas chloroplasts have three. The inner membrane of the mitochondrial envelope is folded into cristae, whereas the thylakoid membranes are not continuous with the inner chloroplast membrane. Instead they enclose a separate space called the **thylakoid lumen** (Figure 2.6). In places the thylakoids are folded into a stack of membranes called a granum (plural grana), connected with other grana by intergranal thylakoids (Figure 2.7).

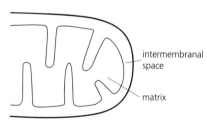

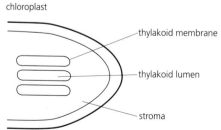

**Figure 2.6** *Comparison of the arrangement of membranes in a mitochondrion and a chloroplast*

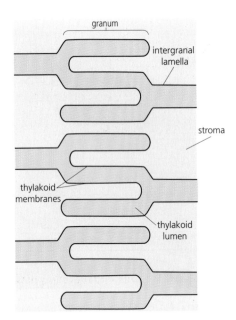

*Figure 2.7 Diagram showing how thylakoid membranes are folded to form stacks (grana). Note that the thylakoid lumen is always separated from the stroma of the chloroplast by the membranes*

The thylakoid membranes are composed of lipid and protein, and some of the membrane proteins are associated with the chlorophyll molecules, forming complexes known as **photosystems** (see Section 2.5, page 39). It is in these photosystems that the transduction of light energy to chemical energy takes place during the light-dependent stage of photosynthesis. The light-independent stage, where carbon dioxide is reduced to carbohydrate, takes place in the stroma.

Chloroplasts contain a circular loop of DNA and their own small (70S) ribosomes, and are able to synthesise their own proteins. You will remember that these facts, together with the size of chloroplasts and their double outer membrane, have suggested to many biologists that they have evolved from photosynthetic bacteria which became incorporated into eukaryotic cells: the **endosymbiont theory** (see *AS Biology*, Box 1.3, page 28).

### 2.4 Chlorophyll

A leaf looks green because it *reflects* mainly green light and *absorbs* other colours of the spectrum. Similarly, if you hold a leaf up to the light it *transmits* green light. The leaf, or more particularly the chlorophyll in the leaf, absorbs light of wavelengths other than green. This can be shown by an **absorption spectrum**, which is a graph showing the absorption by chlorophyll of light of different wavelengths (Figure 2.8).

The graph shows that chlorophyll absorbs light mainly in the blue and red parts of the spectrum, and has a low absorbance at the wavelength of green light, as predicted. In fact, chlorophyll is not a single substance, but a mixture of chemicals. In flowering plants, there are two classes of pigments: the **chlorophylls** (*a* and *b*) and **carotenoids**.

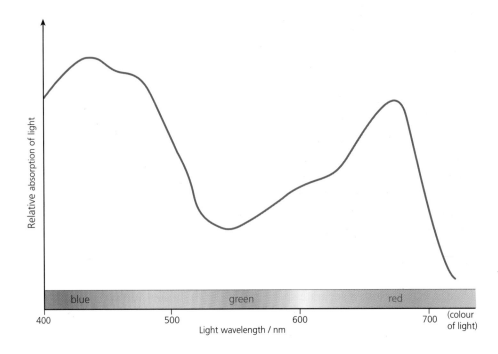

*Figure 2.8 Absorption spectrum of chlorophyll pigments*

**Remember**

The structures of chlorophyll and β-carotene are shown for information only. You do not need to remember them!

Chlorophyll *a* is the most abundant photosynthetic pigment. The molecule of chlorophyll *a* has a hydrophilic 'head' containing a magnesium atom at its centre, and a long hydrophobic hydrocarbon 'tail' (Figure 2.9). The 'head' is a structure known as a porphyrin ring, which is found in several other important biological molecules.

The hydrophobic tail anchors the chlorophyll molecule in the thylakoid membrane. The head region of the molecule is responsible for absorbing and transducing the light energy into chemical energy. Chlorophyll *b* is a very similar molecule to chlorophyll *a* (Figure 2.9).

Carotenoids, on the other hand, are very different chemicals (Figure 2.10). They are either **carotenes** or **xanthophylls**. They may be yellow, orange, red or brown pigments (for example the familiar orange pigment β-carotene, found in carrots).

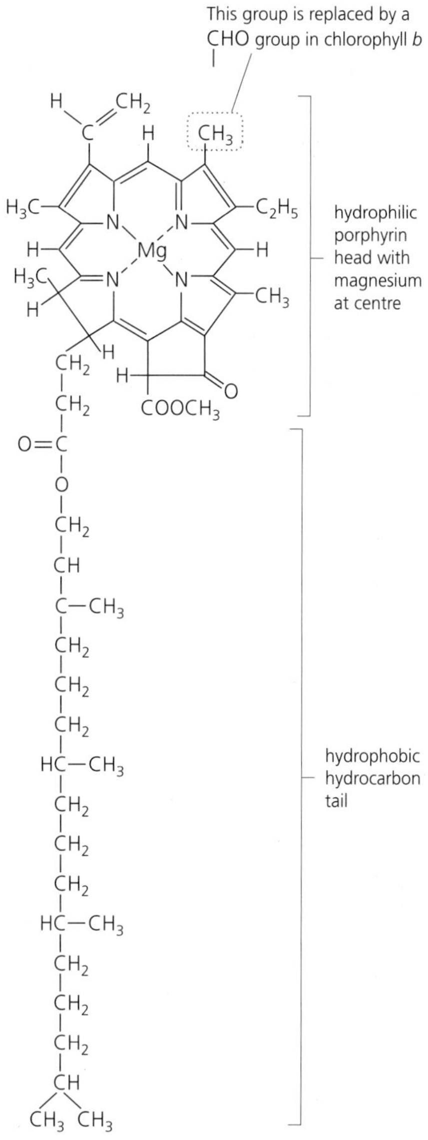

This group is replaced by a CHO group in chlorophyll *b*

hydrophilic porphyrin head with magnesium at centre

hydrophobic hydrocarbon tail

*Figure 2.9 Structure of the chlorophyll a molecule*

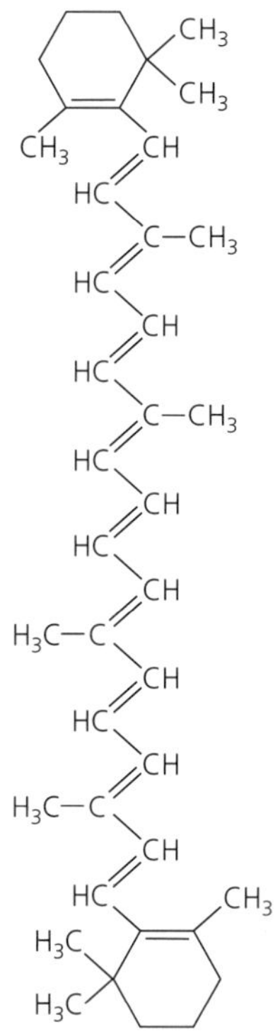

*Figure 2.10 Structure of the accessory pigment ß-carotene, a carotenoid*

*Figure 2.11 Absorption spectra of the different photosynthetic pigments*

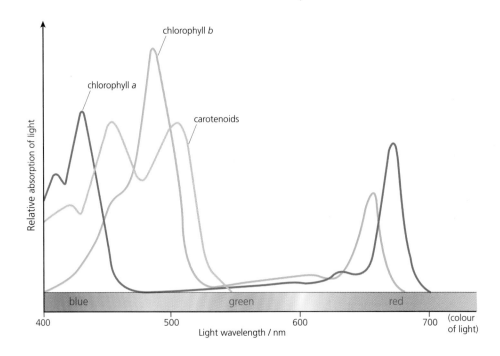

Carotenoids are known as **accessory pigments**. They absorb light of different wavelengths from chlorophyll, and then pass the energy to the chlorophyll molecules for photosynthesis.

The absorption spectra of the carotenoids and the chlorophylls are different (Figure 2.11) and it is their combined absorption characteristics which produce the net absorption spectrum for the mixture of pigments we call 'chlorophyll' shown in Figure 2.8.

**?**

3 Chlorophyll, being green, doesn't absorb green light. What would be a 'better' colour for chlorophyll, which would allow it to absorb light of all wavelengths? Can you suggest a major disadvantage of this colour?

*(2 marks)*

Since different wavelengths of light are absorbed to a different degree, it follows that different wavelengths will produce different rates of photosynthesis. If, instead of light absorbance, we plot the rate of photosynthesis (of a plant) against wavelength, this produces a graph called an **action spectrum** (Figure 2.12). You can see that the shape of the curve of the action spectrum is very similar to that of the absorption spectrum for the combined photosynthetic pigments, as might be expected.

### Separation of pigments in chlorophyll

The pigments that make up chlorophyll can be easily separated by the technique called **chromatography** (see *AS Biology*, Section 3.22, page 92). The chlorophyll is extracted from leaves by grinding them in a suitable solvent, and

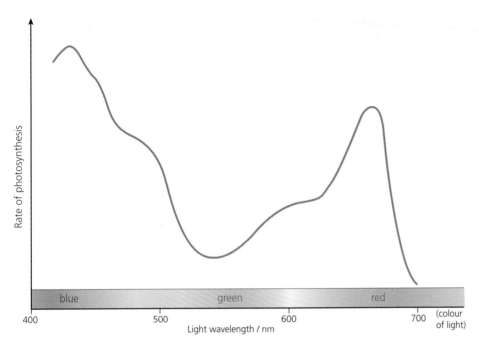

*Figure 2.12* Action spectrum for a plant

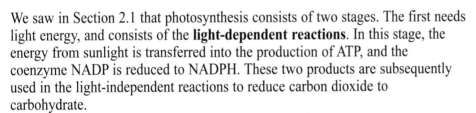

the dissolved pigments separated by allowing the solution to be absorbed (or 'run') through a suitable medium, such as filter paper or silica gel (see Box 2.1). It is an advantage that the pigments are coloured, since it means that they do not need to be stained to make them visible in the chromatogram.

## 2.5 The light-dependent reactions

We saw in Section 2.1 that photosynthesis consists of two stages. The first needs light energy, and consists of the **light-dependent reactions**. In this stage, the energy from sunlight is transferred into the production of ATP, and the coenzyme NADP is reduced to NADPH. These two products are subsequently used in the light-independent reactions to reduce carbon dioxide to carbohydrate.

The absorption of light is carried out by the photosynthetic pigment molecules in two different units or **photosystems**, called **photosystem I (PSI)** and **photosystem II (PSII)**. These can be seen by electron microscopy as different-sized particles attached to the thylakoid membranes. PSI particles are mainly present on the intergranal lamellae, whereas those of PSII are on the granal membranes. Each unit contains several hundred chlorophyll and carotenoid molecules. One particular chlorophyll molecule, called a **primary pigment**, acts as a **reaction centre** for the photosystem, while the other pigment molecules 'catch' the photons of light and transfer their energy to the central molecule. These other molecules are called **accessory pigments**. Because of their role, the accessory pigments are often described as acting as an '**antenna complex**' for light. There are two primary pigments, each a form of chlorophyll *a*. PSI contains the primary pigment **P700**, with an absorption peak at a (light) wavelength of 700 nm, and PSII contains **P680**, with its peak at 680 nm.

### Remember

The **absorption** spectrum shows the amount of light absorbed by chlorophyll, or its component pigments, at different wavelengths of light. The **action** spectrum shows the rate of photosynthesis of a plant at different wavelengths. They are related because the rate of photosynthesis depends on the effectiveness of different pigments in absorbing light.

tall beaker or chromatography jar

lid

solvent front

separated pigments

origin

solvent

thin-layer chromatography strips (front and side view)

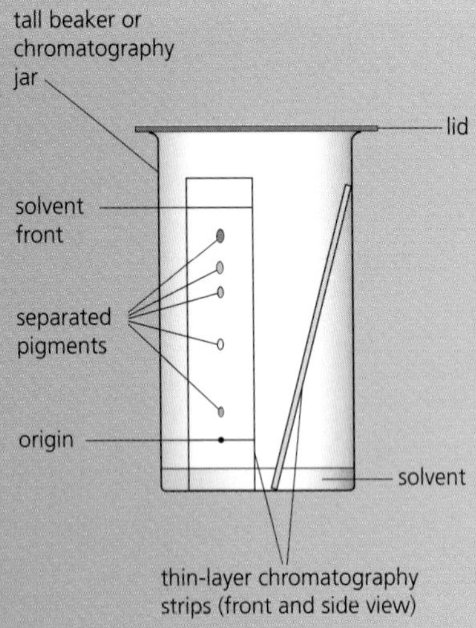

*Figure 2.13 Apparatus for separating chlorophyll pigments by thin-layer chromatography*

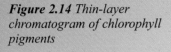

*Figure 2.14 Thin-layer chromatogram of chlorophyll pigments*

## Box 2.1  Separation of photosynthetic pigments by thin-layer chromatography

The method is as follows:

Make up a solvent consisting of 55% cyclohexane and 45% ethyl ethanoate (by volume). Place this to a depth of 0.5 cm in a tall beaker or chromatography jar, and cover the beaker to allow the air to become saturated with solvent vapour. Remove some leaves from a suitable plant, such as spinach, carrot or grass. Cut the leaves into small pieces and remove any stalks. Grind the leaves with a few cm³ of propanone in a mortar to produce a concentrated green solution. Filter this solution to remove plant debris.

Take a ready-prepared strip of thin-layer chromatography sheet coated with silica gel. Mark a line in pencil 1.5 cm from one end of the strip. Using a fine glass tube or pipette, place a small drop of the chlorophyll extract in the middle of the pencil line. Allow this to dry and repeat the 'spotting' several times, allowing to dry between each application, in order to get a concentrated spot of extract. This position is called the **origin**.

Place the strip in the beaker, leaning it against the side, making sure that the origin is above the level of the solvent (Figure 2.13). Cover and leave for 1–2 hours, until the solvent has travelled up the silica gel to a point near the end of the strip, making sure that the solvent does not run off the end. Remove the strip and mark this position (the **solvent front**). Allow the strip to dry.

A typical result is shown in Figure 2.14. The position of the various pigments can be measured, and their **$R_f$ values** calculated:

$$R_f = \text{distance moved by pigment} \div \text{distance moved by solvent}$$

'$R_f$' stands for 'relative front'. These values should be constant for a particular pigment if the same solvent and the same absorbent medium are used. They can be used to identify unknown pigments from different samples. However, it is important to realise that the $R_f$ values may be completely different for a different solvent or medium.

❓

4  What would be the $R_f$ value for a pigment which travelled 75 mm, when the solvent front had travelled 112 mm?                    *(1 mark)*

Energy reaching the reaction centre raises an electron in the primary chlorophyll molecule to a higher energy level. This so-called **excited** electron is then emitted by the chlorophyll (i.e. the chlorophyll is oxidised) and taken up by an **electron acceptor** or **carrier** molecule:

$$chlorophyll \xrightarrow{\text{light energy}} chlorophyll^+ + e^-$$

(electron passed to acceptor)

This is effectively the point where light energy is converted into chemical energy. The light has raised the energy level of the electron enough to make the chlorophyll able to donate the electron to the acceptor (this would not happen with the 'unexcited' electron). In other words, the chlorophyll has become a very strong reducing agent.

The electron acceptor is therefore reduced by the gaining of an electron from the reaction centre. The electron is then passed along a series of electron carriers. As it moves from one carrier to the next, each carrier becomes reduced and then reoxidised in a series of **redox** reactions. Each carrier in the sequence has a lower energy level than the preceding one, so that as the electrons are passed along, enough energy is released to synthesise ATP from ADP and inorganic phosphate. This principle should be familiar to you: it is a very similar mechanism to the one acting in the electron transfer chain of mitochondria (Section 1.9, page 14). However, the process is 'driven' by light energy rather than chemical energy, so it is known as **photophosphorylation** rather than the oxidative phosphorylation of respiration.

**Extension**

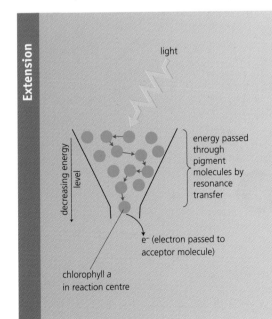

light

decreasing energy level

energy passed through pigment molecules by resonance transfer

e⁻ (electron passed to acceptor molecule)

chlorophyll a in reaction centre

### Box 2.2 Energy transfer in the antenna system

In the photosystems, photons of light are not emitted by one pigment molecule and absorbed by another. Instead the energy is transferred by a physical process called **resonance transfer**. This process can be likened to the way a vibrating tuning fork, when placed near another, will pass its energy to the second fork and start it vibrating. Eventually most of the energy is passed to the reaction centre, which loses an electron. The energy level of the pigments decreases as they get nearer to the reaction centre, which ensures a favourable energy gradient, transferring the energy in the direction of the reaction centre. An easy way to visualise this antenna system is as a funnel, with the accessory pigments in the top of the funnel and the reaction centre at the neck (Figure 2.15).

***Figure 2.15** Funnel model of antenna system*

**5** List two other differences and two similarities between
photophosphorylation and oxidative phosphorylation. *(4 marks)*

Both P680 and P700 lose 'excited' electrons in this way, leaving the primary
pigments of the photosystems oxidised. In order for photosynthesis to continue,
they must be reduced again by replacement of the missing electrons. This takes
place in a different way in the two photosystems, as will become clear when we
consider what happens to the lost electrons.

### The fate of the 'excited' electrons

The fate of the electrons lost from PSI and PSII is shown in Figure 2.16. This
diagram is called the **Z-scheme** and is commonly shown in textbooks of
photosynthesis biochemistry. In fact, the original diagram was drawn rotated by

*Figure 2.16 The Z-scheme of electron
transfer, showing how the light-
dependent reaction produces ATP and
NADPH*

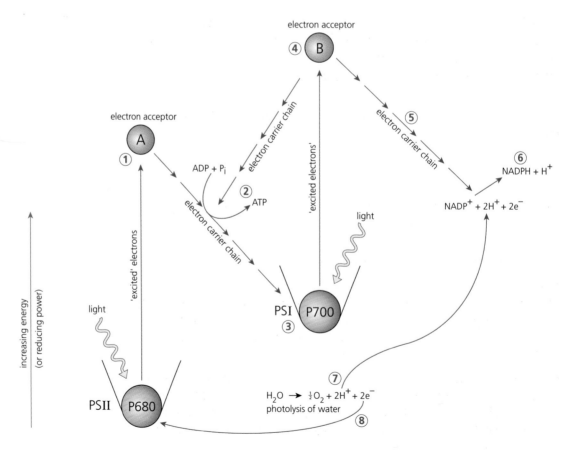

90°, so would have been a Z shape – nowadays it would be more correct to call it the N-scheme! The numbers in the following account refer to stages labelled 1 to 9 in Figure 2.16.

Electrons from the two photosystems are captured by two different acceptor molecules (for simplicity, called A and B on Figure 2.16). The electrons then pass through a chain of electron carriers via a series of redox reactions. Electrons from PSII are taken up by acceptor A (1). Their transfer from acceptor A through the carrier chain is coupled to the synthesis of ATP (2). This uses a mechanism very similar to that of ATP generation in mitochondria, involving proton pumping (see Box 2.3).

At the end of this carrier system, the electrons are used to replace those lost from the P700 chlorophyll molecules of PSI (3).

Electrons from PSI are captured by acceptor molecule B (4) and then passed through a different chain of carriers (5). At the end of the chain, they combine with hydrogen ions and the coenzyme NADP (which strictly speaking is a charged compound NADP$^+$). This forms the reduced form of the coenzyme, NADPH (6):

$$NADP^+ + 2H^+ + 2e^- \rightarrow NADPH + H^+$$

The hydrogen ions for this reaction are supplied by the photolysis of water (7):

$$2H_2O \rightarrow O_2 + 4H^+ + 4e^-$$

Associated with PSII is an enzyme which catalyses this oxidation reaction. As the word 'photolysis' suggests, light (photo) is needed for the splitting (lysis) of the water, but little is known about the mechanism of the reaction. It is a reaction which has great significance for most organisms, since it is the source of oxygen for aerobic respiration.

The electrons from the photolysis of water are used to replace those lost from the P680 of PSII (8).

6  What is the advantage of having two photosystems rather than one? (Look at the Z-scheme of Figure 2.16).          *(2 marks)*

### Non-cyclic and cyclic photophosphorylation

Production of ATP by this Z-scheme is known as **non-cyclic photophosphorylation**. This is because the electrons lost from the primary chlorophyll pigments of PSI and PSII are not used to reduce the pigment again. In other words, the electrons are not 'recycled'. The end-products of the process are ATP and NADPH.

Electrons from PSI may also be passed straight back to P700, via a chain of carriers linked to those from electron acceptor A (9). This is called **cyclic photophosphorylation**, and results in ATP synthesis, but no production of NADPH. The differences between these two processes are summarised in Table 2.1.

| Non-cyclic photophosphorylation | Cyclic photophosphorylation |
|---|---|
| • involves both PSI and PSII | • involves only PSI |
| • electrons used to reduce chlorophyll derived from oxidation of water | • electrons used to reduce chlorophyll derived from, and returned back to P700 |
| • electrons from photosystems used to reduce NADP | • electrons returned back to P700, so no NADP reduced |
| • ATP, NADPH and oxygen formed | • only ATP formed |

*Table 2.1* Comparison of non-cyclic and cyclic photophosphorylation

**Extension**

### Box 2.3  ATP synthesis in the chloroplast

The source of energy for the synthesis of ATP is a gradient of protons (i.e. hydrogen ions, $H^+$). Protons are maintained at a high concentration inside the thylakoid lumen (Figure 2.17).

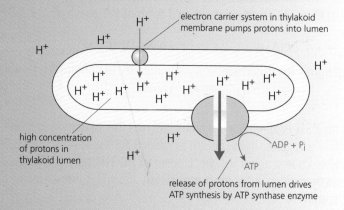

*Figure 2.17  Proton gradients across the thylakoid membrane drive ATP synthesis*

This high concentration of protons is set up partly as a result of the photolysis of water and partly by pumping of protons into the lumen of the thylakoids as the electrons pass along the carrier molecules in the membrane. The high concentration of protons inside the thylakoid lumen establishes both a concentration and an electrical gradient. This **electrochemical** gradient provides the potential energy to synthesise ATP. The protons are released from the lumen through special protein pores associated with an ATP synthase enzyme on the stroma side of the pore. A useful analogy is to compare the proton gradient with a dam of water. Opening the dam (allowing the protons out through the pore) releases the potential energy stored in the system. This is very similar to the process occurring in the mitochondrion which generates ATP (see Section 1.9, page 14).

**?**

**7** What will be the difference in pH between the inside and outside of the thylakoid membrane?                    *(1 mark)*

## 2.6  The light-independent reactions

The second stage of photosynthesis consists of the **light-independent reactions**. It takes place in the stroma of the chloroplast, and uses the products of the light-dependent reactions, NADPH and ATP, to fix carbon dioxide to carbohydrate. NADPH supplies the reducing power, and ATP the energy source.

The light-independent reactions incorporate a cyclical metabolic pathway called the **Calvin cycle** (Figure 2.18). This is named after Calvin and his co-workers, who established the sequence of enzyme-controlled reactions in the cycle (see Box 2.4).

Carbon dioxide diffuses through the air spaces of the leaf and dissolves in the watery layer surrounding the photosynthetic cells. It diffuses into the cells and enters the chloroplasts down a concentration gradient. In the stroma of the chloroplasts it combines with a 5C sugar called **ribulose 1,5 bisphosphate** (**RuBP**). This compound is known as the **carbon dioxide acceptor molecule**. The reaction is catalysed by the enzyme **ribulose bisphosphate carboxylase** (known by the acronym **rubisco**).

The product of this reaction is an unstable six-carbon (6C) compound which immediately breaks down into two 3C molecules of **glycerate 3-phosphate** or **GP** (sometimes called phosphoglyceric acid, or PG). This **carboxylation** reaction is the point at which carbon dioxide is actually fixed into an organic molecule. Both GP molecules are then phosphorylated using ATP, then reduced using NADPH, forming two molecules of **glyceraldehyde 3-phosphate**, a 3C or **triose phosphate**.

Some of the molecules of triose phosphate are used, along with more ATP, to regenerate the RuBP by a complex series of reactions, providing more of the carbon dioxide acceptor so that the cycle can continue.

*Figure 2.18  The Calvin cycle*

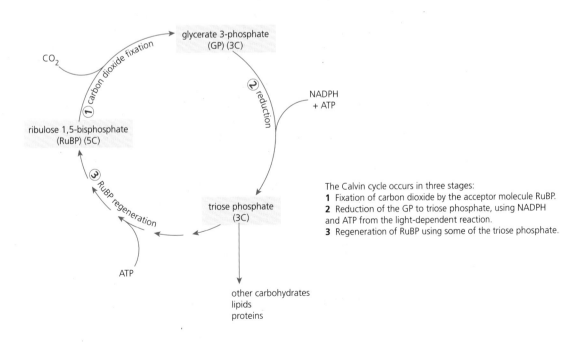

The Calvin cycle occurs in three stages:
1 Fixation of carbon dioxide by the acceptor molecule RuBP.
2 Reduction of the GP to triose phosphate, using NADPH and ATP from the light-dependent reaction.
3 Regeneration of RuBP using some of the triose phosphate.

❓

**8** Apart from triose phosphate, what is also needed to regenerate RuBP (see Figure 2.18)? *(1 mark)*

The rest of the triose phosphate is immediately converted into other organic compounds (see Section 2.7). For every six molecules of triose phosphate made, only one is used to make products of photosynthesis, while the other five are recycled to produce more RuBP.

## Box 2.4  Calvin's experiments

**Extension**

The reactions of the Calvin cycle were discovered by Melvin Calvin and his colleagues at Berkeley University in the United States, between 1946 and 1953. They grew cultures of the unicellular green alga *Chlorella* in a glass flask. This flask was flat in cross-section, which has led to it being nicknamed the 'lollipop' apparatus (Figure 2.19).

The algae were supplied with radioactive $^{14}C$ in the form of hydrogencarbonate ions. The cells were illuminated for short periods of time (of the order of seconds to minutes) and then killed by dropping them quickly into boiling methanol. This solvent extracted and preserved the radioactively labelled intermediate compounds, which were then separated and identified by two-way paper chromatography (see *AS Biology*, Section 3.22, page 92). Their positions on the chromatogram were identified by autoradiography. This technique involves

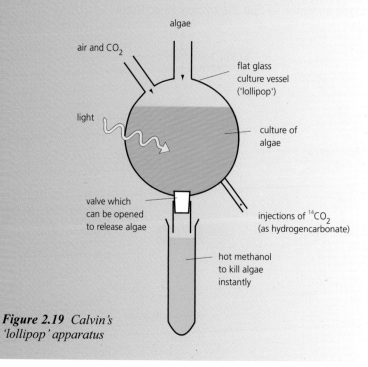

*Figure 2.19* Calvin's 'lollipop' apparatus

placing the chromatograms on an X-ray film for several days. The radioactive compounds show up as dark spots on the film.

Using this method, Calvin was able to show that after 2 seconds exposure to $^{14}$C, most of the radioactivity was associated with glycerate 3-phosphate. After 5 to 10 seconds hexose phosphates and organic acids were formed, and after 1 minute a wide range of compounds, including many organic acids, sugars and amino acids had been synthesised.

Calvin went on to identify all the intermediate stages in the cycle, and for this work was awarded a Nobel Prize in 1961.

## 2.7  Conversion of triose phosphate into other organic compounds

Triose phosphate is the starting point for many of the other organic compounds that the plant can produce as a result of photosynthesis (Figure 2.20).

By reversing some of the reactions of the glycolysis pathway of respiration (see Section 1.6, page 10) triose phosphate can be converted into 6C sugars such as glucose and fructose. Such monosaccharides can then be combined to form disaccharides, such as sucrose, which is the main sugar transported through the phloem of flowering plants. Monosaccharides can also be polymerised to form storage carbohydrates such as starch, or structural carbohydrates such as cellulose.

Triose phosphate can also be converted into pyruvate, from which it can be used to form fatty acids, one of the components of lipids such as triglycerides. Glycerol, the other component of triglycerides, is also easily formed from triose phosphate.

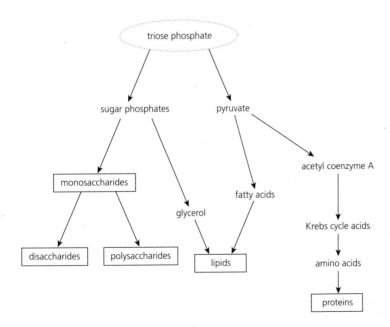

*Figure 2.20  Conversion of triose phosphate into other products of photosynthesis*

All of the above lipids and carbohydrates contain only the elements carbon, hydrogen and oxygen. The manufacture of amino acids, the building blocks of proteins, also needs the element nitrogen, and for some amino acids, sulphur. These are obtained by the plant from soil water in the form of the mineral salts nitrate and sulphate (see Section 2.8). The starting point for amino acid synthesis is the formation of acetyl coenzyme A from the triose phosphate. Acetyl coenzyme A enters the Krebs cycle, where it forms organic acids. These combine with ammonia, formed from the reduction of nitrates, to produce amino acids.

### 2.8 Plant mineral requirements

Plants require the element nitrogen for protein synthesis, as well as the formation of nucleic acids, coenzymes, vitamins and chlorophyll. They must therefore take up large quantities of nitrate from the soil, or from the surrounding water in the case of aquatic species. For this reason nitrate is known as a **macronutrient**. Without sufficient supplies of the mineral, a plant will show poor growth and yellowing of the leaves, a condition called **chlorosis**.

Phosphate is also absorbed by the plant in large quantities from the soil solution, mainly as hydrogen phosphate ($HPO_4^{2-}$) in normal soils or dihydrogen phosphate ($H_2PO_4^-$) in acid soils. It is first assimilated into organic form as ATP, during the metabolic pathways of respiration (see Chapter 1) and the phosphate group is subsequently incorporated into a large number of other compounds, including nucleic acids, sugar phosphates and phospholipids. A deficiency of phosphate in the soil also results in poor plant growth, particularly of the roots.

Other elements are needed in somewhat smaller quantities, for example magnesium, absorbed as ions ($Mg^{2+}$). This is a component of chlorophyll (see Figure 2.9, page 37) so that a magnesium ion deficiency may also lead to chlorosis of the leaves. Magnesium ions are also required as an **activator** of some enzymes, such as ATPase.

Plants also need a number of other mineral ions in very small amounts, such as copper, molybdenum, manganese and zinc. Each has particular functions, often acting as enzyme activators. Because they are needed in minute quantities they are called **micronutrients** or **trace elements**.

### Uptake of mineral ions

Just behind the apex of roots is a region covered by root hairs (Figure 2.21). These are long hair-like extensions of the outer epidermal cells of the root, penetrating between the soil particles. They increase the surface area of the root, and are the main point of entry for both water and mineral ions. Ions enter either passively, by diffusion, or by active transport (see *AS Biology*, Section 2.6, page 48). Which of these is the case for a particular ion depends upon its **electrochemical gradient**. In the case of uncharged solutes, uptake by a cell depends only on the concentration gradient of the solute, but movement of a charged ion depends on its concentration both inside and outside the cell, and also the electrical potential difference across the cell membrane. The electrochemical gradient is a combination of both these factors.

***Figure 2.21*** *Region behind root apex showing root hairs*

All anions such as nitrate, sulphate and phosphate enter root hair cells against their electrochemical gradient, and must therefore be taken up by active transport. Conversely most cations (e.g. sodium, calcium, magnesium) have an electrochemical gradient which favours their entry, which can therefore happen by passive diffusion (in fact they need to be moved *out* of the cells by active transport).

On entering the root, ions may remain in the cytoplasm or vacuoles of cells, or pass out to the cellulose cell walls. They then pass through the outer parts of the root via the **apoplast** or **symplast** pathways, following the routes taken by water (see *AS Biology*, Section 12.7 page 353).

9 Which pathway (apoplast or symplast) occurs through the cytoplasm and which along the cell walls? *(1 mark)*

At the **endodermis** of the root, the presence of the **Casparian strip** ensures that all ions are transported through living cells of the endodermis, mainly by active transport, and prevents them diffusing back out again. The ions then enter the xylem, and are carried to other parts of the plant (see *AS Biology*, Figure 12.17, page 355).

## 2.9 Limiting factors in photosynthesis

The concept of limiting factors was briefly described in *AS Biology* (Box 14.2, page 420) but is worth reviewing at this point, since we can link it with the biochemistry of photosynthesis.

We have seen that the metabolic pathways of photosynthesis consist of a number of reactions, some dependent on light and others light-independent. The rate at which the overall process proceeds depends upon the slowest reaction in the pathway. For example, if the carbon dioxide concentration in the air is very low, its fixation by RuBP will slow down the rest of the reactions. The carbon dioxide is then described as the **limiting factor** for the process. Another factor which might influence the rate is light intensity, through its effects on the light-dependent stage. In the dark this stage cannot occur. As the light intensity increases, more photosynthesis takes place, until some other factor (such as carbon dioxide levels) becomes limiting. Temperature affects the rate of all chemical reactions, and a low temperature may also be a limiting factor (for example on a sunny day in winter). The effects of these three limiting factors can be shown as graphs of photosynthesis rate against light intensity at different temperatures and carbon dioxide concentrations (Figure 2.22).

### The compensation point

Consider the case of a plant which has abundant carbon dioxide and high temperatures to allow photosynthesis. In this case an increase in light intensity will increase the rate of photosynthesis. In the dark, the plant will only respire, producing carbon dioxide. As the light intensity increases, more photosynthesis will take place, using up carbon dioxide. At some point, usually at quite a low light intensity, the amount of carbon dioxide produced by respiration will equal the amount used up by photosynthesis. This level of light is called the **compensation point** (Figure 2.23).

The compensation point of different plant species varies. Some plants have anatomical and physiological adaptations which allow them to photosynthesise efficiently in low light intensities. They are called **shade plants**. A good example is ivy, which is well adapted to low light conditions, and out-competes other plants in dark habitats, such as the floor of deciduous woodlands. Other plants require high light intensities, and are called **sun plants**.

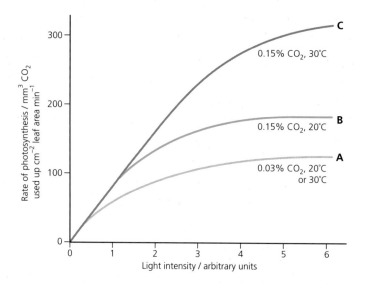

*Figure 2.22  Effects of light intensity on the rate of photosynthesis at different concentrations of carbon dioxide and temperature*

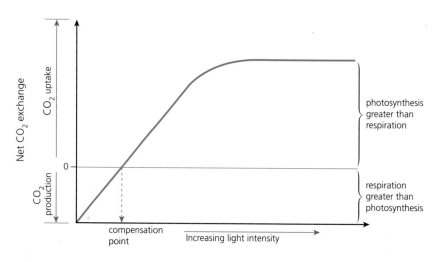

*Figure 2.23 Light intensity and the compensation point*

The compensation point is the light intensity at which there is no net exchange of $CO_2$.

## Summary – (2) Photosynthesis

- Photosynthesis is a metabolic pathway by which the inorganic compounds water and carbon dioxide are converted into carbohydrates, using light energy absorbed by chlorophyll.

- Leaves are adapted to carry out photosynthesis. Their adaptations include a large surface area, a thin cross-section, air spaces in the spongy mesophyll, stomata for gas exchange, palisade mesophyll with many chloroplasts, and xylem and phloem tissues for transport of water and products of photosynthesis.

- Chloroplasts are organelles adapted for photosynthesis. They contain membranes called thylakoids, where the photosynthetic pigments are located and where some of the initial reactions occur, and a matrix called the stroma, where the later stages of photosynthesis take place.

- 'Chlorophyll' is really a mixture of pigments, including chlorophylls and carotenoids. Each pigment absorbs light maximally at different wavelengths (as shown by the absorption spectrum) allowing photosynthesis to take place over a range of wavelengths (called the action spectrum).

- In the light-dependent reactions of photosynthesis, light energy is absorbed by chlorophyll and used to generate ATP and NADPH, using H atoms from water. Oxygen gas is released as a waste product.

- In the light-independent reactions, carbon dioxide is fixed by the acceptor molecule RuBP, and NADPH from the light-independent reactions, along with the ATP, used to reduce the product to triose phosphate.

- Triose phosphate is converted into a number of other substances, including monosaccharides, disaccharides, polysaccharides, lipids and proteins.

- Plants require mineral ions for the synthesis of many organic compounds, including amino acids, nucleic acids and chlorophyll. Some ions are absorbed by passive diffusion and others by active transport, mainly through root hair cells.

- The limiting factor in photosynthesis is the factor which is preventing the rate of the process from increasing. This may be carbon dioxide concentration, light intensity or temperature. The compensation point is the light intensity where the rate of a plant's respiration is equal to its rate of photosynthesis, and so no net loss or gain of carbon dioxide occurs.

**? Answers**

1   Photosynthesis is the reverse of respiration. The products of respiration (carbon dioxide and water) are the reactants in photosynthesis, and the products of photosynthesis (glucose and oxygen) are the reactants of respiration *(1)*.

2   A leaf is adapted for absorbing light. It is likely to absorb infra-red (heat) radiation well too *(1)*.

3   Black would be a 'better' light absorbing colour *(1)* although it would be more likely to overheat *(1)*.

4   The $R_f$ = 75 ÷ 112 = 0.67 *(1)*.

5   Differences:

   Oxidative phosphorylation occurs in mitochondria, photophosphorylation in chloroplasts; During oxidative phosphorylation, oxygen is the terminal H acceptor, producing water, during photophosphorylation water is split into oxygen and hydrogen; NAD is the first H acceptor of oxidative phosphorylation, NADP is the final H acceptor of photophosphorylation *(any 2)*.

   Similarities:

   Both derive energy from the flow of electrons along an electron transport chain; Both result in the synthesis of ATP; Both generate the energy for ATP synthesis from proton pumping *(any 2)*.

6   Two photosystems can raise the electrons to a higher energy level *(1)* for the reduction of NADP to NADPH *(1)*.

7   The inside of the thylakoid has more hydrogen ions, so the pH is lower (more acidic) *(1)*.

8   ATP *(1)*.

9   The apoplast pathway occurs along the cell walls and the symplast pathway through the cytoplasm *(1)*.

# End of Chapter Questions

**1** The photomicrograph below shows the structure of part of a leaf as seen in transverse section.

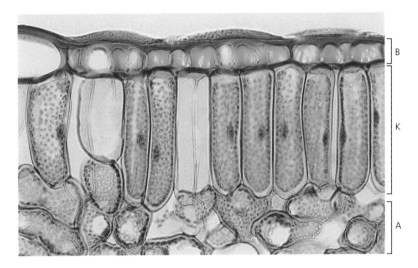

(a) Name the parts labelled A and B. *(2 marks)*

(b) Explain *two* ways in which the cells labelled K are adapted for their function. *(4 marks)*

*Edexcel 1997* *(Total 6 marks)*

**2** T.W. Englemann investigated the effect of different wavelengths of light on photosynthesis. He placed a filamentous green alga into a test tube along with a suspension of motile bacteria to use up the available oxygen and then illuminated the alga with light that had been passed through a prism to form a spectrum. After a short time, he observed the results shown in the diagram. Bacteria, which are indicated by the tiny rectangles, were evenly distributed throughout the test tube at the start of the experiment.

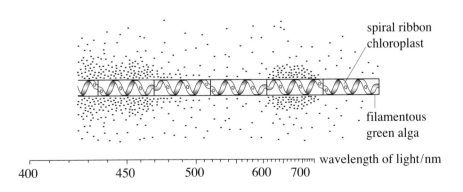

(a) With reference to the passage and to the diagram, sketch a graph to show how the rate of photosynthesis varies with the wavelength of light. Label the axes on your graph. *(2 marks)*

(b) Explain the reasons for the results observed in the diagram. *(4 marks)*

(c) Explain the role of the Calvin cycle in fixing carbon dioxide to form carbohydrates. *(8 marks)*

(In this question, 1 mark is available for the quality of written communication.)

*OCR 2000*                                                    *(Total 14 marks)*

**3** (a) The pigments present in a leaf may be separated by chromatography. The diagram shows a chromatogram of leaf pigments.

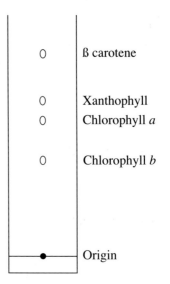

What further information, missing from the diagram, would you need in order to calculate the $R_f$ value of chlorophyll $a$? *(1 mark)*

(b) (i) During the light-dependent reaction of photosynthesis, ATP is produced. Describe one way in which this ATP is used in the light-independent reaction of photosynthesis. *(2 marks)*

(ii) When a solution of pure chlorophyll is illuminated, it gives off heat. No heat is released when the chlorophyll solution is kept in the dark. Suggest a reason for these observations. *(2 marks)*

*AEB 1998*                                                    *(Total 5 marks)*

**4** The diagram below summarises the biochemical pathways involved in photosynthesis.

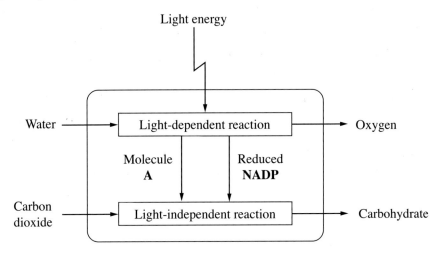

**(a)** Name molecule **A**. *(1 mark)*

**(b)** (i) Describe how NADP is reduced in the light-dependent reaction. *(2 marks)*

(ii) Describe the role played by reduced NADP in the light-independent reaction. *(2 marks)*

*AQA(B)* *(Total 5 marks)*

**5** The diagram shows the main stages in the light-independent reactions in photosynthesis.

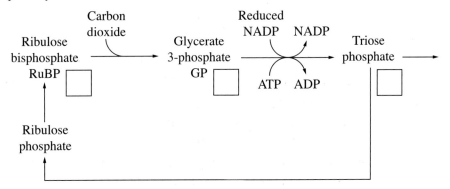

**(a)** Write in the boxes in the diagram the number of carbon atoms in each of the relevant substances. *(1 mark)*

**(b)** What is the role of ATP in the conversion of:

(i) glycerate 3-phosphate to triose phosphate; *(1 mark)*

(ii) ribulose phosphate to ribulose bisphosphate? *(1 mark)*

**(c)** A plant was allowed to photosynthesise normally. The light was then switched off. Explain why there was a rise in the amount of glycerate 3-phosphate present in the chloroplasts of this plant. *(2 marks)*

*AQA(A)*                                                  *(Total 5 marks)*

**N3.1**  **6**  The diagram shows the rate of carbon dioxide exchange (uptake or release) for two species of plant, A and B. The horizontal line X–X shows the carbon dioxide production due to respiration.

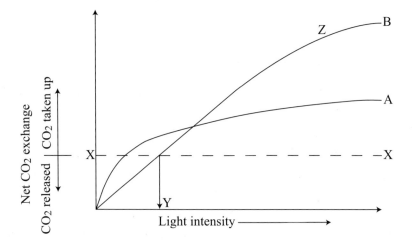

**(a)** For species B, what is the point on the graph labelled Y called? *(1 mark)*

**(b)** What is the significance of this point? *(1 mark)*

**(c)** Explain why the rate of uptake of carbon dioxide for species B is decreasing at point Z on the graph. *(2 marks)*

**(d)** Which plant, A or B, is likely to be a shade plant? *(1 mark)*

**(e)** Explain your answer to (d). *(1 mark)*

*(Total 6 marks)*

Genetics is the study of inheritance. It explains why organisms inherit certain features from their parents and what features may be passed on to their offspring. **Genes** from parents determine the genetic makeup or **genotype** of their offspring.

Genes may exist as different **alleles**. A gene is found at a particular site on the chromosome called the **locus.** As chromosomes normally exist in pairs, there are two gene loci and therefore two alleles determining most features. The characteristics we display, our **phenotype**, depends both on our genes and the environment. Features like blood group, tongue-rolling ability and eye colour largely depend on alleles, but other features like height result from the interaction of genes with the environment.

**This chapter includes:**
- monohybrid crosses
- human genetics
- codominance
- dihybrid crosses
- multiple alleles
- sex linkage
- gene interaction
- genes versus the environment
- chi-squared test.

## 3.1 Terms used in genetics

There are many terms used in genetics which you need to know. Several will already be familiar to you:

- **Genetics** – the study of inheritance, controlled by genes.
- **Gene** – a section of DNA which codes for a particular protein.
- **Alleles** – different forms of the same gene, e.g. height in pea plants is controlled by the two alleles **T** and **t**.
- **Genome** – the total genetic make up of an organism.
- **Dominant alleles** – alleles that always appear in the phenotype (they are expressed), shown by capital letters, e.g. **TT** and **Tt** both display the dominant feature.
- **Recessive alleles** – alleles that are only expressed if both alleles of a pair are recessive, shown by lower case, e.g. **tt**.
- **Genotype** – the genetic makeup of an organism, made up of alleles which determine a particular feature.
- **Phenotype** – the features resulting from expression of the genes and their interaction with the environment.
- **Homozygote** – two alleles of a pair are the same, e.g. **TT** or **tt**.
- **Heterozygote** – two alleles of a pair are different, e.g. **Tt**.
- **Codominance** – two different alleles of a heterozygote are both expressed in the phenotype, e.g. a cat with a black allele (**B**) and a yellow allele (**Y**) for coat colour is tortoiseshell (a mixture of black and yellow).
- **Multiple alleles** – there are more than two types of allele for a particular gene, e.g. blood groups involve three alleles, **A**, **B** and **O**.
- **Sex-linked genes** – genes found on the sex chromosomes, usually on the X chromosome.
- **Autosomal genes** – genes found on all chromosomes apart from the X and Y chromosomes.

- **Polygenes** – a single feature determined by several genes, e.g. height in humans.
- **Gene mapping** – a technique used to locate the position of a gene on a chromosome.
- **Chi-squared** – a statistical test used to find out whether offspring phenotypes fit an expected ratio.

**Box 3.1  Johann Gregor Mendel**

**Figure 3.1** *Johann Gregor Mendel, 1822–1884*

The Austrian monk Johann Gregor Mendel, 1822–1884, is known as the father of genetics. Mendel was born in Moravia and, after becoming a monk, studied mathematics and natural history for 2 years at the University of Vienna. Mendel joined an Augustinian monastery at Brno, now in the Czech Republic, as he needed financial support. He studied theology at Brno and became a priest, but he was not successful in the parish. Mendel attempted but failed the exams necessary to be a science teacher and returned to the monastery in 1853 to concentrate on his breeding experiments. Over the next 8 years, Mendel carried out his now famous experiments on inheritance in the garden pea, *Pisum sativum*.

As a result of his painstaking and thorough work, we now understand the laws by which genes are inherited. The garden pea was an excellent choice for study as there are many varieties of peas with clear differences. Fortunately the features he studied have a genetic basis with alleles that are either dominant or recessive, which simplified the interpretation. As pea plants are normally self-pollinating, with the petals completely enclosing the reproductive parts, Mendel could control the breeding process with little or no interference from wind or insects. The process involved a lot of dissection of the flowers and removal of the anthers to prevent self-pollination, enabling Mendel largely to control the crosses that took place. The garden pea is easy to grow and large numbers of seeds are produced, so increasing the reliability of the findings.

Mendel presented his findings in 1865 and a year later they were published. Unfortunately his results were largely ignored until the early 1900s, when his findings were re-discovered, confirmed and extended.

**Extension**

**3.2  Monohybrid inheritance**

This is the inheritance of *one* character, e.g. seed type in peas. Pea seeds are either round or wrinkled, depending on the alleles present. The gene locus for a particular characteristic is found at the same position on homologous chromosomes, so a diploid organism will possess two alleles for each character in each diploid cell.

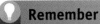

**Remember**

**Summary of meiosis and variation**

A knowledge of meiosis is essential to an understanding of genetics (see *AS Biology*, Section 6.11, page 169). Meiosis is the nuclear and cell division that results in the production of gametes. Two divisions take place. In the first division, separation of homologous pairs of chromosomes occurs and in the second division, the chromatids are separated. Consequently, in the gametes only one allele is present from each pair. The gametes are haploid – they contain only one chromosome from each homologous pair. When gametes fuse at fertilisation, the diploid number is restored by the homologous chromosomes pairing up.

Meiosis increases variation due to crossing over and independent assortment. (A further increase in variation, not directly related to meiosis, includes mutations, which may take place, and the random fusion of unique gametes at fertilisation which follows meiosis.) Genetic variation increases the chance of survival of a population if the environment changes. For example, the appearance of a new virus may kill some organisms but others may be genetically resistant to the virus and survive. If all the organisms were genetically identical, they might all be killed. The new conditions select those variants of a species most adapted to the conditions. Some individuals are better suited to their environment due to their differing genes. Those individuals are described as 'fitter' and are more likely to survive and pass on their useful genes. This is called 'survival of the fittest'. As a result of this selection, the remaining individuals will be different from those suited to previous conditions. The change in a species over time is called evolution. Species that are not suited to the conditions will not survive, their numbers will decrease and they may die out and become extinct.

The allele for round seed, **R**, is dominant to the allele for wrinkled seed, **r**. There are three possible genotypes for seed type (Table 3.1 and Figure 3.2).

| Genotype | RR | Rr | rr |
|---|---|---|---|
| **Description** | homozygous dominant | heterozygous | homozygous recessive |
| **Phenotype** | round | round | wrinkled |

*Table 3.1  Genotypes and phenotypes for seed type*

*Figure 3.2 One pair of autosomes to show the different alleles possible for seed type in pea plants*

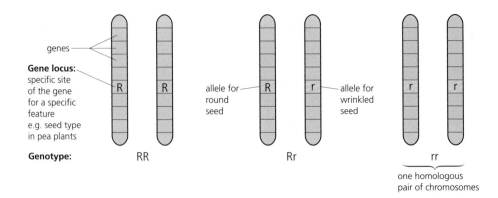

If the two alleles of a pair are the same, the organism is said to be **pure breeding** for that feature. When pure breeding pea plants with round seeds, **RR**, are crossed, they produce the same round seeds generation after generation.

A genetic diagram can be drawn to show the inheritance of round seeds in pure breeding varieties (Figure 3.3). This type of diagram is called a Punnett square (after a geneticist called Punnett).

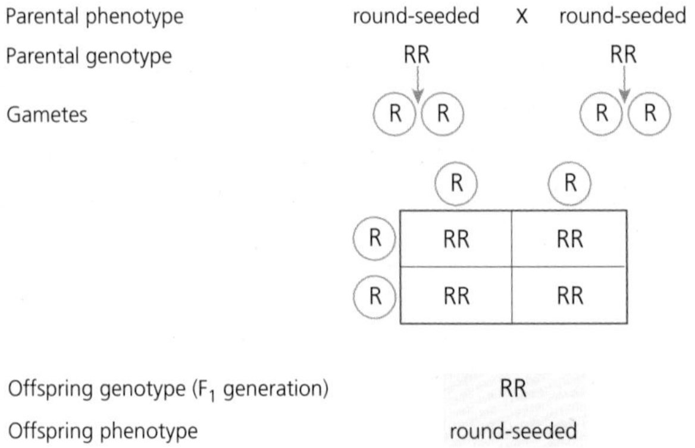

*Figure 3.3 Genetic diagram to show the inheritance of round seeds in pure breeding (homozygous) varieties*

| Parental phenotype | round-seeded | X | round-seeded |
| Parental genotype | RR | | RR |

Gametes

| | R | | R |

|  | R | R |
|---|---|---|
| R | RR | RR |
| R | RR | RR |

| Offspring genotype (F₁ generation) | RR |
| Offspring phenotype | round-seeded |

### Remember

The term $F_1$ refers to the offspring of homozygous parents, an example clearly shown in Figure 3.4. $F_2$ refers to the offspring produced by crossing the $F_1$ (see Figure 3.5).

However, if we cross a pure breeding (homozygous) round-seeded plant with a pure breeding wrinkled-seeded variety, differences will appear later (Figure 3.4).

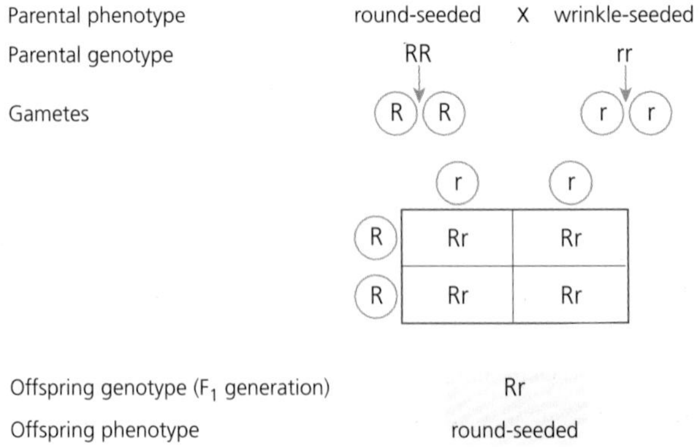

| Parental phenotype | round-seeded | X | wrinkle-seeded |
| Parental genotype | RR | | rr |

Gametes

|  | r | r |
|---|---|---|
| R | Rr | Rr |
| R | Rr | Rr |

| Offspring genotype (F₁ generation) | Rr |
| Offspring phenotype | round-seeded |

*Figure 3.4 Crossing a homozygous round-seeded with a homozygous wrinkled-seeded variety*

The offspring ($F_1$) from Figure 3.4 all have round seeds but each plant is **Rr**, i.e. it carries both a dominant and a recessive allele. In other words the plants are heterozygous for seed type.

If two of these offspring are crossed, the results will be as shown in Figure 3.5.

*Figure 3.5* *Crossing the F$_1$* *(also known as selfing the F$_1$)*

| F$_1$ phenotype | round-seeded | X | round-seeded |
|---|---|---|---|
| F$_1$ genotype | Rr | | Rr |

Gametes (R)(r)   (R)(r)

|  | R | r |
|---|---|---|
| R | RR | Rr |
| r | Rr | rr |

Offspring genotypes (F$_2$ generation)   RR : 2Rr : rr

Offspring phenotypes   3 round-seeded : 1 wrinkle-seeded

**Remember**

Numbers rarely fit a ratio precisely as fertilisation is a random event and the fusion of gametes is down to chance. Also some organisms fail to grow, while others may die or be eaten. To be confident that numbers fit a particular ratio, it is necessary to use a statistical test called a **chi-squared** test to check whether numbers are significantly different from those expected (see Section 3.11, page 94).

The classic **3:1 ratio** is found in the F$_2$ of a monohybrid cross when two heterozygotes are crossed ($\frac{3}{4}$ of the total are round seeds, $\frac{1}{4}$ of the total are wrinkled seeds). In fact Mendel reported counting 5474 round seeds and 1850 wrinkled seeds, which approximates to a 3:1 ratio.

This shows that the allele for wrinkled peas must be present in the F$_1$ generation. It does not show in the phenotype as the allele is recessive to the round allele present. This demonstrated to Mendel that there must be two alleles determining each feature as the recessive allele must have been present in the F$_1$ to subsequently appear in the F$_2$ as the homozygote.

If we now cross a heterozygote with a recessive homozygote, a different ratio occurs (Figure 3.6). This is called a backcross or **test cross**. A test cross can be used to find out unknown genotypes.

*Figure 3.6* *Crossing a heterozygous round-seeded plant with a wrinkled-seeded variety*

| Parental phenotype | round-seeded | X | wrinkle-seeded |
|---|---|---|---|
| Parental genotype | Rr | | rr |

Gametes (R)(r)   (r)(r)

|  | r | r |
|---|---|---|
| R | Rr | Rr |
| r | rr | rr |

Offspring genotypes   Rr : rr

Offspring phenotypes   1 round : 1 wrinkled

A **1:1 ratio** is found when a heterozygote is crossed with the recessive homozygote in a monohybrid cross. If the original round-seeded parent had been RR, then all the offspring would have round seeds. The 3:1 ratio in the F$_2$ of a monohybrid cross and the 1:1 ratio in the backcross are characteristic of monohybrid crosses.

Extension

### Box 3.2  Gregor Mendel and his wrinkled peas

Wrinkled peas are the result of a mutation in the gene which normally produces the enzyme responsible for converting sugar to starch. As a result, these mutant peas make less starch, and more carbohydrate remains as sugar instead. The extra sugar lowers the water potential and more water is taken into the seed by osmosis as it develops. When the seed is dried, this extra water is lost and wrinkling occurs. Normal peas which have less sugar and therefore less water in them, initially lose less water when dried and therefore do not wrinkle (Figure 3.7).

*Figure 3.7  Round and wrinkled peas*

### 3.3  Human genetics

In the study of human genetics, we cannot decide which individuals should mate in the cause of research, so we must rely on family trees to investigate the patterns of inheritance. Fortunately, records are extensive and it is possible to trace the inheritance of certain characteristics over several generations. Dominant alleles are immediately obvious as they normally appear in each generation, while recessive alleles become apparent when two 'normal' parents have a child that displays an 'abnormal' characteristic, thus indicating that both parents are carriers of a particular feature (see Box 3.3).

Extension

### Box 3.3  Dominant and recessive human alleles

| Dominant | Recessive |
| --- | --- |
| brown eyes | blue eyes |
| pigmented skin | albinism (no melanin) |
| Roman nose | straight nose |
| curly hair | straight hair |
| non-red hair | red hair |
| free ear lobes | attached ear lobes |
| achondroplasia (dwarfism) | normal height |
| tongue rolling | tongue non-rolling |

## Cystic fibrosis

Cystic fibrosis is an inherited, autosomal (not sex-linked) recessive disorder affecting one in 2000 white people in Britain. Sufferers therefore have the recessive homozygous genotype. Using **D** to represent the allele for the healthy condition, i.e. not suffering, and **d** to represent the cystic fibrosis allele, there are three possible genotypes (Table 3.2).

| Genotype | DD | Dd | dd |
|---|---|---|---|
| **Phenotype** | healthy | healthy but a carrier | sufferer |

*Table 3.2  Possible genotypes and phenotypes in the inheritance of cystic fibrosis*

Carriers are healthy as they have a dominant allele, but carry the recessive allele and may pass it on to their children.

Sufferers of cystic fibrosis produce excessive amounts of mucus, in their bronchi and bronchioles, (see *AS Biology*, Section 9.5, page 239), in the pancreatic duct, and in the testes in males. This leads to blocked airways and infection in the lungs, and incomplete digestion in the duodenum as pancreatic enzymes are unable to reach the food. Males with excess mucus in the testes may be sterile.

This genetic disorder is caused by a deletion mutation on chromosome number 7. As the affected section of DNA is shorter than the healthy section, this condition can be detected by electrophoresis, as the sufferer's DNA produces a smaller fragment which travels further along the gel (see *AS Biology*, Box 16.5, page 499).

Figure 3.8 shows the inheritance of cystic fibrosis in one family. The appearance of the disorder in a child of unaffected parents is typical of autosomal recessive conditions and indicates that both the parents must be carriers of the condition. To suffer from cystic fibrosis, a child must receive a recessive allele from both parents, but the parents do not display symptoms as the dominant allele is also present. Therefore the parents must be heterozygous for the condition. This shows that the allele must be recessive or it would be expressed in the parents.

It is therefore possible for two parents without the condition to have a child with cystic fibrosis.

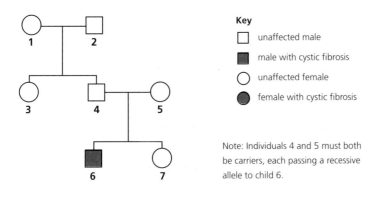

**Key**

☐ unaffected male

■ male with cystic fibrosis

○ unaffected female

● female with cystic fibrosis

Note: Individuals 4 and 5 must both be carriers, each passing a recessive allele to child 6.

 **Did You Know**

**'Scientists finish first draft of DNA blueprint' June 26 2000**

Scientists have just finished deciphering the human genome. The three billion letter DNA blueprint, only involving four different letters, will be posted on the internet to enable researchers to unravel the causes of disease. Scientists have before them a book of letters from which genes may be deciphered. The sequences of thousands of genes are already known, but there are many more that are not. Identifying the genes for inherited diseases will increase our understanding and hopefully lead to treatment, if not elimination, of some inherited disorders. However, relatively few diseases are determined solely by one gene and there are already tests for some of these. Some inherited disorders may also be affected by the environment – our knowledge is far from complete.

*Figure 3.8  A family tree showing the inheritance of cystic fibrosis*

*Figure 3.9* Result of a cross between two heterozygous (or carrier) parents (parents 4 and 5 from Figure 3.8)

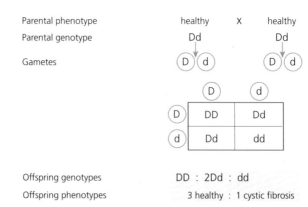

| | | |
|---|---|---|
| Parental phenotype | healthy X | healthy |
| Parental genotype | Dd | Dd |

Offspring genotypes     DD : 2Dd : dd
Offspring phenotypes     3 healthy : 1 cystic fibrosis

Two healthy parents who are carriers therefore have a 1 in 4 or 25% chance of having a child with cystic fibrosis (Figure 3.9). Again this is a classic 3:1 ratio found in the $F_2$ of a monohybrid cross.

If two heterozygous parents have three children who are healthy, it does not mean that a fourth child will definitely be a sufferer, as the fusion of gametes is a random event. Even if a couple have six children, it does not mean that they are bound to have an affected child, but every time they have a child there is a 25% chance that their child will have cystic fibrosis. It is interesting that the incidence of cystic fibrosis is higher in some populations than in others. In Britain, it is particularly common amongst the white population.

It has recently been discovered that carriers of cystic fibrosis, **Dd**, who clearly do not suffer from cystic fibrosis, have resistance to typhoid. Typhoid is an intestinal disease caused by the bacterium, *Salmonella* sp. Therefore there is a selective advantage in being a carrier for cystic fibrosis in areas of the world where typhoid is present. This is called **heterozygous advantage**. (see Box 4.4, page 128). Resistance will increase the proportion of carriers and lead to an increase in the frequency of the harmful cystic fibrosis allele in the population. This may account for the high incidence of cystic fibrosis in certain areas.

There is significant research in the treatment of this condition using somatic cell gene therapy to produce an aerosol spray containing the healthy gene (see *AS Biology*, Section 16.11, page 493).

### Huntington's disease

This autosomal mutation is caused by a dominant allele. Only one dominant allele needs to be present to be expressed, therefore sufferers who are heterozygous for the condition (having a dominant and a recessive allele) will be affected.

Huntington's disease (previously known as Huntingdon's chorea) affects 1 in 10 000 people and is characterised by progressive lack of muscular coordination and mental deterioration. In fact the brain shrinks by about a quarter. Uncontrollable shaking and dance-like movements combined with dementia, slurring of speech and personality changes are typical symptoms. The symptoms do not usually appear until middle age, when sufferers already may have children who have inherited the condition. As it is caused by a dominant allele, the chance of children inheriting the condition from an affected parent is 50% or 1 in 2.

## Box 3.4 Maple syrup urine disease

Maple syrup urine disease (MSUD) is an autosomal recessive disorder. The gene mutation results in a defective enzyme unable to carry out the normal function of deamination and transamination of the essential amino acids leucine, isoleucine and valine. The concentration of these amino acids builds up in sufferers and interferes with other metabolic pathways, leading to progressive degeneration of nerves and early death.

The name comes from the burnt sugar smell of the urine of sufferers, like maple syrup. If untreated, apparently healthy infants die soon after birth. Early detection, preferably before birth, involves testing the blood of the possible sufferer for the amino acids leucine, isoleucine and valine. High concentrations of up to 10 times the normal level indicates MSUD. The condition can be treated successfully with a controlled diet, limiting the amount of proteins containing the damaging amino acids.

MSUD is rare in the general population, affecting only about one in 20 000 births, but it is much more common in particular communities. In the Old Order Mennonite communities of Pennsylvania, USA, the incidence is closer to one in 150 births. It is likely that in this community, all the sufferers and carriers of MSUD inherited the allele from one common ancestor and that they are in fact related. The high incidence is due to their lifestyle and religion, which isolates them from the outside world. This increases the chance of one carrier marrying another from the community, which would be rare in the outside world. This is an example of the **founder effect**, with an original or founding member of the community carrying the recessive allele, combined with inbreeding, leading to a high incidence of the harmful allele. The chance of two parents, both carriers, having an affected child is 1 in 4 or 25%.

Figure 3.10 shows the inheritance of Huntington's disease in one family.

Typical of the pattern of an autosomal dominant allele is the appearance of the condition in every generation. Every affected person has an affected parent (this is not the case with recessive alleles like that for cystic fibrosis). The age of onset varies, but symptoms most commonly start between 40–50 years of age and from then onwards life expectancy averages 15 years.

The gene responsible was found on chromosome 4 in 1993. The mutation codes for a protein called a 'huntingin' whose function is unknown. Non-sufferers have a segment of DNA with a triplet base sequence of CAG repeated 11–35 times. In sufferers, this is repeated 39–100 times, and is referred to as a 'stutter' in the gene. The greater the number of repeats, the earlier the disease appears. The age at which the symptoms appear depends on the number of repeats. Forty repeats lead to symptoms starting at the age of 59, with 42 repeats leading to symptoms at 37 years of age and 50 repeats leading to symptoms at about 27 years of age.

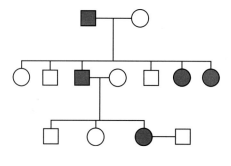

**Key**

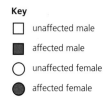

unaffected male

affected male

unaffected female

affected female

***Figure 3.10*** *The inheritance of Huntington's disease*

> **Remember**
>
> In autosomal *recessive* features:
>
> - the condition appears rarely unless it confers selective advantage
> - the condition may not appear in every generation
> - affected individuals may not have an affected parent.
>
> In autosomal *dominant* features:
>
> - the condition appears in every generation
> - if a generation does not show the condition, it will not appear in later generations
> - all affected individuals have an affected parent.

A healthy lifestyle has no effect on this certain outcome. In a study of Huntington's disease in Long Island, USA, 12 generations of a family were investigated and more than a thousand members were affected. They all descended from two brothers who emigrated to America from Suffolk in 1630.

Genetic testing can identify those who will develop the disease, but many children of affected parents prefer not to know whether they will also develop the condition. Genetic counselling both before and after testing is essential.

**1** There is clearly no biological advantage in having Huntington's disease. Suggest why the number of people with this condition has remained fairly constant. *(2 marks)*

## 3.4 Codominance

When two different alleles of a single gene are both expressed in the same phenotype, they are called **codominant**. If only one gene is involved it is monohybrid inheritance.

This is quite different from the dominant–recessive situation described so far. One example is found in **Andalusian fowl**, whose feathers can be black, 'splashed white' or blue (Figure 3.11). The colour of the feathers is controlled by two alleles which are codominant. Breeders favour the blue colour.

The allele for black feathers is represented by **B**, and that for 'splashed white' feathers by **W**. (In this case capital letters are used for both alleles to indicate that the alleles are equally dominant.) This produces three phenotypes:

**BB** = black feathers
**WW** = 'splashed white' feathers
**BW** = blue feathers.

*Figure 3.11 The three types of Andalusian fowl*

Figure 3.12 shows the possible offspring resulting from a mating between a black and a 'splashed white' bird.

*Figure 3.12*  *A cross between two Andalusian fowl*

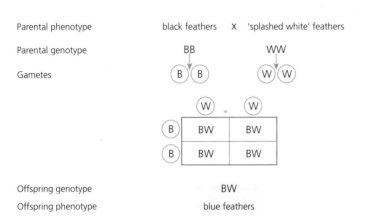

| | | |
|---|---|---|
| Parental phenotype | black feathers    X | 'splashed white' feathers |
| Parental genotype | BB | WW |
| Gametes | B  B | W  W |

| | W | W |
|---|---|---|
| B | BW | BW |
| B | BW | BW |

Offspring genotype                     BW
Offspring phenotype                blue feathers

Whenever a third phenotype appears in offspring, which is different from both parents, codominant alleles may be involved.

**2** What offspring would you expect and in what ratio, if you mated two birds with blue feathers?

*(1 mark)*

*Figure 3.13*  *Variation in colour of snapdragons*

**Snapdragon plants** can have red, white or pink flowers (Figure 3.13). If a red plant is crossed with a white plant, all the offspring are pink. This third phenotype indicates codominance between the two alleles for red and white.

The alleles can be expressed by the following symbols:

**R** = allele for red flower

**W** = allele for white flower.

Possible genotypes:

**RR** = red flower

**WW** = white flower

**RW** = pink flower.

If a red flowered snapdragon (**RR**), is crossed with a snapdragon with white flowers (**WW**), all the offspring have pink flowers (**RW**).

The interesting results obtained when two pink snapdragons are crossed are illustrated in Figure 3.14.

*Figure 3.14 Crossing two pink snapdragons*

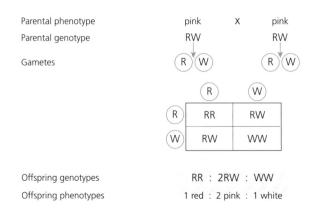

| Parental phenotype | pink | X | pink |
|---|---|---|---|
| Parental genotype | RW | | RW |

Gametes   (R)(W)     (R)(W)

| | R | W |
|---|---|---|
| R | RR | RW |
| W | RW | WW |

Offspring genotypes          RR : 2RW : WW
Offspring phenotypes        1 red : 2 pink : 1 white

The cross produces a **1 to 2 to 1 ratio** in the offspring – 50% of the offspring should be pink, 25% red and 25% white.

**3** What genotypes and phenotypes would you expect if you crossed a pink snapdragon with a white one?        *(2 marks)*

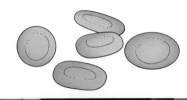

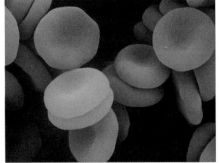

*(a)*

### Sickle cell anaemia

Another example of codominance is seen in the inheritance of sickle cell anaemia. This is caused by a gene mutation (see *AS Biology*, Section 5.11, page 138) which alters the structure of haemoglobin. The faulty gene is recessive and in the homozygous state causes 'sickling' of the red blood cells (making them sickle shaped).

Sickle cell anaemia usually leads to early death, particularly in the first two years of life. About 4% of all babies born in East and West Africa suffer from the disease. In Britain, an estimated 5000 people suffer from sickle cell anaemia, most of whom are of Afro-Caribbean origin. Figure 3.15 shows the appearance of normal and sickled red blood cells.

The abnormal haemoglobin has reduced solubility which causes it to crystallise and distort the red blood cell into a crescent or sickle shape, instead of the familiar biconcave disc shape. Many red blood cells are destroyed. In addition, the sickle shape reduces the amount of oxygen that is carried to the tissues, causing anaemia.

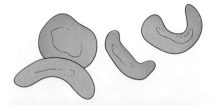

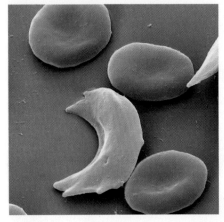

*(b)*

*Figure 3.15 Diagrams and photomicrographs of*
*(a) normal red blood cells*
*(b) sickled and normal red blood cells*

The distorted red blood cells stick together, causing blockages in the narrow capillaries and severe pain, particularly in the joints (see *AS Biology*, Section 5.13, page 144).

Table 3.3 shows possible genotypes and phenotypes. **Hb^A** represents the allele for normal haemoglobin; **Hb^S** represents the allele for abnormal haemoglobin.

**Table 3.3** *Possible genotypes and phenotypes for haemoglobin*

| Genotype | Hb^AHb^A | Hb^AHb^S | Hb^SHb^S |
|---|---|---|---|
| Description | normal red blood cells | normal and slightly sickled red blood cells | sickled red blood cells |
| Phenotype | healthy | sickle cell trait – mild anaemia | sickle cell anaemia – severe anaemia |

The heterozygous condition, **Hb^AHb^S**, has both normal and slightly distorted red blood cells. It is therefore an example of codominance, as the influence of both alleles is evident in the phenotype of the heterozygote, with about 40% of the haemoglobin affected. The person is generally healthy and only suffers from relatively mild anaemia under normal conditions. The main advantage of the heterogous condition is the resistance to malaria which it confers. Therefore in malarial parts of the world the proportion of those with sickle cell anaemia and those heterozygous for the condition are far higher than in non-malarial regions. Unfortunately, this means that it is likely that two individuals with the genotype **Hb^AHb^S** may have a child, each with a 1 in 4 chance of suffering from sickle cell anaemia (Figure 3.16).

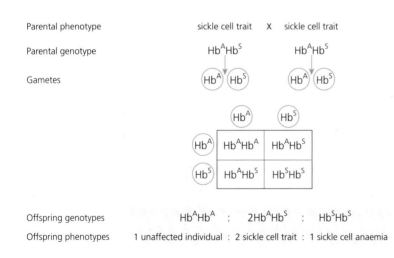

**Figure 3.16** *A cross between two heterozygous parents, both carriers of the sickle cell allele (sickle cell trait)*

From the cross, 50% of the offspring are likely to have the sickle cell trait and be resistant to malaria. They have the genotype **Hb^AHb^S**. Twenty-five percent will be homozygous for the normal allele, **Hb^AHb^A**, and 25% will suffer from sickle cell anaemia, with the genotype **Hb^SHb^S**. Although it is not part of a continuous variable, the selective advantage of the heterozygote can be seen as a form of stabilising selection (see Section 4.7, page 126).

## 3.5 Dihybrid inheritance

This is the inheritance of two characteristics. If the genes are found on different chromosomes, we describe the two genes as unlinked. To understand the mechanism of dihybrid inheritance it is easier to consider a specific example.

Pea plants vary in height and in the type of seeds they produce. These features are genetically determined. The allele for tall pea plants, **T**, is dominant to the allele for short pea plants, **t**. Round seeds, **R**, are dominant to wrinkled seeds, **r**. The possible genotypes and phenotypes for height are:

| Genotype: | **TT** | **Tt** | **tt** |
|---|---|---|---|
| Phenotype: | tall | tall | short |

Pea plants also have one of the following genotypes and phenotypes for seed type:

| Genotype: | **RR** | **Rr** | **rr** |
|---|---|---|---|
| Phenotype: | round | round | wrinkled |

The genes determining these features are carried on different chromosomes. In each case there are two alleles determining the feature. There is one allele at each gene locus on each chromosome of a homologous pair (Figure 3.17).

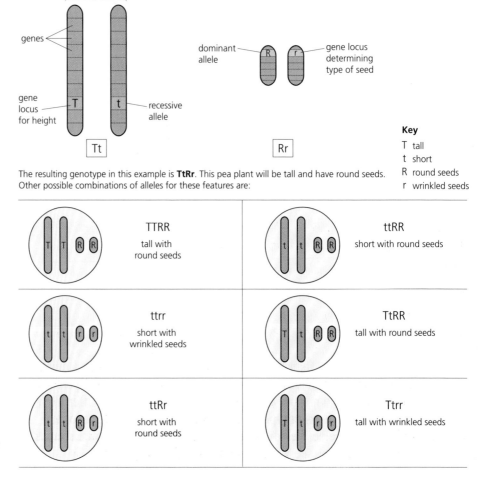

*Figure 3.17 Two pairs of homologous chromosomes and their alleles*

Therefore a tall pea plant with round seeds, could have the genotype **TTRR**, **TtRR**, **TTRr** or **TtRr**. In each example, at least one dominant allele for each feature is present, so both dominant features are expressed.

A short pea plant with round seeds could have the genotype **ttRr** or **ttRR**, as the recessive feature is determined by the homozygous recessive **tt**, and the dominant allele, **R**, results in round seeds. Similarly, a tall pea plant with wrinkled seeds could have the genotype **TTrr** or **Ttrr**.

The only certain genotype is that of a short plant with wrinkled seeds which must be double recessive – **ttrr**, as no dominant alleles can be present.

## A dihybrid cross

If one parent is homozygous for tall and round seeds and the other is short with wrinkled seeds, all the offspring will have the dominant features and be tall with round seeds. Figure 3.18 shows a cross between a pure breeding parent with the dominant features and one with the recessive features.

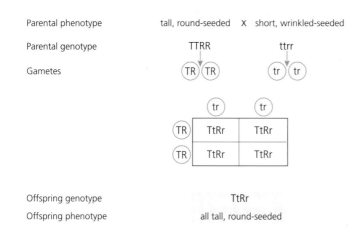

*Figure 3.18 A dihybrid cross between a pure breeding parent with dominant features and a homozygous recessive parent*

All the offspring from this cross will be heterozygous for both conditions (**TtRr**), as both dominant and recessive alleles of each gene are present. They will be tall with round seeds.

## Gamete genotypes

Remember that in meiosis the chromosome number is halved so the gametes only receive one chromosome from each of the homologous pairs. In pea plants, therefore, the gametes will only receive one chromosome with one allele for height (**T** or **t**), and one chromosome with one allele for type of seed, (**R** or **r**). The four possible gamete types from the $F_1$ genotype (**TtRr**) are **TR**, **Tr**, **tR** or **tr**. The allele combination depends on independent assortment, (see *AS Biology*, Section 6.11, page 174), during prophase of meiosis. Statistically the gametes should appear in equal ratios, 1:1:1:1. However, independent assortment is a random event and this can lead to unequal numbers of each type of gamete.

At fertilisation, homologous pairs of chromosomes are restored, so that the cells of the zygote have pairs of alleles again, e.g. **TR** and **TR** combine to form **TTRR**. If a **TR** gamete joins with a **tr** gamete, then the resulting zygote will have the genotype **TtRr**, displaying both dominant characteristics, and indistinguishable from the homozygous dominant individual. Figure 3.19 shows the results of a cross between two of the heterozygous F$_1$ individuals (selfing).

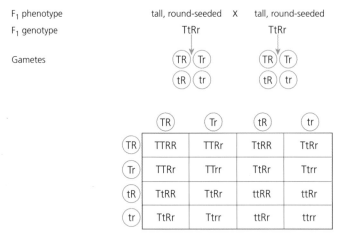

*Figure 3.19  Crossing two of the F$_1$ individuals from Figure 3.18*

The phenotypes show the classic **9:3:3:1 ratio** expected with unlinked genes in a dihybrid cross of two double heterozygotes. As with the monohybrid cross, a **backcross** or **test cross** is a means of establishing the genotype when it is unknown. This involves crossing the unknown genotype with the double recessive, whose genotype is certain. As we cannot cross humans of specific genotypes to order, we have to restrict genetic experiments to other animals or plants. So, back to peas!

### Use of a test cross to establish genotype

Pea plants may be tall or short and have round or wrinkled seeds. To determine the genotype of a tall, round-seeded pea plant, a test cross is carried out with the double recessive, the short, wrinkled-seeded plant. A tall, round-seeded pea plant may have four possible genotypes – **TTRR**, **TTRr**, **TtRR** or **TtRr**. A short, wrinkled-seeded plant can only have one possible genotype – **ttrr**. In each of the crosses shown in Figure 3.20 a tall, round-seeded plant is crossed with a short, wrinkled-seeded variety, but the genotype of the tall, round-seeded plant differs in each case.

The only possible gamete type from **ttrr** is **tr** (one of each type of allele). From the tall, round seeded plant, the gamete types vary depending on the genotype of the parent. Therefore, from the phenotypes of the offspring it is possible to establish the original parent's genotype.

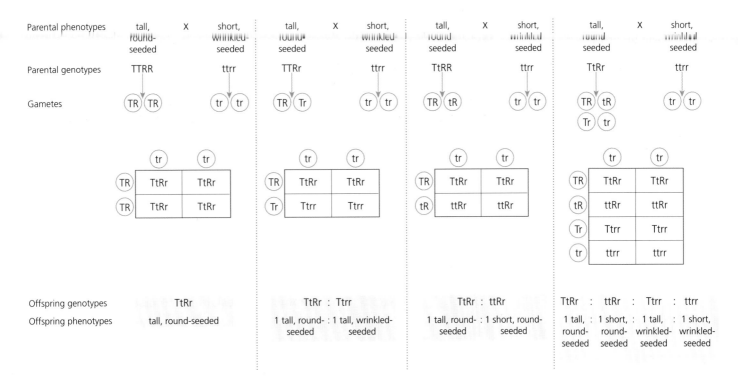

Figure 3.20 *Test crosses in peas*

## 3.6  Multiple alleles

Many genes have more than two alleles, although only two can occupy the gene locus on a pair of homologous chromosomes at any one time in any one individual.

### Human ABO blood groups

Human ABO blood groups are determined by multiple alleles. There are three possible alleles available, **A**, **B** and **O**, represented by the symbols $I^A$, $I^B$ and $I^O$.

The alleles **A** and **B** are codominant, so both are expressed in the genotype $I^AI^B$. Alleles **A** and **B** are both dominant to the allele **O**, which is recessive. There are four possible blood groups as shown in Table 3.4.

It is possible for parents, one of whom is blood group A and the other blood group B, to produce any of the four blood groups in their offspring if both parents are heterozygotes (Figure 3.21).

| Blood group phenotype | Possible genotypes |
|---|---|
| A | $I^AI^A$, $I^AI^O$ |
| B | $I^BI^B$, $I^BI^O$ |
| AB | $I^AI^B$ |
| O | $I^OI^O$ |

Table 3.4 *Genotypes for human ABO blood groups*

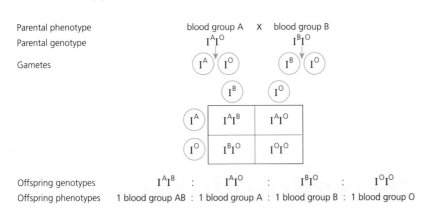

Fig 3.21 *Results of a cross between a person who is heterozygous for blood group A with another who is heterozygous for blood group B*

**?**

4 Four babies were born in the same hospital on the same day to different parents. Owing to a power cut their identities were confused before they could be given to their parents. (Nowadays DNA testing would be used to establish parentage.) The babies' blood groups were A, B, AB and O. The blood groups of the parents were:

AB  Mr and Mrs Darwin – A and B

B  Mr and Mrs Pasteur – B and O

O  Mr and Mrs Mendel – O and O

A  Mr and Mrs Fleming – AB and O.

Which baby should each couple take home? Explain your answer. (*4 marks*)

**Extension**

**Box 3.5  Blood groups and resistance to disease**

Human blood groups seem to be connected to susceptibility to cholera. People with type O blood are most susceptible to infection from cholera, whereas there is a dominance hierarchy of likely infection for the other blood types. The blood type AB confers the greatest resistance, then group A followed by group B, with group O having the least resistance. People of blood type AB are so resistant to cholera that they are virtually immune to the disease. Cholera is caused by the *Vibrio* bacterium, which settles in the gut causing diarrhoea except in those fortunate people who are blood type AB. Therefore in areas with cholera, individuals with type AB blood are more likely to survive to pass on their alleles, **A** and **B**. This maintains the **B** allele in the population as **AA** is more resistant than **BB** but the heterozygote, **AB**, is the most cholera resistant.

It is interesting that the **O** allele survives even though it provides almost no resistance to cholera. It appears that people who are type O blood seem to be a little more resistant to malaria than the other blood groups and this is probably responsible for keeping the **O** allele in the population.

**Multiple alleles and dominance hierarchy**

Multiple alleles can also form a dominance hierarchy as in this example of dog coat colour. Three alleles exist for determining coat colour: $A^s$ produces a uniform dark colour; $a^y$ produces a tan coat colour; $a^t$ produces a spotted coat.

$A^s$ is dominant to $a^y$, which is dominant to $a^t$, i.e. $A^s > a^y > a^t$. The possible genotypes and phenotypes for coat colour are as follows:

- $A^sA^s$, $A^sa^y$, $A^sa^t$ – all with *dark* coat as $A^s$ is dominant
- $a^ya^y$, $a^ya^t$ – all with *tan* coat, as $a^y$ is dominant to $a^t$
- $a^ta^t$ – *spotted* coat.

A dog with dark fur is mated with a dog with tan fur. Some of their puppies have dark fur, some have tan fur and some have spotted fur (Figure 3.22).

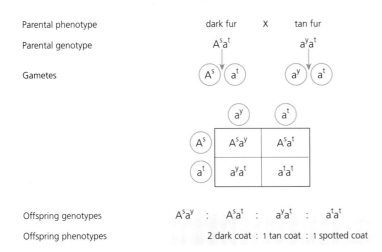

| Parental phenotype | dark fur | X | tan fur |
|---|---|---|---|
| Parental genotype | $A^s a^t$ | | $a^y a^t$ |

Offspring genotypes $A^s a^y$ : $A^s a^t$ : $a^y a^t$ : $a^t a^t$

Offspring phenotypes    2 dark coat : 1 tan coat : 1 spotted coat

*Figure 3.22  Crossing a dark-coated dog with a tan-coated dog*

It is possible to deduce the genotype of the parents from their offspring. This particular cross produces a 1:2:1 ratio, one tan to two dark to one spotted.

**5** Figure 3.23 shows a family tree of dog coat colour. The phenotypes are given. Work out the genotypes for each dog. *(3 marks)*

In rabbits there is a series of alleles for coat colour, again with a dominance hierarchy. The top dominant allele is $E^D$ and gives a black coat. The allele $E$ gives a brown coat and the recessive allele $e$, a yellow coat. Figure 3.24 shows the result of a cross between a black rabbit of genotype $E^D e$ and a brown rabbit of genotype $Ee$.

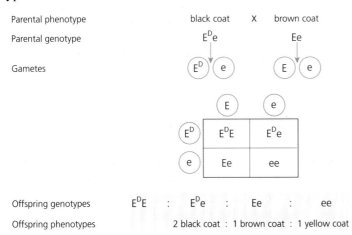

| Parental phenotype | black coat | X | brown coat |
|---|---|---|---|
| Parental genotype | $E^D e$ | | $Ee$ |

Offspring genotypes $E^D E$ : $E^D e$ : $Ee$ : $ee$

Offspring phenotypes    2 black coat : 1 brown coat : 1 yellow coat

This produces a phenotypic ratio in the offspring of 2 black, 1 brown and 1 yellow, i.e. 2:1:1.

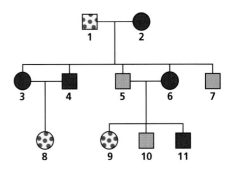

**Key**
- dark male
- dark female
- tan male
- tan female
- spotted male
- spotted female

*Figure 3.23  A family tree of coat colour in dogs (not sex linked)*

*Figure 3.24  Coat colour in rabbits*

**?**

**6** In repeated crosses between a black and a brown rabbit, the following offspring were obtained.

| Cross | Black | Brown | Yellow |
|-------|-------|-------|--------|
| 1 | 18 | 0 | 0 |
| 2 | 13 | 14 | 0 |
| 3 | 15 | 7 | 8 |

For each cross, state the possible genotypes of the parents. *(3 marks)*

## 3.7 Sex chromosomes

In humans there are 23 pairs of chromosomes, of which one pair is known as the sex chromosomes. The other 22 pairs are known as autosomes. Males have an X and a shorter Y chromosome making up their sex chromosomes, whilst females have two X chromosomes. Whether we are male or female depends on the sex chromosomes we inherit from our father.

A **karyotype** displays all the chromosomes present. Karyotype analysis involves collecting a cell sample and stimulating the process of mitosis. Once the cells have been allowed to divide several times, colchicine is added. This chemical inhibits spindle formation and the process of mitosis is stopped. Without a spindle the chromosomes remain at metaphase, as there are no spindle fibres to pull the chromatids apart. A hypotonic saline solution is added, causing the chromosomes to swell and become more visible. A chromosome stain such as acetic orcein is then added, making the chromosomes clearly visible. The stained chromosomes are then photographed under the microscope.

Individual chromosomes are cut from the photograph and paired up according to size. The longest pair of homologous chromosomes are pair number one. The chromosomes are paired and numbered in order of decreasing size to produce a karyotype. The 23rd pair are the sex chromosomes.

Figure 3.25 shows karyotypes from three different individuals; a normal male and female, and a female with Down's syndrome. Note the sex chromosomes of the male – XY, and female – XX. The condition known as Down's syndrome is caused by the presence of an extra chromosome number 21, which can be observed by studying a karyotype (see Box 4.1, page 107 for more details).

The homologous pairs of chromosomes have come from each parent, one of each pair from the mother and one from the father. All gametes contain one of the two sex chromosomes which pair up at fertilisation to form a pair of sex chromosomes, either XX or XY. The X chromosome does look X-shaped but the Y chromosome looks more like a squashed X.

**Remember**

**Summary of stages involved in producing a karyotype**

- collect human cells and stimulate mitosis
- add colchicine to stop spindle formation
- add hypotonic saline to make the chromosomes swell
- use acetic orcein to stain the chromosomes
- photograph the chromosomes, which should be stuck at metaphase
- cut the photograph to pair up the chromosomes and number them in order of decreasing size.

**?**

**7** Why are human red blood cells unsuitable for producing a karyotype?

*(1 mark)*

### Gender determination in humans

Females produce ova by meiosis, each ovum containing half the chromosome number of the parent (23 instead of 46 in humans). Only one chromosome from each homologous pair is present in each ovum, including one X chromosome. Females, XX, are called the **homogametic sex** as they produce only one kind of gamete, all with an X.

Males also produce sperm by meiosis. This leads to two types of sperm, half with an X chromosome and half with a Y chromosome. Males, XY, are therefore called the **heterogametic sex** as they produce two types of gametes. (*Homo*gametic means that gametes are the *same*, all with an X chromosome; *hetero*gametic means that gametes are *different*, some have an X chromosome and others a Y chromosome.) Figure 3.26 shows sex determination in humans.

Statistically the ratio of males to females produced is 1:1, but this is only a theoretical ratio as the fusion of gametes is random. However, every time a woman gives birth, there is a 50% chance that the child will be a girl and a 50% chance it will be a boy.

Figure 3.26 shows that the sex of the child depends on the type of sperm. If the X chromosome of the ovum joins a sperm with an X chromosome, a girl is produced, whereas if it joins a sperm with a Y chromosome, a boy results.

### Sex-linked characteristics

Genes which are found on the sex chromosomes are described as sex linked. The small Y chromosome has very few alleles on it, so most sex-linked genes are located on the X chromosome. If alleles are present on the X chromosome which are unmatched by alleles on the Y chromosome, then the phenotype in males will be determined by these alleles. Whatever unmatched alleles are present on the X chromosome in males will be expressed, whether dominant or recessive.

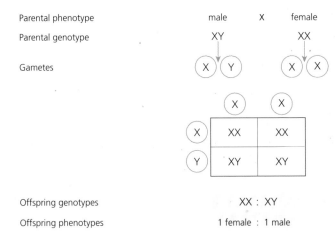

**Figure 3.26** *Sex determination in humans*

*Figure 3.25 Human karyotypes*
*(a) normal female,*
*(b) normal male,*
*(c) female with Down's syndrome*

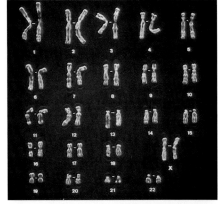

*(a)*

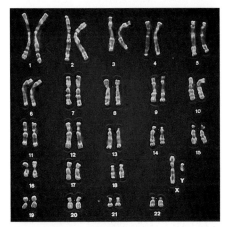

*(b)*

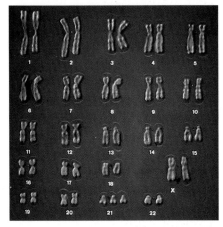

*(c)*

# ③ Genetics

In females, with two X chromosomes, the phenotype is normally dependent on dominance. A recessive feature will only be expressed in females if recessive alleles are present on both X chromosomes.

## Haemophilia

Haemophilia is an example of a recessive, sex-linked, genetic disorder. About one in 20 000 of the population of Europe suffer from this disease, including about 0.1% of the males in Britain. Haemophilia affects the ability of the blood to clot. In healthy people, clotting prevents the loss of blood and the entry of pathogens such as bacteria, and is part of the body's defence mechanism (see *AS Biology*, Section 11.8, page 324).

Haemophilia is caused by a recessive gene mutation which prevents the formation of the protein involved in blood clotting and so leads to prolonged bleeding in those affected. This harmful recessive allele is found on the X chromosome.

Bleeding both externally and internally into joints and muscles leads to severe blood loss. People suffering from haemophilia may need blood transfusions for minor injuries to replace the excessive blood lost, and injections to replace the missing clotting factor. Unfortunately, contaminated blood has caused the spread of HIV to some sufferers. As the allele is found on the X chromosome, haemophilia can theoretically affect both males and females, although it is far more common in males.

Males need only inherit one recessive allele to suffer, $X^hY$. Females inheriting just one harmful recessive allele, $X^HX^h$, are unaffected and are therefore carriers as the other allele will be the dominant allele for normal blood clotting.

Boys normally inherit the recessive allele from an unaffected mother who is a carrier, and a healthy father. It is therefore possible for two healthy parents to have a child with haemophilia (Figure 3.27).

> ### Did You Know
>
> Even before Mendel, people recognised some patterns of inheritance. The Talmud, which is the main source of Jewish law, specifies that if a boy dies of bleeding following circumcision, his younger brothers and male cousins on his mother's side should not undergo the ritual. An understanding that 'bleeding' runs in families and is passed on from mothers was therefore recognised long before Mendel.

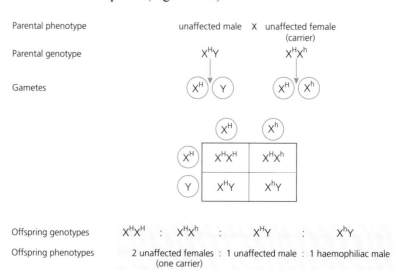

*Fig 3.27 Inheritance of haemophilia from two healthy parents*

Females are much less likely to suffer from haemophilia as a dominant allele will prevent the condition. To inherit haemophilia, a girl would have to inherit a recessive allele from both parents, i.e. she would have to be the daughter of a

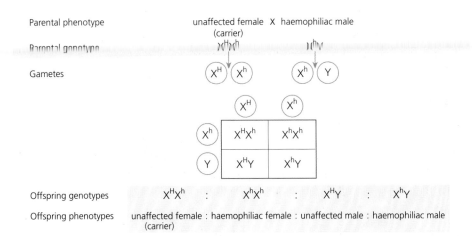

**Figure 3.28** *Inheritance of haemophilia resulting in a haemophiliac female*

haemophiliac father and a carrier mother. This is possible but very unlikely (Figure 3.28). Only females with two recessive alleles will be sufferers, $X^hX^h$. This condition is usually fatal in utero and results in a miscarriage. Regular menstruation would also lead to major problems and in reality female haemophiliacs are extremely rare.

Queen Victoria (1819–1901) was a carrier for haemophilia and therefore unaffected by the condition herself. Figure 3.29 shows the family tree of Queen Victoria, mapping the inheritance of haemophilia in six of her nine children.

**Figure 3.29** *Family tree of Queen Victoria and the inheritance of haemophilia in the royal families of Europe*

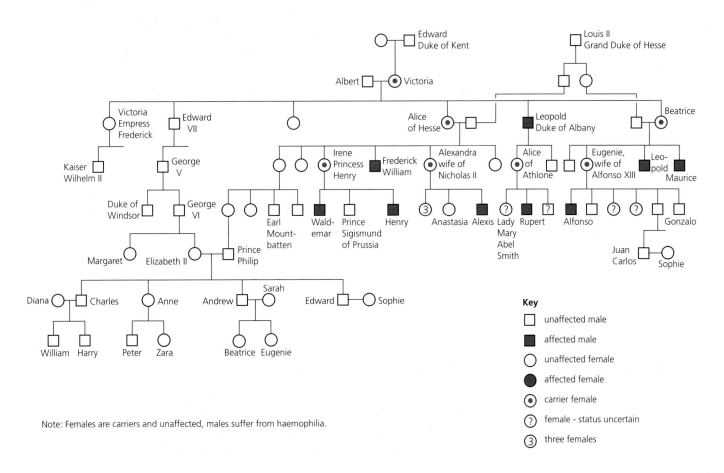

Note: Females are carriers and unaffected, males suffer from haemophilia.

*Figure 3.30* Queen Victoria (seated) with her husband, prince Albert (left) and their nine children (from left) Alfred, Helena, Alice, Arthur, Beatrice (baby), Victoria, Louise, Leopold and Edward. Alice and Beatrice are known to have been carriers of haemophilia, and Leopold suffered from the disorder

In the 19th century Britain was a superpower and British princesses were sought after for marriage. Unfortunately Queen Victoria, who had nine children, passed on the harmful recessive allele to her descendants, so spreading the recessive allele around the royal families of Europe.

The family tree in Figure 3.29 shows how the disease mainly affects males but is transmitted by female carriers. This pattern is typical of sex-linked, recessive disorders. Of Queen Victoria's nine children, one son, Leopold, suffered from haemophilia and two daughters were carriers. Leopold died after a minor fall led to a major haemorrhage. His daughter was inevitably a carrier while his son was clear of the condition, having inherited a Y chromosome from his haemophiliac father.

Queen Victoria's daughter Alice was also a carrier, with three children inheriting the recessive allele. Alice's son Frederick had haemophilia and died at the age of three after falling out of a window and bleeding to death. Alice's daughter Alexandra married Nicholas, who became the last Tsar of Russia. Unfortunately, their son Alexis inherited the recessive allele from his mother and suffered from haemophilia. Victoria's son Edward VII was unaffected by the condition, and as a result the present British royal family are clear of the disease.

Although haemophilia often leads to early death, the frequency of the harmful allele has not decreased in the population as it causes no harm in the heterozygous state and is a recurring mutation.

## Box 3.6 The fate of the last Russian royal family, the Romanovs

The Russian royal family, Nicholas, Alexandra and their children, were sent to Siberia in July 1918 at the start of the Russian revolution. After the Communists had seized power the Romanovs, together with some of their servants, were shot and buried in a nearby forest. In July 1991 nine skeletons – six adults and three children – were found in a shallow grave in the Urals. The gold teeth and porcelain fillings in their teeth suggested that these were aristocrats. Forensic scientists examined the skeletons and noted the signs of violence to which they had been subjected. It was suspected that these were the remains of the Romanovs, Tsar Nicholas II, his wife Alexandra, three of their daughters and some servants.

*Figure 3.31 The Romanovs, Tsar Nicholas II, his wife Alexandra (granddaughter of Queen Victoria) and their children*

Later, with the collapse of communism, it was decided that a state funeral would be appropriate but their identities had to be confirmed first. Analysis of the DNA in the ancient bones made this identification possible. From the DNA, the sex of the person can be established. Using gel electrophoresis, a genetic fingerprint can be produced to show whether the skeletons are related to each other and to any living members of the same family. The genetic fingerprint showed that the skeletons included a mother, father and their three children. The other skeletons were unrelated. To establish their precise identity involved finding any possible living relations of the Tsar. The Duke of Edinburgh, Prince Philip, who is related to the Russian royal family through his mother (see family tree) agreed to provide some DNA for testing.

Mitochondrial DNA sequences from the skeletons and from Prince Philip were compared (mitochondrial DNA is only inherited through the mother). If two people have the same sequence then they are probably related. The DNA sequences from the mother and her three children were identical to that from Prince Philip, so confirming their relationship. The sequence from the Tsar matched matrilineal descendants from his grandmother.

As a result of these tests the identity of the Romanovs was confirmed and they were given a state funeral in July 1998 in St Petersburg, more than 80 years after their death. The fate of the other children is uncertain, but may be confirmed by DNA analysis in due course, if further skeletons are unearthed. In August 2000 Nicholas was canonised, becoming Saint Nicholas.

**?**

**8** Explain why the sons of a haemophiliac man and a healthy (non-carrier) woman would never inherit haemophilia? *(1 mark)*

Other human sex-linked conditions found on the X chromosome include red-green colour blindness and Duchenne's muscular dystrophy. Red-green colour blindness is caused by a recessive allele found on the X chromosome. Figure 3.32 shows how red-green colour blindness affects vision.

*Figure 3.32 Photographs demonstrating colour blindness. Individuals with normal colour vision see colours as in the left-hand photograph. Colour-blind individuals see colours as in the right-hand photograph*

Duchenne's muscular dystrophy is far more common than haemophilia, affecting 0.25% of boys. This sex-linked, recessive condition affects muscle development and, like haemophilia, mainly affects boys, who inherit the allele from their carrier mothers.

### Y-linked alleles

Rarely, sex-linked genes may be found on the Y chromosome so only males are affected, although there are no absolute examples of Y-linked inheritance. Hairy ears are thought to be determined by a sex-linked allele found on the Y chromosome but this has not been confirmed. This trait is common in certain populations, such as parts of India, Pakistan and Israel. If the gene is found only on the Y chromosome, then it should only be found in men and must be inherited by all their sons, who inherit the Y chromosome from their father and the X chromosome from their mothers. A study of hairy ears in Israel showed that 76% of sons inherited the trait from their fathers and none of the daughters. This seems to confirm that this is a Y-linked gene, but why didn't all the sons develop hairy ears?

It is possibly an example of **epistasis** (see Section 3.8, page 85), where one gene is inhibited or suppressed by another.

9  Explain why sons will always inherit the hairy ear allele from their
   fathers?                                                    *(1 mark)*

10 Why would females be unaffected by sex-linked genes found on the Y
   chromosome?                                                 *(1 mark)*

## Coat colour in cats – codominance and sex-linkage

A particularly interesting example of sex linkage controls coat colour in cats.
This involves both sex linkage and codominance. One of the genes that controls
coat colour in cats has its locus on the X chromosome. In cats, as in humans,
females are the homogametic sex (XX) and males are the heterogametic sex
(XY).

As the allele is only found on the X chromosome, females have two alleles for
coat colour, one on each X chromosome, whilst males only have one.

The allele $X^o$ produces orange fur while the allele $X^b$ produces black fur. These
two alleles exhibit codominance, as both are expressed in the heterozygote,
which produces tortoiseshell cats having patches of orange fur and black fur
(Figure 3.33).

Table 3.5 lists all the possible genotypes and phenotypes for coat colour.

|              | Genotype | Phenotype |
|--------------|----------|-----------|
| **Female cats** | $X^oX^o$ | orange fur |
|              | $X^oX^b$ | orange and black fur (tortoiseshell) |
|              | $X^bX^b$ | black fur |
| **Male cats** | $X^oY$ | orange fur |
|              | $X^bY$ | black fur |

**Table 3.5** *Possible genotypes and phenotypes for coat colour in cats*

To be tortoiseshell, cats need to have both $X^o$ and $X^b$ alleles present. This is
possible in females which have two XX chromosomes, but not in males with
only one X chromosome. Therefore all tortoiseshell cats are females.

**Figure 3.33** *Coat colour in cats*

## Fruit flies

In the fruit fly, *Drosophila melanogaster*, the gene for eye colour is sex linked. The allele for white eyes (**r**) is recessive to the allele for the normal red eye colour (**R**). Sex-linked genes are carried on the X chromosome. In fruit flies, the female has two X chromosomes (XX) and males have an X and a Y chromosome (XY).

When a red-eyed female is crossed with a white-eyed male, red- and white-eyed flies of both sexes are produced. However, when a white-eyed female is crossed with a red-eyed male the offspring produced are quite different. All the females have red eyes and all the males have white eyes. This is explained in Figure 3.34.

*Figure 3.34 Sex linkage in* Drosophila

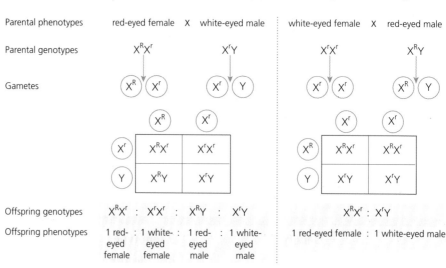

| | red-eyed female X white-eyed male | white-eyed female X red-eyed male |
|---|---|---|
| Parental phenotypes | red-eyed female X white-eyed male | white-eyed female X red-eyed male |
| Parental genotypes | $X^R X^r$ — $X^r Y$ | $X^r X^r$ — $X^R Y$ |
| Gametes | $X^R$ $X^r$ — $X^r$ $Y$ | $X^r$ $X^r$ — $X^R$ $Y$ |

Cross 1:
| | $X^R$ | $X^r$ |
|---|---|---|
| $X^r$ | $X^R X^r$ | $X^r X^r$ |
| $Y$ | $X^R Y$ | $X^r Y$ |

Cross 2:
| | $X^r$ | $X^r$ |
|---|---|---|
| $X^R$ | $X^R X^r$ | $X^R X^r$ |
| $Y$ | $X^r Y$ | $X^r Y$ |

Offspring genotypes: $X^R X^r$ : $X^r X^r$ : $X^R Y$ : $X^r Y$ | $X^R X^r$ : $X^r Y$

Offspring phenotypes: 1 red-eyed female : 1 white-eyed female : 1 red-eyed male : 1 white-eyed male | 1 red-eyed female : 1 white-eyed male

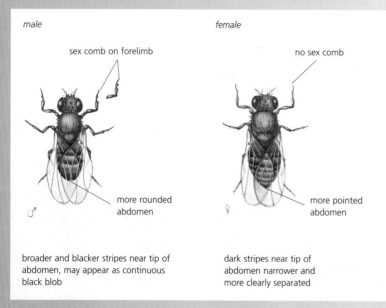

*male*

sex comb on forelimb

more rounded abdomen

♂

broader and blacker stripes near tip of abdomen, may appear as continuous black blob

*female*

no sex comb

more pointed abdomen

♀

dark stripes near tip of abdomen narrower and more clearly separated

*Figure 3.35 Male and female* Drosophila, *showing distinguishing features*

### Box 3.7 Choice of species for genetic crosses

*Drosophila* is an ideal choice for genetic study. This insect is easy to breed and keep due to its small size and short life cycle, about 10 days at 25 °C. Female *Drosophila* produce large numbers of offspring – about 100 eggs at a time. This allows statistical analysis of phenotypes to show significant differences. In addition, the fruit flies have clear recognisable differences with a genetic basis and males can easily be distinguished from females. Females can therefore be separated from males while still virgins to ensure that only the required matings take place (Figure 3.35).

*Figure 3.36* *The temperature of the eggs determines the sex of the young crocodiles*

## 3.8  Gene interaction

Sometimes a phenotype depends on the interaction of two or more different genes. There are two main types of gene interaction:

- epistasis
- polygenic inheritance.

### Epistasis

Here, one gene masks or suppresses the action of another gene. Epistatic genes are commonly called **inhibiting genes**. The genes involved are found at different gene loci, they are not alleles of the same gene.

A classic example of epistasis is found to determine coat colour in mice. Coat colour is not sex linked. Most wild mice have a grey coat with a banded appearance described as **agouti**. Other mice have a black or white coat (Figure 3.37).

*Figure 3.37* *The three types of coat colour in mice – agouti, black and white*

| Genotype | Phenotype |
|---|---|
| AAMM, AAMm, AaMM, AaMm | agouti |
| aaMM, aaMm | black |
| aamm, Aamm, AAmm | albino |

*Table 3.6 Possible genotypes and phenotypes for coat colour in mice*

One gene controls the distribution of melanin pigment. The **A** allele produces an uneven distribution of melanin, resulting in a banded appearance (agouti). The **a** allele produces an evenly black coat. A second gene, with the alleles **M** and **m**, determines whether melanin is produced at all.

The dominant allele **M** is needed to form any melanin, while the allele **m** prevents the production of melanin, leading to albino mice (white):

**A** = agouti, **a** = black

**M** = pigment (melanin), **m** = no pigment (albino).

Therefore only mice with the **M** allele will produce a pigment; those with the **m** allele will be albinos, regardless of the presence of the **A** or **a** allele. The possible genotypes and phenotypes are shown in Table 3.6.

**11** If two agouti mice, both with the genotype **AaMm**, are crossed, what genotypes and phenotypes would you expect in their offspring, and in what ratio? *(4 marks)*

### Did You Know

Mink are small mammals with high-quality fur. The breeding of mink used to be highly profitable as mink coats were very popular. Twelve different pairs of alleles all at different gene loci interact to influence fur colour and patterns. Mink farmers bred for desired combinations of these alleles in the mink to increase their profits. The great variety of pelts was possible due to gene interaction.

*Figure 3.38 Mink*

### Polygenic inheritance

Polygenic means *many genes*. A polygenic feature like height in humans, is determined by many genes acting together. The genes are found at different gene loci but have a cumulative effect on the phenotype.

If a person inherits many height genes all coding for tall, then the person will be very tall. Another person with many height genes all coding for short will be very short. Most people, however, will inherit a mixture of 'tall' and 'short' genes leading to an average height. The many possible combinations of alleles explains how a range of heights is possible in the population from very short to very tall. The feature is described as showing a **continuous variation** (see Section 4.1, page 105).

In addition, height is also affected by the environment. A starving child will not grow as tall as her better fed identical twin even though they have the same genes.

## 3.9  Genes interacting with the environment

The main effect on the phenotype is the genotype. However, external environmental factors can influence the phenotype coded by genes, but the result of environmental influence is not inheritable. For example, two plants with identical genotypes may end up with dissimilar appearances if their environments are different. If plant A receives more light than genetically identical plant B, it will grow taller and stronger due to increased photosynthesis. The difference in size will not be inherited and therefore the size difference is not directly related to evolution.

The relative influence of the environment on the phenotype differs from one characteristic to another. For example, human height is more affected by the genotype, whilst human weight will be more affected by the environment. This is further complicated by the polygenic basis of human height (see Section 4.1, page 106).

Experiments with *Potentilla glandulosa*, a small plant which grows at a range of altitudes in the mountains of California, have also contributed to our understanding of the interaction between genes and the environment (Figure 3.39).

Two sets of plants were used. In one group genetically identical plants were grown in different conditions and in the other group, non-identical plants were grown in identical conditions. Individual plants were collected from three locations all at different altitudes above sea level – plant A from high altitude, plant B from medium altitude and plant C from a low level. The different plants varied due to their differing genotypes and experienced different environmental conditions at each location.

A plant from each location was separated into three cuttings, all genetically identical, and a cutting was planted at each of the three altitudes, high, medium and low levels, so experiencing different environmental conditions.

Each location therefore had a cutting from plant A, B and C, so the plants were genetically different but experienced the same environmental conditions. Figure 3.40 Compares the effect of genotype and the environment on the growth of *Potentilla glandulosa*. Each vertical column has the same environmental conditions and plants in each horizontal row have the same genotype.

In a particular environment, for example at high altitude, different genotypes produce different phenotypes. Plants in the same row with the same genotype display differing phenotypes in the different environments. This is called **phenotypic plasticity** and is particularly apparent in plants.

All these results confirm that the phenotype is the result of interaction between genes and the environment, with the genotype determining the range of phenotypic possibilities.

*Figure 3.39* Potentilla glandulosa

**Did You Know**

Rabbits and Siamese cats develop dark tips to their nose, ears, paws and tail only at low temperatures. This 'Himalayan' colouring is caused by an allele which is only activated by low temperatures, found at the extremities of the animal. In other words, it is an environmental trigger.

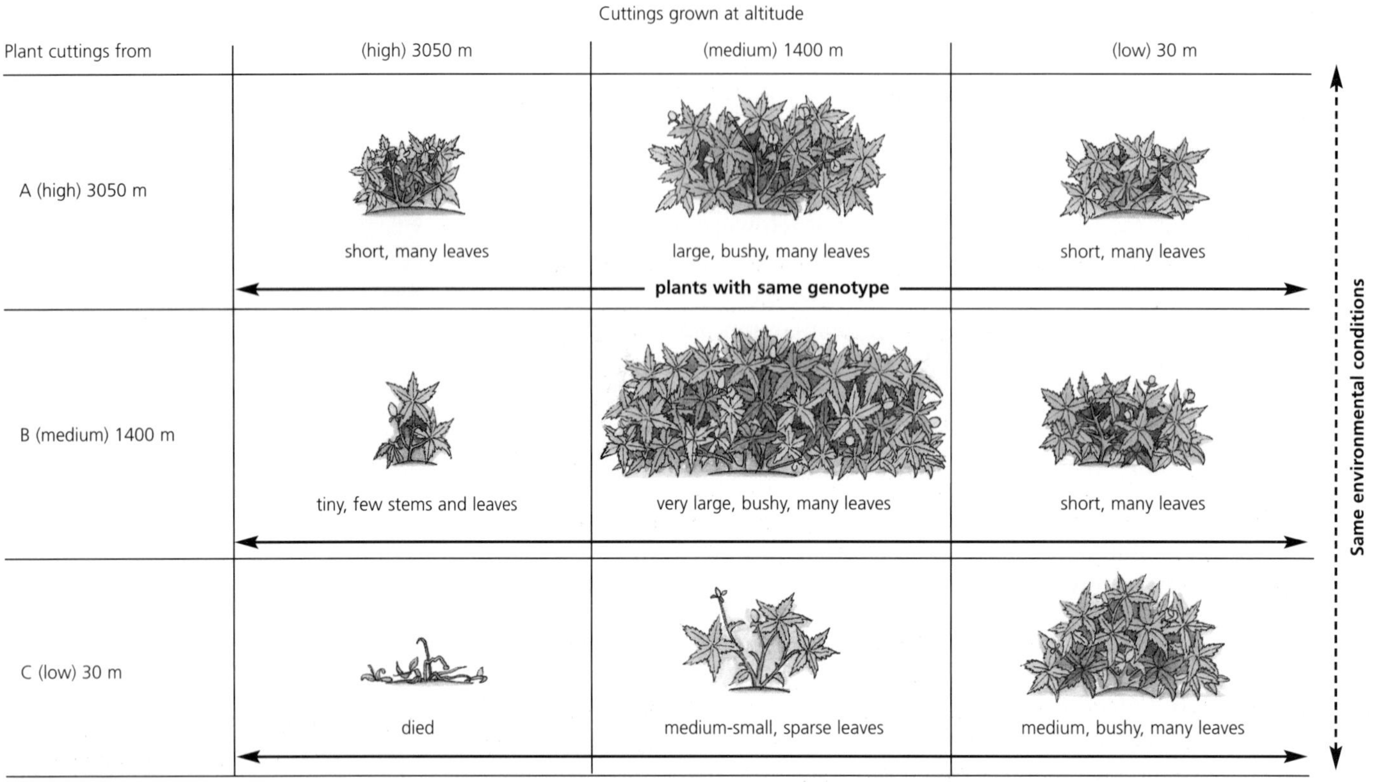

| Plant cuttings from | Cuttings grown at altitude | | |
|---|---|---|---|
| | (high) 3050 m | (medium) 1400 m | (low) 30 m |
| A (high) 3050 m | short, many leaves | large, bushy, many leaves | short, many leaves |
| B (medium) 1400 m | tiny, few stems and leaves | very large, bushy, many leaves | short, many leaves |
| C (low) 30 m | died | medium-small, sparse leaves | medium, bushy, many leaves |

plants with same genotype

Same environmental conditions

*Figure 3.40 Comparing the effect of genotype and the environment on the growth of* Potentilla glandulosa

## 3.10 Location of genes on chromosomes

Genes can be described as:

- unlinked
- sex linked
- showing autosomal linkage.

In humans, all the genes determining thousands of characteristics are found on just 23 pairs of chromosomes. Inevitably, each chromosome must carry many different genes, in other words each pair of chromosomes possesses many gene **loci**.

### Unlinked genes

Genes which are found on different pairs of chromosomes are referred to as **unlinked**. During prophase I in meiosis (see *AS Biology*, Section 6.11, page 170), each homologous pair of chromosomes arranges itself independently along the equator. Whether the maternal or paternal chromosome is facing a particular pole is quite random in each pair. This is **independent assortment** and is one of the causes of variation in the gametes. Therefore unlinked genes are transmitted independently of each other, leading to a random distribution of genes.

## Sex-linked genes

Genes found on the same chromosome are called linked genes. If they are found on one of the sex chromosomes they are said to be **sex linked** (see Section 3.7, page 77).

## Autosomal linkage

If genes are located together on a non-sex chromosome (an autosome), the linkage is described as **autosomal linkage**. Two or more genes located on the same chromosome are described as linked. All the genes on a particular chromosome comprise a **linkage group** and tend to be transmitted together (Figure 3.41).

Experiments have shown that the number of linkage groups corresponds to the number of chromosomes. *Drosophila*, with four pairs of chromosomes, has four linkage groups. Two alleles on the same chromatid will tend to be inherited together. When the chromatids are pulled to opposite poles during anaphase, both will go into the same daughter cell. See *AS Biology*, Chapter 6, for a reminder of the events of meiosis.

It is possible for these alleles to be separated, however, if **crossing over** occurs at prophase I of meiosis (Figure 3.42). The paternal and maternal chromatids of a homologous pair of chromosomes intertwine and cross over each other at points called **chiasmata**. The chromatids break at these points and rejoin so that the alleles from one chromatid join the alleles on the other. An exchange of alleles takes place between chromatids.

**Figure 3.41** *Linkage groups – the fruit fly* Drosophila melanogaster *has four linkage groups corresponding to the four pairs of chromosomes present*

One nucleus from the fruit fly showing the four pairs of chromosomes present.

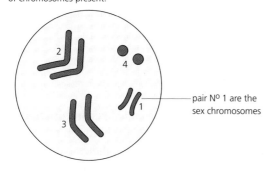

pair N⁰ 1 are the sex chromosomes

Linkage groups – genes found on the same chromosome pair (only a few of the many gene loci known are included here).

Chromosome pair N⁰

genes linked together as on same chromosomes

| 1 | 2 | 3 | 4 |
|---|---|---|---|
| yellow body | dumpy body | sepia eyes | bent wings |
| white eyes | jammed wings | hairy body | eyeless |
| ruby eyes | black body | scarlet eyes | |
| cut wings | reduced bristles | pink eyes | |
| singed bristles | purple eyes | curled wings | |
| lozenge eyes | cinnabar eyes | stubble bristles | |
| miniature wings | vestigial wings | striped thorax | |
| forked bristles | curved wings | hairless bristles | |
| carnation eyes | humpy body | ebony body | |
| | brown eyes | rough eyes | |
| | balloon wings | claret eyes | |
| | | minute bristles | |

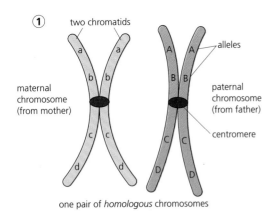

① two chromatids

alleles

maternal chromosome (from mother)

paternal chromosome (from father)

centromere

one pair of *homologous* chromosomes

*Prophase I meiosis*
Homologous chromosomes pair up.
(**Note:** each chromosome has replicated to form two *chromatids*.)

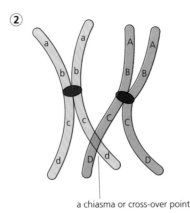

②

a chiasma or cross-over point

The chromosomes intertwine to form **chiasmata**. The chromatids break off at the chiasma and recombine, so an exchange of alleles takes place between maternal and paternal chromatids.

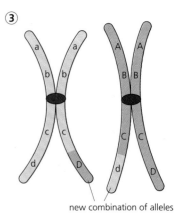

③

new combination of alleles

The resulting chromatids have a new combination of alleles. These chromatids then separate during prophase II and each ends up in a different gamete.

④ *Gametes produced*

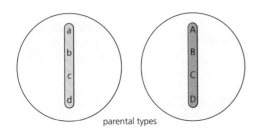

parental types

These gametes have the parental combination of alleles. They are produced whether crossovers occur or not. Most gametes will therefore have the parental combination of alleles.

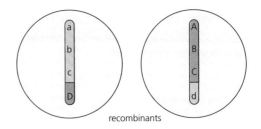

recombinants

These gametes have a new combination of alleles and are called **recombinants**. They are the result of crossing over, separating alleles on a chromatid. If there are no crossovers, no recombinants are formed.

*Figure 3.42 Crossing over*

The new combination of alleles that are produced are called **recombinants** and are another source of variation. If crossing over takes place between two particular genes, they will be separated. The chance of two alleles being separated by a crossover depends on their relative position on the chromosome. The closer they are, the less likely they are to be separated and, conversely, the greater their distance apart on the same chromosome the more likely they are to be separated. It is possible to calculate how often crossovers occur during meiosis by counting the percentage of offspring with recombinant features. From this knowledge, **chromosome maps** can be produced to show the position of alleles along a chromosome. Ten per cent crossovers indicate that genes are 10 map units apart, 40% crossovers indicate they are 40 map units apart – the greater distance increases the frequency of crossovers.

Extensive study of the fruit fly *Drosophila* has allowed the construction of detailed chromosome maps, showing the location of genes on each chromosome.

## Box 3.8  Autosomal linkage

In *Drosophila* the genes for **wing length** and **abdomen width** are on the same chromosome. The allele for long wing (**L**) is dominant to that for vestigial (short) wing (**I**). The allele for broad abdomen (**B**) is dominant to that for narrow abdomen (**b**). If a homozygous long winged fly with a broad abdomen (**LLBB**) is crossed with a vestigial-winged fly with narrow abdomen (**llbb**), all the offspring have the dominant features as we would expect. However, when these F₁ flies are crossed, unexpected results are found in the F₂ (Figure 3.43).

*Figure 3.43  Linkage in* Drosophila

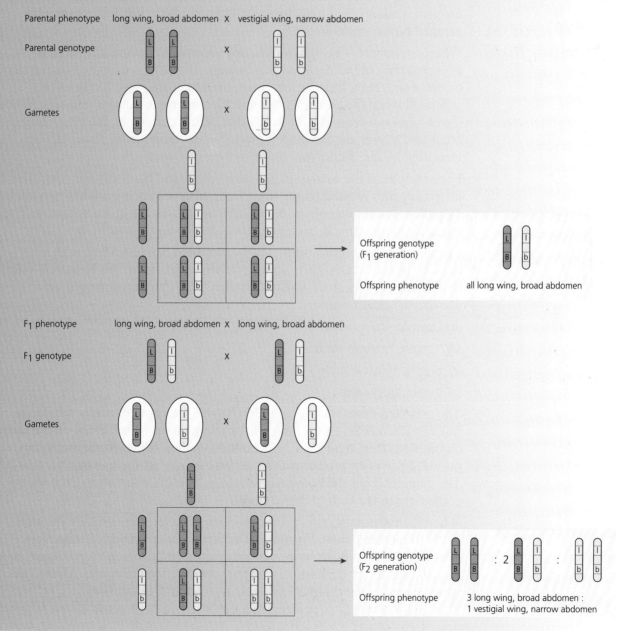

## Box 3.8 Autosomal linkage continued

In a normal dihybrid cross, a 9:3:3:1 ratio would be expected in the $F_2$. These unexpected results can be explained by linkage – the alleles for wing length and abdomen size are both found on the same chromosome.

There are no flies with a mixture of parental features, i.e. no long-winged flies with a narrow abdomen or vestigial-winged flies with a broad abdomen. This indicates that the alleles for long wing and broad abdomen are *linked* and carried together on the same chromosome. Similarly, vestigial wing and narrow abdomen are linked. As there are no flies showing the result of crossing over between the two alleles, we can conclude that these alleles are not only linked but situated very close together, making chiasmata between them unlikely. For this reason only parental types appear in the $F_2$.

When crossing two $F_1$ individuals, any significant deviation from the expected ratio of 9:3:3:1 indicates linkage. If alleles are located on the same chromosome but a little further apart, then they are more likely to be separated when the parental chromosomes intertwine during meiosis. Where the chromosomes cross over each other, at the chiasmata, the chromatids break and alleles are exchanged producing recombinants. These can be identified by the phenotypic ratios in the $F_2$.

For example, in the tomato plant, stem colour and hairiness are controlled by different genes located on the same chromosome. Each gene has two alleles. The allele for purple colour, **A**, is dominant to that for green colour, **a**. The allele for hairy stem, **B**, is dominant to that for hairless stem, **b**.

Crossing a homozygous purple, hairy stemmed plant, **AABB**, with a homozygous green, hairless variety, **aabb**, results in all the $F_1$ having purple, hairy stems as expected, **AaBb**. If two of the $F_1$ are now crossed, an unexpected ratio results:

300 purple, hairy stem

102 green, hairless stem

12 purple, hairless stem

14 green, hairy stem

The expected ratio in the $F_2$ is 9:3:3:1 – the unexpected results are due to linkage. The alleles controlling colour and hairiness are located on the same chromosome and tend to be transmitted together. When the homozygous purple, hairy plant produces gametes by meiosis, all the gametes receive **AB**. The gametes from the homozygous green, hairless plant all receive **ab**, so combining in the $F_1$ to produce **AaBb**.

When the heterozygous $F_1$ plants produce gametes, the gametes are more likely to receive parental genotypes which do not depend on crossovers. Therefore, there will be more gametes that have the parental genotypes **AB** and **ab**, and fewer gametes having the recombinant genotypes **Ab** and **aB** which depend on crossovers for their formation (Figure 3.44).

## Box 3.8 Autosomal linkage continued

This results in a 3:1 ratio of purple, hairy stems to green, hairless stems (the actual figures were 300 purple, hairy stems to 102 green, hairless stems). The few recombinant gamete genotypes Ab and aB, produce the genotypes shown in the table on the right at fertilisation.

This results in a 2:1:1 ratio of two purple, hairy stem, one purple, hairless stem and one green, hairy stem. As there are far fewer of these gamete genotypes, they will be a smaller proportion of the total (the actual figures were 12 purple, hairless plants and 14 green, hairy varieties).

| gametes | | Ab | aB |
|---|---|---|---|
| gametes | Ab | AAbb | AaBb |
| | aB | AaBb | aaBB |

*Figure 3.44 Linkage in the tomato plant – the genes controlling colour and hairiness are located on the same chromosome*

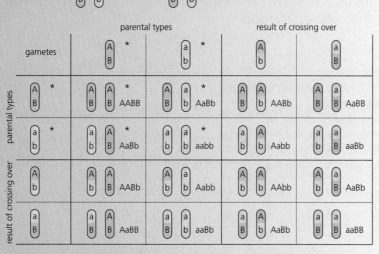

Large numbers of *parental* gametes are produced as they do not depend on crossing over. This leads to large numbers of F₂ offspring with the genotypes AABB, AaBb (purple, hairy stem - 300) and aabb (green, hairless stem - 102).

**Formation of gametes**

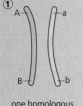

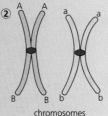

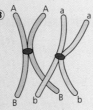

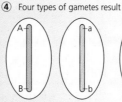

one homologous pair of chromosomes

chromosomes replicate

exchange of chromatids

parental types

recombinants

## 3.11 Chi-squared test

Breeding experiments involve an element of chance. The fusion of gametes at fertilisation is a random event and the resulting offspring may only approximate to expected ratios. To find out if numbers fit an expected ratio a statistical test called a **chi-squared** test is carried out. Chi-squared is given the Greek letter $\chi^2$ (pronounced kye) or chi-squared. The results of this test indicate whether numbers fit the expected ratio or differ significantly from those expected. The difference between the observed and the expected ratios is the deviation from the expected. The frequencies that are observed are compared with those expected, based on a prediction which is in the form of a hypothesis. The purpose of the chi-squared test is to find out whether the difference between the observed and expected ratios is significantly different from those expected or merely due to chance.

There are two types of hypotheses, one is from genetics ratios; the second is the null hypothesis.

### Genetics ratios

It is obvious that 90 tall pea plants and 30 short pea plants exactly fits a 3:1 ratio. But what about 100 tall and 27 short? Is this still a 3:1 ratio? Do the numbers differ *significantly* from those expected or are the differences just down to chance?

To find out if observed numbers fit a 3:1 ratio, a chi-squared test is carried out using the following formula.

$$\chi^2 = \sum \frac{(O-E)^2}{E}$$

O = observed numbers obtained
E = expected numbers
$\Sigma$ = sum of

The hypothesis being tested here is that the numbers approximate to a 3:1 ratio. In this example, if it is a 3:1 ratio, then we would expect $\frac{3}{4}$ of the total to be tall:

$$\frac{3}{4} \text{ of } (100 + 27) = 95.25$$

and we would expect $\frac{1}{4}$ of the total to be short:

$$\frac{1}{4} \text{ of } 127 = 31.75$$

Using the formula above, the observed and expected numbers are inserted, taking one feature at a time (Table 3.7). This table is often used in a chi-squared test.

***Table 3.7*** *Calculating chi-squared*

| Class | (O) | (E) | (O – E) | (O – E)² | $\frac{(O-E)^2}{E}$ |
|---|---|---|---|---|---|
| tall | 100 | 95.25 | 4.75 | 22.56 | 0.237 |
| short | 27 | 31.75 | 4.75 | 22.56 | 0.710 |
| | | | | sum of figures | 0.947 |

| Degrees of freedom | Probability (p) as % | | | | | | |
|---|---|---|---|---|---|---|---|
| | 90% | 50% | 20% | 10% | 5% | 2% | 1% |
| | as a decimal | | | | | | |
| | 0.90 | 0.50 | 0.20 | 0.10 | 0.05 | 0.02 | 0.01 |
| 1 | 0.02 | 0.46 | 1.64 | 2.71 | 3.84 | 5.41 | 6.64 |
| 2 | 0.21 | 1.39 | 3.22 | 4.61 | 5.99 | 7.82 | 9.21 |
| 3 | 0.58 | 2.37 | 4.64 | 6.25 | 7.82 | 9.84 | 11.34 |
| 4 | 1.06 | 3.36 | 5.99 | 7.78 | 9.49 | 11.67 | 13.28 |

*Table 3.8 Part of table used to find probability (values are given as decimals and percentages)*

Adding together the two values for $(O - E)^2 \div E$ (0.237 + 0.710) = 0.947 (0.95 to two decimal places). Therefore $\chi^2 = 0.95$. To find out what this number indicates, it is necessary to use a probability table (Table 3.8).

**Degrees of freedom** are calculated by counting the number of classes minus one (tall and short classes = 2 – 1 = 1). In this case there is one degree of freedom. Looking at one d.f. (degrees of freedom) on the left-hand side of the table, the numbers are read horizontally to find where the chi-squared number fits. A chi-squared of 0.95 is found between a probability of 0.2 and 0.5 (between 20 and 50%).

**Five per cent is the critical value**. This example indicates a probability above 0.05 or 5%, so the ratio being tested is accepted – this is still a 3:1 ratio.

*If p is greater than 5% (0.05), the hypothesis is supported and accepted*

*If p is less than 5% (0.05), the hypothesis is not supported and is rejected*

Other genotypic ratios can be tested by the chi-squared test for 'goodness of fit' to expected ratios. For example, when pure bred tall, round-seeded pea plants are crossed with short, wrinkled-seeded pea plants, all the offspring are tall with round peas. This tells us that the alleles for tall and round are dominant to the alleles for short and wrinkled. However, when two of the $F_1$ are selfed or interbred, four types of plants are found in the $F_2$:

P    tall, round seeds × short, wrinkled seeds

↓

$F_1$    all tall, round seeds (interbred)

↓

$F_2$    118 tall round, 30 tall wrinkled, 24 short round, 6 short wrinkled
(total = 178)

As this is a normal, unlinked dihybrid cross, the ratio expected is 9:3:3:1. To find out whether the observed numbers differ significantly from those expected, a chi-squared test is carried out. The expected numbers can be calculated from the expected ratio:

● tall, round – the expected number is $\frac{9}{16}$ of the total (178) =100

● tall, wrinkled – the expected number is $\frac{3}{16}$ of the total (178) = 33.375

**Remember**

With a chi-squared test, when the probability is *greater than 5%*, numbers fit the expected ratio and the hypothesis is accepted, i.e. there is no significant difference between the observed numbers and the numbers expected from the genetics ratios.

| Phenotype | (O) | (E) | (O – E) | (O – E)$^2$ | $\dfrac{(O – E)^2}{E}$ |
|---|---|---|---|---|---|
| tall, round | 118 | 100 | 18 | 324 | 3.24 |
| tall, wrinkled | 30 | 33.375 | 3.375 | 11.39 | 0.34 |
| short, round | 24 | 33.375 | 9.375 | 87.89 | 2.63 |
| short, wrinkled | 6 | 11.125 | 5.125 | 26.27 | 2.36 |
| | | | | sum of figures | 8.57 |

***Table 3.9*** *Calculating chi-squared*

- short, round – the expected number is $\frac{3}{16}$ of the total (178) = 33.375
- short, wrinkled – the expected number is $\frac{1}{16}$ of the total (178) = 11.125.

Table 3.9 shows how chi-squared is calculated. The numbers in the end column added together = 8.57. Therefore chi-squared = 8.57 (no units).

The degrees of freedom are the *number of classes minus one.* Here there are four classes, tall round, tall wrinkled, short round and short wrinkled. Therefore the degrees of freedom are 4 – 1 = **3**.

Using Table 3.8, at three degrees of freedom and moving across the table, the chi-squared number of 8.57 is found between 0.02 and 0.05 (between 2% and 5%). As the probability is less than the critical value of 5%, the difference between the observed and expected results is significant, so the deviation from the expected results is too great. The hypothesis is therefore rejected, it does not fit a 9:3:3:1 ratio.

12 In *Drosophila,* the alleles for ebony body and curled wings are recessive to the alleles for grey body and normal wings. A heterozygous grey-bodied, normal-winged fly was crossed with an ebony bodied, curled-winged fly. The following offspring were obtained:

| Phenotype | Number |
|---|---|
| grey body and normal wings | 32 |
| grey body and curled wings | 22 |
| ebony body and curled wings | 29 |
| ebony body and normal wings | 21 |

Are these numbers consistent with the expected 1:1:1:1 ratio?

*(4 marks)*

### The null hypothesis

Sometimes it is necessary to compare observed and expected frequencies based on a prediction in the form of a **null hypothesis**. A null hypothesis assumes that there is no difference expected in numbers.

For example, lichen distribution can be compared on the North-, South- and West-facing sides of a wall. To find out if aspect affects distribution, a null hypothesis can be produced. For this investigation it would be 'The aspect of the wall does not affect the abundance of lichen'. If aspect does not affect abundance, then equal numbers would be expected on each side of the wall.

The number of colonies observed are as follows:

- north – 20
- south – 32
- west – 53.

If no difference is expected, then the numbers should be the same on all three aspects. The total number of observations is 20 + 32 + 53 = 105. 105 ÷ 3 = 35. Therefore 35 colonies are expected on each side of the wall based on a null hypothesis.

Using the formula on page 94, chi-squared is calculated as 15.942 (see Table 3.10 for calculations). There are two degrees of freedom (3 – 1 = 2). From Table 3.8 the probability is found to be less than 0.01 (less than 1%). As the probability is less than the critical value of 5%, the null hypothesis is rejected. Therefore aspect does affect abundance.

Thus chi-squared can be used to compare the observed and expected frequencies based on both a null hypothesis and numbers expected from genetics ratios.

*Table 3.10* Calculating chi-squared

| Aspect | (O) | (E) | (O – E) | $(O - E)^2$ | $\dfrac{(O - E)^2}{E}$ |
|---|---|---|---|---|---|
| north | 20 | 35 | 15 | 225 | 6.428 |
| south | 32 | 35 | 3 | 9 | 0.257 |
| west | 53 | 35 | 18 | 324 | 9.257 |
| | | | | sum of figures | 15.942 |

## Summary – ③ Genetics

- Offspring inherit genes from their parents.
- A gene is a section of DNA that codes for a particular polypeptide.
- Pairs of homologous chromosomes contain pairs of alleles making up a gene.
- Dominant alleles will be expressed even if only one is present.
- Two recessive alleles are needed for a recessive feature to be expressed.
- Mendel contributed to the understanding of genetics through his work on pea plants.
- A monohybrid cross involves the inheritance of one feature.
- Numbers rarely fit a ratio precisely as the fusion of gametes is a random event.
- Cystic fibrosis is an inherited, autosomal, recessive disorder caused by a deletion mutation. It will only be expressed if both alleles of a pair are recessive.
- Huntington's disease is an inherited, autosomal, dominant disorder which only needs one allele to be expressed.
- A dihybrid cross is the inheritance of two characteristics. If the two genes are found on different chromosomes the genes are said to be unlinked.
- If two organisms, both heterozygous for two genes are crossed, the resulting phenotypes display a 9:3:3:1 ratio if the genes are unlinked.
- A test cross is used to discover the genotype of an organism showing the dominant trait. The organism of unknown genotype is crossed with the relevant homozygous recessive one.
- When two different alleles of a gene are both expressed in the phenotype of a heterozygote, they are called codominant.
- Sickle cell anaemia is an example of codominance caused by a base substitution. Resistance to malaria is conferred by the heterozygous state.
- If there are more than two alleles of a gene, it is described as having multiple alleles.
- A karyotype displays the chromosomes present in a nucleus, paired according to size.
- Sex determination in humans depends on which sex chromosomes are inherited, XX producing a girl and XY a boy. The statistical ratio is 1:1.
- Genes found on the sex chromosomes are said to be sex linked. Most sex-linked alleles are found on the X chromosome and can therefore affect both males and females. Recessive sex-linked disorders are more common in males, who inherit the allele from unaffected carrier females.
- Haemophilia is an example of a recessive, sex-linked disorder found on the X chromosome.
- Gene interaction is found when the phenotype depends on the interaction of two or more different genes. Epistasis is a type of gene interaction in which one gene inhibits the action of another. A feature determined by many genes acting together is described as polygenic.
- The phenotype of an organism depends on the interaction between genes and the environment.
- Autosomal linkage occurs if two genes are located on the same chromosome and are transmitted together. An unexpected genetics ratio indicates that genes may be linked.
- Chi-squared is a statistical test used to find the significance of any differences between observed and expected ratios. It can be used to find out if observed numbers fit an expected ratio.

# ? Answers

1  Huntington's disease does not cause early death *(1)* and sufferers live long enough to pass on the allele to their children before symptoms appear *(1)*.

2  BW × BW

|   | B  | W  |
|---|----|----|
| B | BB | BW |
| W | BW | WW |

1 black : 2 blue : 1 white *(1)*.

3  RW × WW produces pink (RW) and white (WW) flowers in a 1 to 1 ratio *(2)*.

4  The Darwins should take baby AB, the Pasteurs baby B, the Mendels baby O and the Flemings baby A *(2)*. Explanation *(2)*.

5  1 = $a^t a^t$, 2 = $A^s a^y$, 3 = $A^s a^t$, 4= $A^s a^t$, 5 = $a^y a^t$, 6 = $A^s a^t$, 7 = $a^y a^t$, 8 = $a^t a^t$, 9 = $a^t a^t$, 10 = $a^y a^t$, 11 = $A^s a^t$ or $A^s a^y$ *(3)*.

6  Cross 1: $E^D E^D$ × EE (or Ee) *(1)*

Cross 2: $E^D E$ × EE (or Ee) or $E^D e$ × EE *(1)*

Cross 3: $E^D e$ × Ee *(1)*.

7  Human red blood cells do not contain a nucleus and therefore have no chromosomes *(1)*.

8  The son of a haemophiliac father must inherit a Y chromosome from his father making him male. He does not inherit the X chromosome with the harmful allele as this would make him female! Therefore a son cannot inherit haemophilia from his father *(1)*.

9  A Y-linked allele will inevitably pass from a father to his son as it is the Y allele that makes him male. Any alleles on the Y chromosome must pass to the son *(1)*.

10 As females inherit the X chromosome from their father, they cannot be affected by alleles on the Y chromosome. Their sex chromosomes are XX only *(1)*.

11 AaMm × AaMm

|    | AM   | Am   | aM   | am   |
|----|------|------|------|------|
| AM | AAMM | AAMm | AaMM | AaMm |
| Am | AAMm | AAmm | AaMm | Aamm |
| aM | AaMM | AaMm | aaMM | aaMm |
| am | AaMm | Aamm | aaMm | aamm |

*(2)*

This produces nine agouti, four albino and three black coats *(2)*.

12 Chi-squared = 3.3 *(1)*. There are three degrees of freedom (4 – 1 types of *Drosophila* = 3). This gives a probability between 0.2 and 0.5 ( 20–50%) *(1)*. As this is above the critical value of 5%, the difference in numbers from those expected is not significant and is due to chance. Therefore the numbers do conform to the expected 1 to 1 to 1 to 1 ratio *(2)*.

### End of Chapter Questions

1   In human beings the ability to taste phenylthiocarbamide (PTC) is determined by a single gene with two alleles. For people with the allele **T** a dilute solution of PTC has a bitter taste while people who have only the allele **t** cannot taste the solution. Another gene, at a different locus, determines eye colour so that individuals with the allele **B** have brown eyes whilst those with only **b** have blue eyes.

   **(a)** What is meant by the term locus?                                    *(1 mark)*

   **(b)** Distinguish between the terms genotype and phenotype.        *(1 mark)*

   **(c)** A woman with brown eyes, who could taste PTC (a taster), married a man with blue eyes who was a non-taster. List all the possible genotypes and phenotypes of the children they could produce.        *(2 marks)*

   **(d)** Which of the following men could be excluded as the biological father of a child with brown eyes who is a non-taster and whose mother is a taster with blue eyes? Give your reasoning.

   (i)   Brown-eyed taster.

   (ii)  Brown-eyed non-taster.

   (iii) Blue-eyed non-taster.

   (iv)  Blue-eyed taster.                                                *(2 marks)*

   **(e)** How could you determine whether the genes for eye colour and tasting are linked on the same pair of homologous chromosomes?    *(2 marks)*

   *Oxford 1998*                                                   *(Total 8 marks)*

2   The inheritance of ABO blood groups is controlled by three alleles of the same gene, $I^A$, $I^B$ and $I^O$. The alleles $I^A$ and $I^B$ are codominant. Both $I^A$ and $I^B$ are dominant to the allele $I^O$.

   **(a)** Explain what is meant by an allele.                                  *(1 mark)*

   **(b)** (i) Complete the table below to show the missing genotypes.

| Blood group phenotype | Possible genotype |
|:---:|:---:|
| A | $I^A I^A$, ........... |
| B | $I^B I^B$, ........... |
| AB | ..................... |
| O | ..................... |

                                                                      *(2 marks)*

   (ii) Children of blood groups A and O were born to parents of blood groups A and B. Complete the genetic diagram to show the possible ABO blood group phenotypes of the children which could be produced from these parents.

| Parental phenotypes | Blood group A | Blood group B |
|---|---|---|
| Parental genotypes | | |
| Genotypes of gametes | | |
| | | |
| Genotypes of children | | |
| Phenotypes of children | | *(3 marks)* |

AQA 1999                              *(Total 6 marks)*

**3** In tomatoes, the allele for red fruit, **R**, is dominant to that for yellow fruit, **r**. The allele for tall plant, **T**, is dominant to that for short plant, **t**. The two genes concerned are on different chromosomes.

   **(a)** A tomato plant is homozygous for allele **R**. Giving a reason for your answer in each case, how many copies of this allele would be found in

   (i)  a male gamete produced by this plant,

   (ii) a leaf cell from this plant?                              *(2 marks)*

   **(b)** A cross was made between two tomato plants.

   (i) The possible genotypes of the gametes of the plant chosen as the male parent were **RT**, **Rt**, **rT** and **rt**. What was the genotype of this plant?                              *(1 mark)*

   (ii) The possible genotypes of the gametes of the plant chosen as the female parent were **rt** and **rT**. What was the phenotype of this plant?                              *(1 mark)*

   (iii) What proportion of the offspring of this cross would you expect to have red fruit? Use a genetic diagram to explain your answer.
                                   *(3 marks)*

NEAB 1998                              *(Total 7 marks)*

**4** The diagram shows the inheritance of cystic fibrosis in a human family. The allele for cystic fibrosis is recessive to the normal allele.

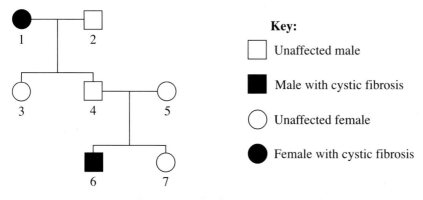

**Key:**

☐ Unaffected male

■ Male with cystic fibrosis

◯ Unaffected female

● Female with cystic fibrosis

   **(a)** What is the possibility that the next child born to individuals 4 and 5 will:

   (i)  be a male with cystic fibrosis;                              *(1 mark)*

   (ii) have at least one allele for cystic fibrosis?                              *(1 mark)*

N2.2 **(b)** In Britain, the frequency of the cystic fibrosis allele is 0.02. Calculate the proportion of people you would expect to be born with cystic fibrosis. *(1 mark)*

**(c)** In the past, people born with cystic fibrosis usually died before reaching adulthood. Suggest an explanation for the fact that the cystic fibrosis allele remained at a very high frequency in the population in spite of this. *(2 marks)*

*AEB 1997* *(Total 5 marks)*

**5** The fruit fly is a useful animal for studying genetic crosses. The diagram shows the life cycle of the fruit fly.

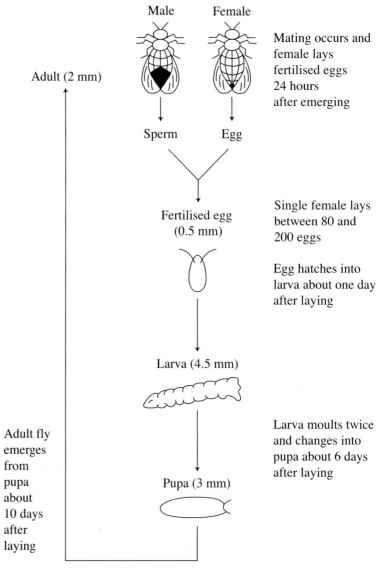

Male  Female

Adult (2 mm)

Mating occurs and female lays fertilised eggs 24 hours after emerging

Sperm  Egg

Fertilised egg (0.5 mm)

Single female lays between 80 and 200 eggs

Egg hatches into larva about one day after laying

Larva (4.5 mm)

Larva moults twice and changes into pupa about 6 days after laying

Adult fly emerges from pupa about 10 days after laying

Pupa (3 mm)

**(a)** Using information from the diagram explain three ways in which the fruit fly is a useful animal for studying genetic crosses. *(3 marks)*

In fruit flies the allele for grey body colour, **G**, is dominant to the allele for black body, **g**, and the allele for normal wings, **N**, is dominant to the allele for vestigial wings, **n**. A cross between a grey-bodied, normal-winged fly and a black-bodied, vestigial-winged fly resulted in the following offspring.

25 grey-bodied, normal-winged
26 grey-bodied, vestigial-winged
24 black-bodied, normal-winged
27 black-bodied, vestigial-winged

**(b)** (i) Give the genotype of the grey-bodied, normal-winged parent.

*(1 mark)*

(ii) Give the genotypes of the gametes which could be produced by one of the grey-bodied, vestigial-winged offspring. *(1 mark)*

**(c)** What ratio would you expect in the offspring produced if the grey-bodied, normal-winged parent had been crossed with a fly of the same genotype?

*(1 mark)*

*AQA 1999* *(Total 6 marks)*

**6** Warfarin is a pesticide which is used to kill rodents such as rats and mice. Some rodents are resistant to warfarin and such resistance was first discovered in wild rats on farms around Welshpool in 1959. Resistance to warfarin is controlled by a gene with two alleles, $W^1$ and $W^2$.

N2.2
C2.2

In 1959, a study was made of the genotypes of 74 rats on farms around Welshpool where warfarin had been used. The genotypes of 74 trapped rats were determined. The results are shown in the table below. The table also shows the expected numbers of each genotype, assuming the frequency of the two alleles, $W^1$ and $W^2$, in the population remains constant.

| Genotype | Phenotype | Observed number | Expected number |
|----------|-----------|-----------------|-----------------|
| $W^1W^1$ | Susceptible to warfarin | 28 | 32 |
| $W^1W^2$ | Resistant to warfarin | 42 | 33 |
| $W^2W^2$ | Resistant to warfarin | 4 | 9 |

**(a)** (i) Determine whether the allele for resistance to warfarin, $W^2$, is dominant or recessive, giving a reason for your answer. *(1 mark)*

(ii) The chi-squared test was used to determine whether the difference between observed and expected numbers is significant. A value of 5.73 was obtained for $\chi^2$.

Using the extract from a table of values, state whether the difference is significant, giving a reason for your answer.

| Probability levels $P$ (%) | 99 | 10 | 5 | 2 | 1 | 0.1 |
|---|---|---|---|---|---|---|
| $\chi^2$ | | 0.00 | 2.71 | 3.84 | 5.41 | 6.64 | 10.83 |

*(2 marks)*

**(b)** Rats with the genotype $W^2W^2$ require much more vitamin K in their diet than those with the other genotypes.

(i) Calculate the observed number of rats of each homozygous genotype as a percentage of the expected number of that genotype. Show your working. *(3 marks)*

(ii) Both homozygous genotypes are at a disadvantage compared with the heterozygous genotype. Using your calculated results from (b) (i), state which of the homozygous genotypes is at the greater disadvantage. *(1 mark)*

(iii) Give a reason why each of these genotypes is at a disadvantage compared with the heterozygous genotype. *(2 marks)*

**(c)** A further study was carried out in a nearby area where warfarin was also used. In 1973, nearly 60% of the rats were resistant to warfarin. The use of warfarin was then discontinued and after two years the number of resistant rats had dropped to less than 40%. Explain why the number of resistant rats dropped after the warfarin was discontinued. *(3 marks)*

**(d)** Explain why the allele for resistance to warfarin is likely to remain in the gene pool when the use of warfarin is discontinued. *(2 marks)*

*London 1999*                                              *(Total 14 marks)*

# 4 Evolution

A **species** is a group of organisms that look alike and can reproduce successfully to produce fertile offspring. Within a species there is variation which can be caused by meiosis, random fertilisation, mutations and the environment (and also artificially, using recombinant DNA technology).

Some organisms have features that will increase their chance of survival in particular conditions, compared to others of the same species. **Natural selection**, or survival of the fittest, explains why these organisms seem to fit into their surroundings while those less well adapted die out. Well-adapted individuals will live long enough to breed and pass on their useful alleles to the next generation. Those less well adapted will not. Selection will increase the proportion of individuals with alleles giving an advantage, leading to an increasing proportion of the population showing the phenotype determined by these useful alleles.

If part of a population becomes isolated from the main group, changes in allele frequency and mutations will take place independently in the two groups and lead to differences developing between them. Eventually they will diverge so much that they can no longer interbreed successfully; they have therefore become different species. This process is called **speciation**. The whole process of producing new species from ones already existing is called **evolution**.

## This chapter includes:
- types of variation
- causes of variation
- Darwin's theory of natural selection
- evidence for natural selection
- evolution in action
- types of selection
- population genetics
- speciation
- artificial selection.

## 4.1 Types of variation

Variation between individuals may be either continuous or discontinuous.

### Continuous variation

When there is a complete range of measurements from one extreme to the other for a named characteristic, it is an example of continuous variation. Although quite different at the extremes, differences between some individuals are slight and grade into each other so individuals often do not fall into distinct categories. Human height shows continuous variation (Figure 4.1). The graph shows the number of people at each height and produces a bell-shaped curve, with the maximum number being of average height (the **mean**) and fewer very tall or short individuals.

In this example, the mean is calculated by adding the heights of all the individuals together and dividing by the number of individuals:

$$\text{mean} = \frac{\text{heights of all the individuals}}{\text{total number of individuals}}$$

When a frequency distribution is plotted against a continuous variable, like height, the resulting bell-shaped curve is called a **normal distribution**. In a normal distribution, the arithmetic mean is also the most commonly occurring value or **mode**. The mode or **modal class** is the most common number or value in a range of values. In this example, the most common height category or modal class is 172–174 cm.

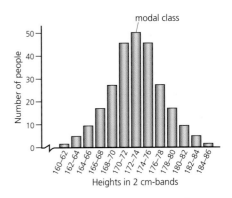

The variation in adult males follows a **normal distribution**. There is a complete gradation from small to tall with no gaps. The majority of adult males fall into the **modal class** 172–174 cm. The modal class is the most commonly occurring set of values.

**Figure 4.1** *Continuous variation shown by variation in human height*

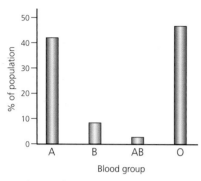

There are four distinct groups in the ABO blood grouping system, with no intermediates, i.e. there is no continuous range of variation. Human blood groups are therefore an example of discontinuous variation, controlled by a single gene with three alleles.

*Figure 4.2* *ABO Human blood groups – distribution in the UK*

Features showing continuous variation can usually be measured and are therefore described as quantitative. A large number of genes is usually involved so they are described as **polygenic** features (see Section 3.8, page 86). In addition, the environment may exert considerable influence. Clearly the environment also affects birth weight – a well-fed mother will generally have a larger baby than if she were starving.

It is fortunate that Mendel did not encounter this type of genetics when first establishing the laws of heredity. Clear, discrete features are easier to separate and analyse statistically, enabling Mendel to deduce his laws.

### Discontinuous variation

A characteristic may fit into separate categories with no intermediates. This is called discontinuous variation. Human ABO blood groups are one example of discontinuous variation. People are either blood group A, B, AB or O. There is nothing in-between. Figure 4.2 shows discontinuous variation in human blood groups. Such features do not give a normal distribution curve and separated bar charts are used to illustrate the discrete nature of the features.

Discontinuous variations are described as **qualitative** (as a feature is either present or it is not) and are normally controlled by a single gene, although two or more alleles may exist for the gene. For example, the ABO blood groups are determined by a single gene with three possible alleles (see Section 3.6, page 73). Another characteristic demonstrating discontinuous variation is tongue rolling, a person can either tongue roll or they cannot, there is no intermediate (Figure 4.3). As with most examples of discontinuous variation, the environment plays a minimal role.

1 Most cats have tails but Manx cats do not. What type of variation is this an example of?
*(1 mark)*

### 4.2 Causes of variation

Variation can be caused both by genes and the environment.

### Genetic causes of variation

Sexual reproduction involves the fusion of gametes produced by meiosis. During the process of meiosis to produce gametes, new combinations of genes are produced, firstly by crossing over and then by independent assortment. Finally, the process of fertilisation randomly combines two unique gametes out of millions.

Although sexual reproduction produces variation it does not actually introduce new alleles. These are only brought in by **mutation**. A mutation is a change in the structure or amount of DNA in an organism, leading to an alteration in the amino acid sequence produced. A mutation in a gamete will be inherited, whereas a mutation in other body cells (**somatic** cells) will not be passed to offspring. Somatic cell mutations may affect the phenotype and, therefore, the survival of the organism but play no direct part in evolution.

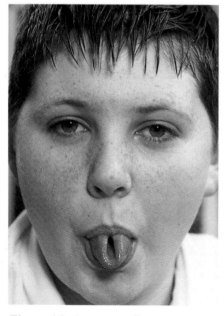

*Figure 4.3* *A tongue-roller*

There are two main types of mutations, gene mutations and chromosome mutations (see *AS Biology*, Chapter 5 for more details on gene mutations).

## Gene and point mutations

A point mutation is an alteration in the base sequence of one or two base pairs, by addition, deletion or substitution (see *AS Biology*, Section 5.11, page 138). A gene mutation may involve just this, or more point mutations. Point mutations occur at a single gene locus on a chromosome, altering the DNA sequence of bases and the amino acid sequence produced. The resulting protein shape may be altered significantly so it does not function as required. This is particularly important if the altered protein is an enzyme which, if changed, may no longer match the shape of its substrate so stopping further enzyme action.

Some mutations are harmful to the organism, others may be neutral and have no effect. However, mutations which are beneficial may increase the chances of survival and be selected for. It is these rare, new, useful alleles that determine the path of evolution. If useful, they are selected for and the frequency of the new allele will increase in the population leading to an increase in the number of individuals with the advantageous phenotype.

## Chromosome mutation

Chromosome mutations involve changes to either the number of chromosomes or to large sections of a single chromosome. Thus many genes are changed. When part of a chromosome breaks off and is lost, it is described as a **deletion mutation**. If a fragment of one chromosome attaches to another non-homologous chromosome, this is a **translocation mutation**, thought to be due to increased 'stickiness' of the chromosomes.

**2** Why are mutations in somatic or body cells unimportant for evolution?

*(1 mark)*

### Box 4.1 Down's syndrome

Down's syndrome is caused by the presence of an extra chromosome number 21 (see Figure 3.25c on page 77 for a karyotype of a female with Down's syndrome). The obvious physical features include a broad forehead, downward sloping eyes, folds in the eyelid, a short nose and protruding tongue. Mental impairment is apparent although there is quite a range of severity. Down's syndrome is more common in the children of older mothers, affecting approximately one in 2300 children of mothers aged 20 years, and about one in 40 children of mothers aged 45 years.

Most cases of Down's syndrome are caused by the failure of the homologous chromosomes number 21 to separate during meiosis.

Mutations are only of evolutionary importance when gametes are affected as the alteration may be passed on to future generations.

### Box 4.1  Down's syndrome continued

This is known as **non-disjunction** and leads to two types of daughter cells, one with two chromosomes number 21 and one with none. (Remember that at meiosis the gametes should receive one chromosome from each homologous pair.) This results in two chromosomes number 21 in an ovum, instead of one. At fertilisation, these combine with the single chromosome number 21 in the sperm, producing three in the zygote. This failure of chromosomes to separate is thought to be due to the increased 'stickiness' of these chromosomes.

After fertilisation, the zygote with three chromosomes number 21, **trisomy 21**, will develop into a child who will suffer from Down's syndrome. There will be a total of 47 chromosomes instead of the normal 46, 2n +1. The ovum without a chromosome number 21 will at fertilisation produce a zygote with only one number 21 from the sperm, a lethal condition known as **monosomy**. Figure 4.4 shows how non-disjunction of chromosome number 21 can lead to Down's syndrome.

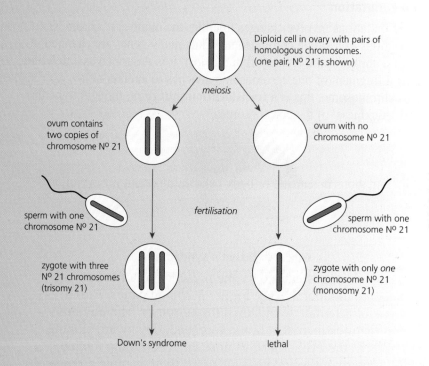

**Figure 4.4** *Non-disjunction of chromosome number 21 leading to Down's syndrome*

Five per cent of people with Down's syndrome have a parent with a translocation mutation. Chromosome 21 joins with chromosome 14 to form a translocation heterozygote. Gametes subsequently produced from this parent have chromosome 14 with the attached number 21 in addition to the normal chromosome number 21.

At fertilisation, there will be three chromosomes number 21 (trisomy 21) in the zygote resulting in Down's syndrome. Mothers over 35 years old, with an increased risk of fetal abnormalities, may choose to have an **amniocentesis** test at about the 15th week of pregnancy. This procedure detects chromosome abnormalities in the fetus. A sample of amniotic fluid containing fetal cells is removed from the uterus using a fine needle. A karyotype is produced and extra chromosomes should be visible (see Section 3.7, page 76). The technique involves a risk to the fetus and the knowledge gained may be used as grounds for termination.

Another technique more recently developed for the same purpose is **chorionic villus sampling**, **CVS**, which involves removing fetal cells from the chorionic villi that grow from the embryo into the mother's uterus wall. The cells are then examined and a karyotype produced. The advantage of CVS is that it can be carried out earlier in the pregnancy when termination is safer.

### Box 4.2  Polysomy and polyploidy

Down's syndrome is an example of a condition known as **polysomy**, when the number of a particular chromosome is more than the normal diploid number (*poly* means many). It is caused by a mutation taking place during meiosis.

Occasionally, a whole set of chromosomes fails to separate and the resulting gametes will either receive pairs of chromosomes, in which case they are diploid, or none. After fertilisation, if the diploid gamete joins with a normal, haploid one the resulting zygote will be triploid, with three sets of chromosomes. The possession of multiple sets of chromosomes is known as **polyploidy**. If a diploid gamete combines with another diploid gamete at fertilisation, the resulting zygote would have four of each type of chromosome instead of the usual two, a condition known as **tetraploid**. In some organisms, particularly plants, there are cases of organisms having three or more sets of chromosomes. Common examples include the bread wheat plant which is hexaploid, 6n (having six of each chromosome) and some potato plants which are tetraploid, 4n (having four of each chromosome). If a tetraploid potato produces gametes with half the parental number of chromosomes, they will each be 2n, with two of each type of chromosome. If two of these gametes fuse, the zygote will be 4n. If a 2n gamete fuses with a normal n gamete, the resulting zygote will be triploid, 3n. In this way a variety of types of polyploidy are produced. Triploid organisms are infertile as their chromosomes cannot pair up during meiosis. Only even numbers can pair up. Sometimes, a mutation during mitosis can result in polyploidy. If the chromosomes double but fail to separate during anaphase, then cells will be have four sets of chromosomes.

### Environmental causes of variation

All phenotypes are the product of the genotype plus the environment. The effect of the environment on variation has already been discussed in Chapter 3. Obviously two plants of identical genotypes will look markedly different if they receive different amounts of light. But the environmental influence will not be inherited. It will not pass into the gametes and will not directly affect evolution. Continuous variation is a product of both polygenes and the environment, whereas discontinuous variation is mainly genetically determined and less affected by the environment. One example of discontinuous variation affected both by environmental and genetic causes, is the yellow leaves found in tobacco plants. Magnesium deficiency in the soil (environmental cause), and mutant genes (genetic cause) can both result in yellow leaves.

One way that the environment can influence evolution is through mutations. **Mutagens** are factors which increase the mutation rate (see *AS Biology*, Chapter 5 for more information on mutations). For example, ultraviolet light and X-rays are mutagens that alter the DNA, and if the resulting mutation is in the gametes, it will be inherited.

The main way that the environment affects evolution is by providing the selection pressure that determines which organisms survive to reproduce and pass on their alleles. For example, a reduction in light, possibly due to long-term pollution, will favour tall plants that can reach sufficient light, so enabling more tall plants to reproduce and pass on the alleles for tallness to the next generation. The constraints exerted upon phenotypes by environmental factors, lead to differential transmission of alleles to the next generation and so contribute to evolution.

### 4.3 Darwin's theory of Natural Selection

Charles Darwin (1809–1882) was the son of a wealthy doctor. He was expected to become a doctor like his father but was unsuccessful at this, and later started studying to be a clergyman. Darwin abandoned these plans also and spent his time at Cambridge University socialising and collecting specimens. His only passion in life was natural history and he seized the opportunity when it arose at the age of 22, to travel around the world. The purpose of the voyage was to survey the South American coast and some of the Pacific Islands. A naturalist was needed to observe, collect and record anything of interest and Darwin was invited to join the voyage. Figure 4.5 shows the route of the 5-year voyage of *HMS Beagle*.

Darwin carefully observed the flora and fauna (plants and animals) during his voyage from 1831–1836, and collected many specimens which he took back to Britain. The finches of the Galapagos Islands in particular caught his attention. The ship made several trips up and down the coast of South America and Darwin discovered giant fossilised skeletons on his trips ashore that belonged to species now extinct.

Darwin did not understand the significance of these finds at the time, but over the next 22 years he developed his theories using evidence from his travels. Another biologist, Alfred Wallace, was independently coming to the same

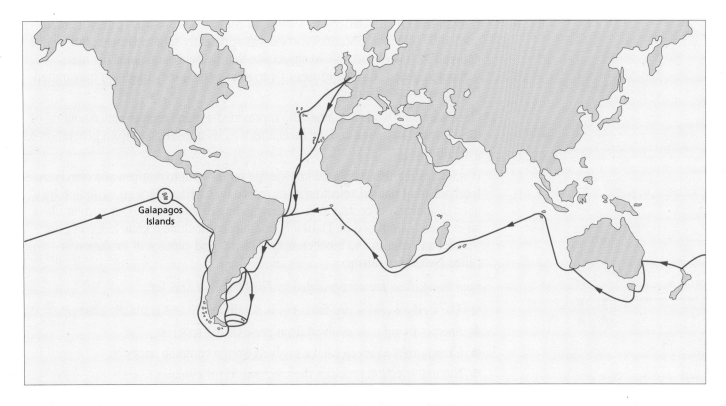

*Figure 4.5  Map showing the route taken by* The Beagle *on its voyage around the world, 1831–1836*

conclusions. In 1858 Charles Darwin, together with Alfred Wallace, published a paper suggesting that natural selection was the main factor causing evolutionary change. A year later in 1859, Darwin published his famous book, *The Origin of Species*.

The main points of Darwin's theory of Natural Selection were based on his observations which were:

- More offspring are produced than actually survive.
- Within a species there is variation, and some individuals are better suited to the conditions than others.

From the above observations Darwin deduced the following:

- There is competition between organisms for resources like food, space and a mate, in other words a struggle for existence.
- The better adapted individuals will live longer and have more offspring than those less well adapted.
- The offspring will inherit the favourable features that helped their parents survive, so leading to organisms better adapted for survival.
- Those less well adapted may die before reaching maturity and therefore may not live long enough to reproduce. The number of individuals with less helpful features will decrease in the population.
- This is 'survival of the fittest' or Natural Selection.
- There will be change in the individuals of a species, with some becoming better adapted to the local conditions.
- Over time this may lead to a new species.

**Remember**

The 'fittest' organisms are those that are best suited to a particular environment. They are not the strongest, but the ones most likely to find food and shelter, attract a mate, breed successfully and avoid predators. The variation found within a species means that some will be better at achieving these than other less fit individuals.

Darwin had no knowledge of genetics, as Mendel's work was largely ignored until 1900. A copy of Mendel's work was apparently sent to Darwin, but he did not read it! It is remarkable to appreciate that Darwin formulated his ideas without knowledge of the causes of variation or how they might be transmitted from parents to offspring.

Once Mendel's work was available it provided a much greater understanding of the causes of variation and the mechanism of its transmission from generation to generation.

It is interesting that increased knowledge serves both to confirm and clarify the mechanism of natural selection as proposed by Charles Darwin. Evolution by natural selection is central to an understanding of all biology. Our current ideas on evolution are based on Darwin's findings but include a knowledge of post Mendelian genetics and biochemistry. This revised concept of evolution is called **Neo-Darwinism**.

Four basic ideas are encapsulated in *The Origin of Species*:

- The living world is not fixed but is continuously and gradually changing.
- Species living now evolved from pre-existing species.
- Closely related species have evolved from a common ancestor.
- Natural selection provides the mechanism for evolution.

### Why have giraffes got such long necks?

Darwin would explain the long neck length in giraffes as an example of natural selection. Variation in neck length in giraffes means that some are better suited to their conditions as they can reach the higher leaves at times of food shortage. Therefore more of the long-necked giraffes will live long enough to breed with other long-necked survivors and pass on this feature to their young. Those with shorter necks will be selected against as they cannot reach the leaves and are less likely to live long enough to breed. Thus the numbers with short necks will decrease. Eventually only long-necked giraffes will remain. They are the fittest.

3 Stick insects are commonly found in hedges. Their stick-like appearance makes them particularly well adapted to their habitat. Using the ideas of natural selection explain how they might have evolved.    *(4 marks)*

### 4.4  Evidence of evolution by natural selection

### Fossils

Evidence from fossils shows how organisms have changed over time. For example, palaeontology, the study of fossils, has shown that modern-day camels and llamas (Camelidae) are closely related and have arisen from a pre-existing camel. The common ancestor is thought to have arisen in North America. Over time the ancestral camels reproduced, increased in number and extended their

### Box 4.3  Other ideas on the origin and development of life

#### Creationism

Various cultures have 'creation' stories. For example, in *The Bible* (the book of Genesis) it states that God created the world and all its species in 6 days. Creationists take this literally and believe that all the species on the Earth appeared together at this time and have remained unchanged since they were individually created by God. Scientific evidence from fossils and carbon dating does not support this view. The earliest simple forms of life seem to have arisen from spontaneous biochemical reactions and, from these simple beginnings, millions of species have evolved. Evidence in the fossil record suggests that some of the species became extinct, while others evolved and continue to this day. It is hard to explain the rise, change and mass extinction of the dinosaurs if you hold creationist views. Darwin's book *The Origin of Species* was not welcomed by many church leaders when it was first published, as it went against the teaching of the church at that time. Today, with the overwhelming scientific evidence, most Christians take a symbolic view of creation and accept Darwin's view of evolution.

#### Lamarck and the theory of acquired characteristics

In 1809, the French biologist Jean-Baptiste Lamarck (1744–1829) proposed a theory of evolution based on the 'use and disuse' of organs. Lamarck proposed that organisms acquire new features which are useful during their life, and these acquired features are then passed on to their young. It is interesting that Erasmus Darwin, Darwin's grandfather, proposed a similar theory some years earlier. Lamarck explained the long neck length in giraffes as being the result of years of stretching. Giraffes constantly stretched their necks to reach leaves at the top of trees so their necks grew longer. Lamarck believed that this acquired feature would be inherited. In the absence of a knowledge of genes, this appeared to make some sense at the time. Like Darwin, Lamarck also believed that those best suited would survive.

Experiments were carried out to investigate these ideas. In one gruesome experiment, mice had their tails cut off to see if their young would be born with shorter tails. They were not. Lamarck's ideas are interesting historically but a modern understanding of genetics completely disproves his idea of inheritance of acquired characteristics.

range, some extending north and others moving south. This occurred during the Pleistocene period about 1–2 million years ago, when there was a land bridge between North America and Russia, across what is now a strip of water called the Bering Straits. Gradually some ancestral camels spread into South America

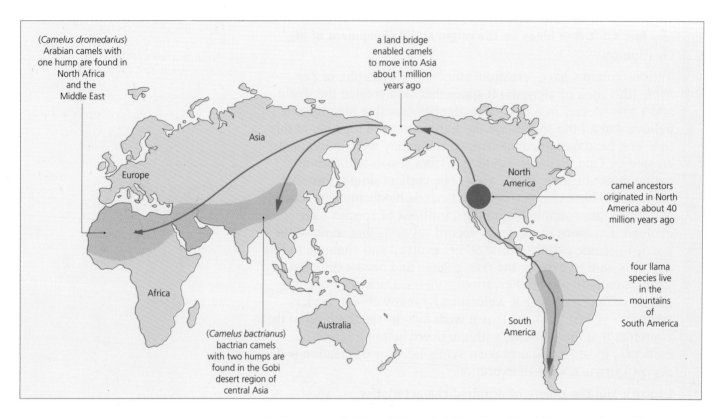

(*Camelus dromedarius*) Arabian camels with one hump are found in North Africa and the Middle East

a land bridge enabled camels to move into Asia about 1 million years ago

Asia

Europe

Africa

North America

camel ancestors originated in North America about 40 million years ago

four llama species live in the mountains of South America

Australia

South America

(*Camelus bactrianus*) bactrian camels with two humps are found in the Gobi desert region of central Asia

**Figure 4.6** *Map showing the origin and dispersal of the camel family (Camelidae)*

and others extended into Asia and Africa. Fossil evidence confirms this movement, which is illustrated in Figure 4.6. Camels and llamas are quite different species, but intermediate forms exist in the fossil record. These forms are now extinct.

As the two groups lived in different places, they were unable to interbreed and natural selection took place in each population, with different features being selected depending on the local conditions. Mutations also took place independently in the two populations leading to further variation. The two separated groups became increasingly different and we now have modern-day camels in Asia and Africa, and llamas in South America. They are clearly different species now, unable to interbreed, but they both evolved from one ancestral type (Figure 4.7).

**Figure 4.7** *A camel and llama, two distinct species that have evolved from a common ancestor*

(a)

*Figure 4.8* Archaeopteryx – this
'missing link' has both reptilian and
bird-like features which suggests that a
common ancestor may have led to both
the reptile and bird classes
*(a) cast of the Berlin fossil of
Archaeopteryx found in rocks from the
upper Jurassic period,
(b) artist's reconstruction of
Archaeopteryx,
(c) comparing bird features with those
of Archaeopteryx (Archaeopteryx
means 'ancient wing')*

(b)

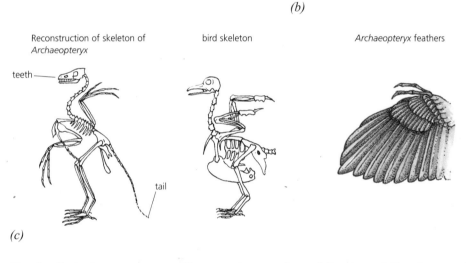

Reconstruction of skeleton of *Archaeopteryx* — teeth — tail    bird skeleton    *Archaeopteryx* feathers    bird feathers

(c)

The fossil *Archaeopteryx* was discovered soon after publication of *The Origin of Species*. The fossil shows both reptilian and bird-like features. *Archaeopteryx* had teeth and a tail with vertebrae, which are reptilian features, and a beak and feathers normally found only in birds. However, it lacked the well-developed keel possessed by birds for the attachment of flight muscles. Having both bird and reptilian features suggests a close evolutionary relationship between reptiles and birds. *Archaeopteryx* is not thought to be an ancestor of either modern birds or reptiles, but is from a 'side branch' of reptile evolution, a cousin of modern birds (Figure 4.8).

The gradual change in a species can be seen from the fossil record showing the evolution of the modern horse. The sequence of fossils is not a linear one but forms part of a 'branching tree' of horse evolution. The fossils are not direct ancestors of the modern horse but they do give some indication of the progression of 'horse' evolution. As more fossils are discovered, direct horse ancestry will become clearer. Figure 4.9 illustrates the evolution of the horse as shown by the fossil record.

Evolution is not always a gradual process. Although there are long periods with very little change, there are occasional, very sudden, large changes in both the structure and number of organisms. This is known as **punctuated equilibrium**.

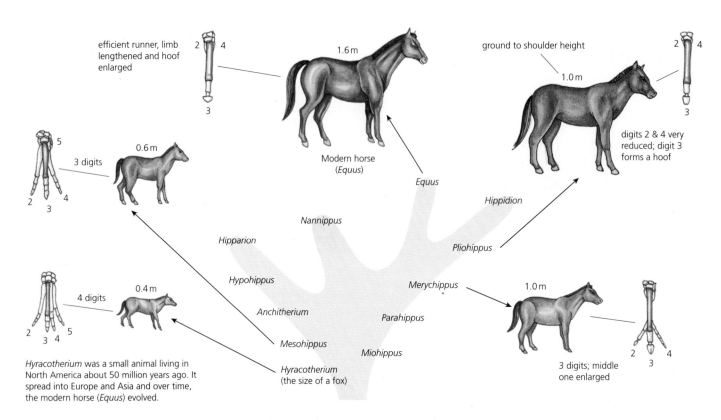

efficient runner, limb lengthened and hoof enlarged

2 4

3

1.6 m

Modern horse (*Equus*)

*Equus*

ground to shoulder height

1.0 m

2 4

3

digits 2 & 4 very reduced; digit 3 forms a hoof

*Hippidion*

5

3 digits

2 3 4

0.6 m

*Nannippus*

*Hipparion*

*Pliohippus*

*Hypohippus*

*Merychippus*

1.0 m

5

4 digits

2 3 4

0.4 m

*Anchitherium*

*Parahippus*

*Mesohippus*

*Miohippus*

*Hyracotherium* (the size of a fox)

2 4

3

3 digits; middle one enlarged

*Hyracotherium* was a small animal living in North America about 50 million years ago. It spread into Europe and Asia and over time, the modern horse (*Equus*) evolved.

***Figure 4.9*** *Evolution of the horse – evidence from fossil record (bones of the right forelimb are shown in each example)*

For example, the mass death of the dinosaurs and the accompanying loss of marine invertebrates was only one of several mass extinctions in the fossil record. These occurred at fairly regular intervals. There are also examples in the fossil record of gradual changes but some species, once present, may not change for millions of years and then may completely disappear. These abrupt changes contrast with Darwin's description of evolution as a slow, gradual process. Punctuated equilibrium is a theory proposed by Gould and Eldridge. This alternative viewpoint to that of gradual evolution is currently being hotly debated and is now accepted by many evolutionary biologists. It is quite possible that both viewpoints are correct at certain times. This is an example of Neo-Darwinism – looking at the theory of evolution by natural selection including new scientific evidence which was not available to Darwin.

**Comparative anatomy**

Comparative anatomy involves comparing structures such as the heart and skeleton in different animals. When the structure of a particular organ is found to be similar in different animals, common ancestry is indicated. The pentadactyl limb is a five-digit limb found in amphibians, reptiles, birds and mammals (see Section 5.3, page 162). It is slightly different in form in the animal groups but all have the same five-digit structure, modified to allow adaptation to different conditions. The most likely explanation for this common structure is that all these different animal groups have evolved from a common ancestor that also had the five-digit structure. A study of embryology among the vertebrate groups shows a remarkable similarity in the early stages of the embryo, again indicating a common ancestor. Figure 4.10 compares embryonic development in the different vertebrate groups.

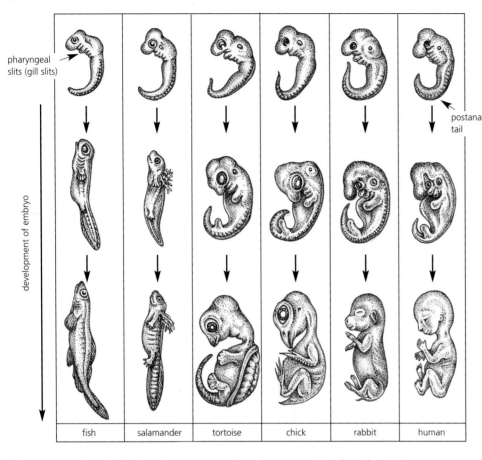

*pharyngeal slits (gill slits)*

*postanal tail*

*development of embryo*

| fish | salamander | tortoise | chick | rabbit | human |

**Figure 4.10** *Comparison of vertebrate embryos*

The greater the similarity between species, the more closely related they are likely to be. All vertebrate embryos have a post-anal tail and pharyngeal (gill) slits, both subsequently lost in humans, but again suggesting common ancestry.

Extending these observations implies a common ancestor for both humans and apes. This initially caused outrage as it led to a misunderstanding that humans actually evolved from apes. Recent DNA studies have confirmed our ancestry and have enabled classification of the rest of the animal and plant kingdom to be more meaningful and truly based on evolutionary relationships (see Section 5.6, page 167).

## Biochemistry

Evidence from biochemistry is probably the most powerful indicator of evolutionary relationships. Not only do all organisms share DNA as the genetic code, but all have ATP as an energy store and follow similar metabolic pathways, for example in respiration. This commonality is unlikely to have arisen so many times by chance.

Immunological research has revealed even more extraordinary relationships between species. Serum is blood plasma without the clotting factors. It contains plasma proteins. If the serum from one animal, A, is injected into another animal, B, the recipient produces antibodies that are specific to the proteins injected. This can be used to investigate the similarity of proteins in different organisms (Figure 4.11).

**Figure 4.11** *Immunological research – procedure used to investigate biochemical relationships (protein similarities)*

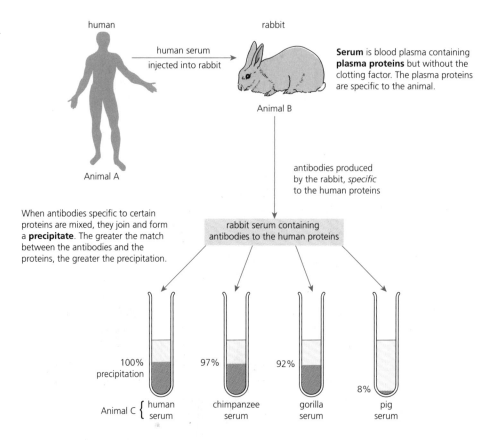

If serum from animal B, now containing the antibodies to animal A's proteins, is mixed with serum from another animal, C, the antibodies in the serum will combine with any proteins similar to that of animal A, forming a precipitate. The greater the number of similar proteins, the greater the precipitate. If the proteins in animal C are the same as in animal A, then 100% precipitation will take place. If there are fewer proteins in common, then less antibody antigen reaction will take place leading to reduced precipitation. The amount of precipitation therefore indicates the closeness of the relationship between animals A and C, as closely related animals will have similar proteins and produce greater precipitation.

Figure 4.11 shows what happens when human serum containing human plasma proteins is injected into a rabbit. The rabbit responds by producing antibodies specific to human proteins. When this rabbit serum containing antibodies to human proteins is then mixed with serum from a variety of animals the following results are obtained:

| Serum: | Human | Chimpanzee | Gorilla | Gibbon | Baboon | Lemur | Hedgehog | Pig |
|---|---|---|---|---|---|---|---|---|
| % precipitate: | 100 | 97 | 92 | 79 | 75 | 37 | 17 | 8 |

The results indicate a high degree of similarity between human proteins and those found in chimpanzees and gorillas, suggesting that we evolved from a common ancestor. By contrast, pigs only have 8% of their proteins in common with humans.

## Did You Know

### The downfall of 'Soapy Sam'

The impact of Darwin's book *The Origin of Species* shook society. On June 30, 1860, six months after publication, a debate took place between Samuel Wilberforce, Bishop of Oxford and T.H. Huxley, a distinguished biologist, at a meeting of the British Association for the Advancement of Science. Wilberforce was known as Soapy Sam because of his patronising manner. He attempted to belittle Darwin and his findings by demanding to know whether Huxley was descended from an ape on his grandmother's or his grandfather's side. In reply, Huxley, an outspoken Darwin supporter known as 'Darwin's bulldog' explained Darwin's theory and stated, 'If the question is put to me, would I rather have a miserable ape for a grandfather or a man highly endowed by nature and possessed of great means and influence, and yet who employs these faculties and that influence for the purpose of introducing ridicule into a grave scientific discussion – I unhesitatingly affirm my preference for the ape'. Soapy Sam sat down defeated. The outcome was not universally welcomed and the wife of the Bishop of Worcester when told of the debate, exclaimed, 'Descended from the apes! My dear, let us hope that it is not true, but if it is, let us pray that it will not become generally known'.

Of course Darwin did not suggest that we are descended from apes, but that we are related by a common ancestor. 'Our ancestor,' he once wrote, 'was an animal which breathed water, had a swim bladder, a great swimming tail, an imperfect skull and undoubtedly was a hermaphrodite.'

Anna Sproule wrote: '*The Origin of Species* was not just a book about biology. It was a formula for revolution. And its painstaking, perfectionist author was one of the world's great revolutionaries'.

*Figure 4.12* *Cartoonists delighted in Darwin's theory of evolution, which was ridiculed at the time by some*

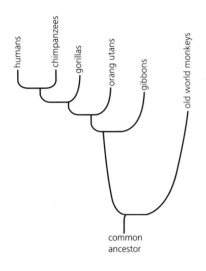

**Figure 4.13** *Hybridisation techniques indicate these relationships between the primates*

## DNA hybridisation

Comparing DNA is another way of discovering how closely two organisms are related. Figure 4.13 shows the evolution of humans, based on DNA hybridisation techniques.

The steps involved in the DNA hybridisation technique are as follows:

1 Collect DNA from two different species.

2 Heat treat to break the double-stranded DNA of both species to form single strands.

3 Use restriction enzymes to cut the DNA into fragments (see *AS Biology*, Section 16.2, page 481).

4 Mix the DNA fragments from the two species together.

5 The DNA fragments pair up if their bases are complementary. Some DNA fragments form a good match as they are from the same species, and hydrogen bonds hold the double strand together firmly. Other DNA fragments join between the different species to form a hybrid, but as fewer bases are complementary, less hydrogen bonds form between them making them less stable.

6 The double-stranded DNA from one species will require a higher temperature to be separated.

7 The hybrid DNA will separate at a lower temperature.

8 The higher the temperature needed to separate a hybrid, the more stable the molecule is, indicating similar DNA chains. The closer the temperature at which the hybrid separates to that of the double-strand DNA of the species to which it is attached, the closer their evolutionary relationship.

### 4.5 Evolution in action

A classic example of natural selection is seen in the peppered moth, *Biston betularia*. One form of the peppered moth is dark and the other light. The dark (melanic) form is due to a single gene mutation resulting in greater production of melanin, hence the term 'melanic'. The moth flies at night and rests on the bark of trees during the day. Before the industrial revolution there was a predominance of light-coloured moths throughout Britain and the darker moths were rarely seen. It was noticed that in the 1950s, the lighter moths were more abundant in rural areas and the darker moths were more common in areas with heavy pollution. In rural areas, the trees are covered with a pale-coloured lichen, camouflaging the light moths. By contrast, the dark moths stand out against the pale bark and are easy prey for the birds that feed on them, resulting in a greater number of light moths surviving in rural areas. In polluted areas, there is no lichen on the trees as lichens cannot tolerate high levels of sulphur dioxide. The trees are much darker in colour and the dark moths are now camouflaged. It is the lighter moths that are more visible to birds such as thrushes, and therefore more light ones will be eaten in polluted areas. Figure 4.14 shows both types of moth photographed on pale bark, and a map of their distribution in Britain. Note

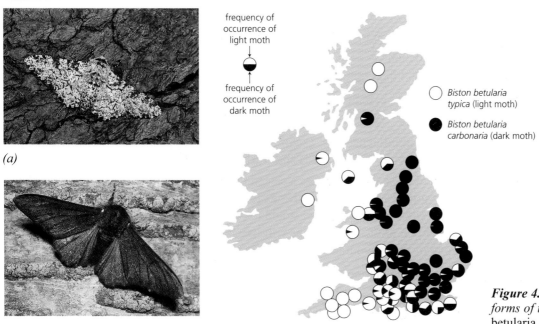

*(a)*

*(b)*                              *(c)*

frequency of occurrence of light moth ↓

↑ frequency of occurrence of dark moth

○ *Biston betularia typica* (light moth)

● *Biston betularia carbonaria* (dark moth)

***Figure 4.14*** *(a) light and (b) dark forms of the peppered moth,* Biston betularia, *(c) distribution in the British Isles*

the greater numbers of light moths in the rural counties of Devon and Cornwall in the South West, where tree trunks are covered with lichen. The dark form is found particularly in the industrial areas of the Midlands where pollution prevents the growth of lichen and darkens the tree bark.

These findings have been confirmed by experiments in which light and dark moths were released into woods in both polluted and unpolluted areas. The proportions of each type recaptured were noted. Observations confirmed that more light ones were eaten in polluted areas and more dark ones were eaten in unpolluted areas, due to their increased visibility. Natural selection therefore favours the light moth in unpolluted areas and the dark moth in polluted areas. Which is 'fittest' depends on the local conditions and this can change. With the Clean Air Act and a reduction in pollution, the numbers of the light moth are increasing in formerly polluted areas as lichen once again covers the tree trunks.

Natural selection causes a change in frequency of particular genotypes and phenotypes in a population, resulting in a change in the proportion of alleles. Examples include the incidence of sickle cell anaemia, warfarin resistance in rats, antibiotic resistance in bacteria and copper tolerance in grasses. All these examples are discussed more fully later in the chapter.

### Genes and natural selection

Darwin's theory of natural selection can now be applied, combined with a modern understanding of genetics. To illustrate the main points of the process we can take as an example the question: Why do hedgehogs have long spines?

Hedgehogs produce more offspring than can survive. A struggle for existence follows with competition for scarce resources, such as food and shelter. This competition represents **selection pressure**. Selection pressure is a factor which causes selection within a population by exerting **environmental resistance**.

***Figure 4.15*** *Why do hedgehogs have long spines?*

Examples of selection pressure include:

- competition for food, territory, and a mate
- predation
- limited light, oxygen or water
- variation in temperature
- disease
- human influence, e.g. use of antibiotics, use of pesticides like DDT (an insecticide, now banned) and warfarin (a rat poison), pollution.

Each of these factors causes a selection pressure within a population, with the fitter organisms having an advantage over others. The best-suited organisms are those whose alleles are best suited to the prevailing conditions. A higher proportion of these organisms will live long enough to breed and so contribute more of their favourable alleles to the next generation, compared to the less fit.

Within a population of hedgehogs there is variation in spine length – some hedgehogs have longer spines than others. The spine size varies depending on the alleles present, in other words it has a genetic basis. The continuous variation in spine length indicates that it is a polygenic feature. New variation will arise by mutation producing new alleles. These alleles and the corresponding phenotype are inherited. Hedgehogs with longer spines are better able to protect themselves and escape from predators. Those with shorter spines will be less successful. Consequently more hedgehogs with long spines will survive to breed with other long-spined survivors. The allele determining long spines will therefore be passed on to their many young. Fewer short-spined hedgehogs will survive to breed and pass on the allele for short spines. As a result there will be an increase in the frequency of alleles causing long spines in the population and a decrease in the frequency of the alleles producing short spines. Natural selection has brought about a change in the allele frequency – this is termed **microevolution**.

The main points of natural selection which can be applied to any example can be summarised as follows:

- there are too many offspring for the available resources
- a struggle for existence follows
- there is variation within a population with a genetic basis
- the best suited survive to breed and pass on the useful alleles; this is survival of the fittest
- those less suited die and their alleles decrease
- a change in the allele frequency takes place due to selection
- the surviving organisms are better suited to the conditions than previous generations.

Natural selection therefore leads to an increase in the number of organisms with favourable alleles and a decrease in the number with less favourable alleles. Thus there is a change in the allele frequency in the gene pool of the next generation.

**Remember**

Natural selection or microevolution brings about a change in allele frequency due to selection.

*Figure 4.16  A peacock displaying his tail*

Another example of selection pressure can be found in the tails of peacocks, which vary both in size and colour (Figure 4.16). The variation is due to different alleles, some arising by mutation. Females choose males with large, colourful tails as mates. Therefore only males with large tails are able to mate and pass on their alleles to the next generation. These are the 'fittest' and this is an example of **sexual selection**. Peacocks with smaller, less spectacular tails will not be selected by females and will be unable to breed and pass on their alleles to the next generation. In other words they are selected against. There will be an increase in the number of peacocks with large, colourful tails and a decrease in those with smaller tails in the next generation due to the change in allele frequency in the population.

A large tail must be a problem for peacocks, making them more vulnerable to predators, but they clearly live long enough to mate and their success in mating compared to short-tailed varieties allows them to pass on the alleles for a large tail. Therefore the reproductive advantages of a large tail outweigh the disadvantage of possible predation, and selection has favoured the larger tail over a smaller one. However, there must come a point where the tail becomes too large and hinders mobility to the extent that selection will favour those with the optimum tail size.

## 4.6  Directional selection

Directional selection involves the selection of organisms whose phenotype, at one end of a range of variation, gives them a selective advantage. In a continuous distribution curve, the phenotype shifts either to the right or the left.

Taking the example of ear length in hares we can see that desert hares have much larger ears than those in milder climates, and hares from the cold tundra have even smaller ears (Figure 4.17). How have these differences come about?

The primary function of ears is to aid hearing, but in this case they are also involved in temperature regulation. The large ears of the antelope jack rabbit (Figure 4.17a) increase the surface area for heat loss by radiation and are useful in the hot, desert conditions to prevent overheating. Similarly, the small ears of the arctic hare (Figure 4.17c), with their reduced surface area, lose less heat by radiation and are useful in the cold, arctic conditions.

*(a) Antelope jack rabbit* (Lepus alleni). *Lives in the Arizona desert (hot climate)*

*(b) Black-tailed jack rabbit* (Lepus californicus). *Lives in Oregon (mild climate)*

*(c) Arctic hare* (Lepus arcticus). *Lives in Arctic tundra (cold climate)*

*Figure 4.17  Ear size in hares from different climates*

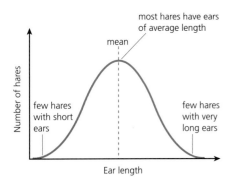

**Figure 4.18** *Graph showing distribution of ear length in a hare population*

Within a population of hares, there will be some variation in the length of ears – some will have smaller ears, others will have larger ears, but most will have ears of average size, the *mean* for that population. This bell-shaped curve is called a normal distribution with a maximum number at the mean and fewer at the extremities. The graph in Figure 4.18 shows the variation in ear length in a hare population. There is a continuous gradation from short to long ears with no gaps. This is an example of continuous variation (see Section 4.1, page 105).

In cold, arctic conditions, hares with small ears will lose less heat than those with larger ears. In excessively cold conditions, small-eared hares therefore have a **selective advantage**. More hares with small ears will survive to breed together and pass on the alleles for short ears to their young, compared with long-eared hares. Consequently, the number of hares with short ears will increase and those with longer ears will decrease, as their long ears give them a **selective disadvantage**. The peak on the graph will shift to the left (towards short ears), moving the mean with it – directional selection.

Similarly, hares living in hot desert conditions will also show some variation in ear length, with a maximum number at the mean. During times of extreme heat, hares with larger ears will be at a selective advantage as they can lose more heat by radiation. More hares with large ears will survive to breed and pass on their alleles, so increasing the frequency of alleles for large ears in the population. This will result in a shift to the right. Hares with large ears are the 'fittest' in excessive heat. Fit individuals pass on more alleles to the next generation than other individuals in the same population. Figure 4.19 shows the effect of extreme cold and heat on ear length in hares.

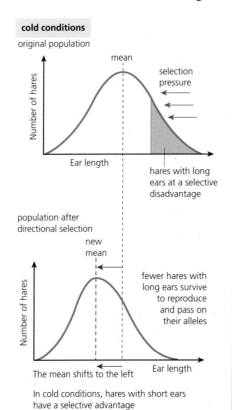
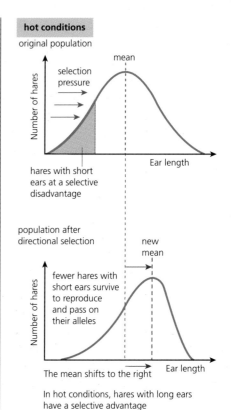

**Figure 4.19** *Graphs showing directional selection in hares*

Directional selection results in one extreme of a range of variation being selected against. This shifts the mean of a feature in one direction. The greater the original variation the greater the possible shift.

### Antibiotic resistance in bacteria

Since the 1940s antibiotics have been widely used to kill bacteria. Although successful at the start, the number of bacteria resistant to antibiotics is increasing, causing a severe problem in tackling infections throughout the world. Resistance to the antibiotic penicillin is now widespread. Bacteria that are resistant to penicillin are not killed by it.

Assuming that a normal population of bacteria shows little resistance to a particular antibiotic, how can a resistant strain come about? Resistance to antibiotics is controlled by genes. A chance mutation may confer penicillin resistance to one bacterium. The mutation produces an allele that results in an enzyme that can inactivate penicillin so providing resistance to the mutant bacterium. Two groups of bacteria result, one resistant and the other not. This is an example of discontinuous variation with two distinct groups. As a result, if a patient is treated with penicillin, non-resistant bacteria will be destroyed and any bacterium resistant to penicillin will survive to reproduce and pass on the resistant allele. Here, the use of penicillin provides the **selection pressure**, selecting the resistant ones and killing the others. Prolonged use of penicillin will result in a large population of bacteria that are resistant to penicillin, making the antibiotic less effective. Taken to the extreme, only penicillin-resistant bacteria will remain and the antibiotic will be useless. This is an example of directional selection favouring the most resistant bacteria. It will only take place while the antibiotic is used. Doctors are therefore reluctant to prescribe antibiotics unless absolutely necessary as it can lead to the emergence of resistant strains.

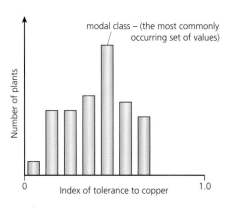

*(a)*

With bacteria, a mutation can give an immediate advantage as each cell only contains one strand of DNA and one allele controlling each gene. (In most organisms chromosomes and alleles are in pairs, with a dominant allele preventing the expression of a recessive allele unless it is in the homozygous state.) The rapid reproductive rate, splitting every 20 minutes in ideal conditions, means a large population of resistant bacteria can therefore emerge very quickly from only one original resistant bacterium.

### Copper tolerance in grasses

The soil around abandoned copper mines is heavily contaminated with copper, which is poisonous to most plants. Grasses vary in their tolerance to copper and this variation has a genetic basis. Some varieties of the grass *Agrostis tenuis* are copper tolerant and can grow in copper-contaminated soil.

*Agrostis tenuis* shows continuous variation with regard to copper tolerance, with a few plants displaying high tolerance to copper, a few having little copper tolerance, and the majority having an average tolerance to copper. Figure 4.20a shows an index of tolerance in *A. tenuis*, ranging from 0 for plants with no tolerance to 1.0 for plants with a high level of copper tolerance. The gradation of tolerance giving a normal distribution suggests that copper tolerance is controlled by polygenes. (Note that the mean is lower then the modal class in this case.)

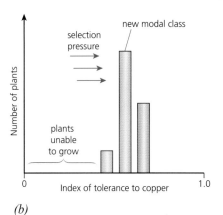

*(b)*

***Figure 4.20*** *Copper tolerance in* Agrostis tenuis
*(a) in soil not contaminated with copper,*
*(b) in soil contaminated with copper (directional selection has taken place)*

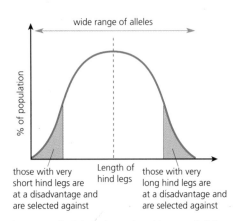

those with very short hind legs are at a disadvantage and are selected against

Length of hind legs

those with very long hind legs are at a disadvantage and are selected against

*Large* standard deviation and a *wide* range of alleles.

*(a)*

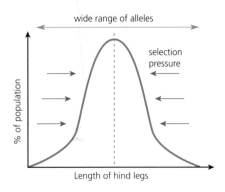

*Lower* standard deviation (the data is clustered around the mean) but the *same* range of alleles as in (a).

*(b)*

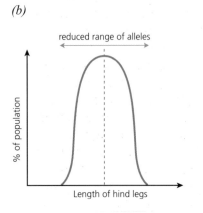

*Lower* standard deviation and a *reduced* range of alleles compared to (a).

*(c)*

When grass seeds from *A. tenuis* were planted on soil contaminated with copper, only a few plants were able to grow. These were the seeds with copper tolerance – the other seeds were unable to tolerate the copper and failed to grow. The surviving plants reproduced and passed on the alleles for copper tolerance to their seeds, which consequently were also able to grow. Over time there was a change in the allele frequency. Eventually the soil was covered by grasses which were all tolerant to copper. There has been directional selection favouring the tolerant grasses. The most commonly occurring phenotype, the modal class, was that of the copper-tolerant plants – they increased in number and the peak shifted towards the more tolerant end of the range, to the right. The selection pressure here was the copper which killed the non-tolerant grasses (Figure 4.20b). In soil not contaminated with copper, there is no selective advantage in being copper tolerant and these plants will not be selected for.

**4.7  Stabilising selection**

In the case of stabilising selection, the extreme phenotypes are selected against. This happens in an unchanging environment. The length of the hind legs in rabbits can be used as an example. The variation in length is due to the different alleles present. It is clearly a disadvantage to have short hind legs, which would reduce the animal's ability to escape from danger. Very long hind legs would also impair mobility. The two extremes will therefore be selected against and the alleles for very long or very short legs will decrease in subsequent generations (Figure 4.21a). The ideal length is somewhere in between the two extremes. The intermediate length is selected for, leading to an increase in the allele frequency for average leg length (Figure 4.21b). As a result there is a decrease in the range of phenotypes for leg length. Continued stabilising selection would lead to loss of the extremes and reduced variation from the mean (Figure 4.21c). The loss of alleles and range of phenotypes reduces the ability of a population to adapt to possible changes in conditions in the future.

**Standard deviation** is a measure of the spread of values from the mean. It can only be applied to measurable phenotypes such as length or weight. The wider the spread in the data, the greater the standard deviation. After stabilising selection, the standard deviation is lower, with a higher proportion of phenotypes clustered around the mean, although the range may be unchanged.

Birth weight in human babies is another example of stabilising selection. Figure 4.22 shows birth weight and human infant mortality (death) rates over a 12-year period at University College Hospital, London. The graph shows that babies of low and high birth weight have a higher mortality rate than babies of average

*Figure 4.21   Stabilising selection in rabbits (a) normal distribution with a wide range of alleles, (b) after stabilising selection, a change in the standard deviation but not in the range of alleles, (c) a change in the standard deviation and in the range of alleles*

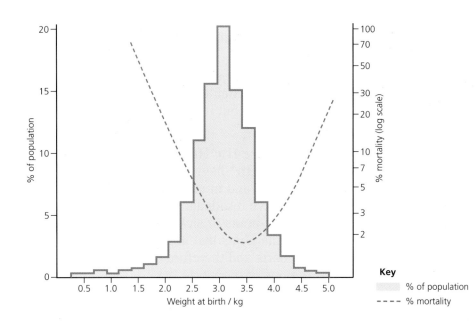

*Figure 4.22* *Human infant mortality and birth weight*

birth weight. In other words, babies of average birth weight are selected for, survive and pass on the alleles for average birth weight. This is stabilising selection – favouring the mean and selecting against the extremes. It is interesting to note that the highest death rate is amongst low birth weight babies and therefore slight directional selection is also taking place to the right. Recent evidence suggests a link between high birth weight and reduced coronary heart disease. As people usually develop this disease after the age at which they become parents, it is unlikely to alter the pattern of selection as their genes will already have been passed on.

*Figure 4.23* *A coelacanth*

## Did You Know

In 1938, a young museum curator from South Africa made the extraordinary discovery of a fish thought to have been extinct for millions of years. Amongst fish caught in a net was a large, strange specimen, 1.6 metres long with hard scales and limb-like fins. It was a **coelacanth** (Figure 4.23), and was given the name *Latimeria chalumnae* (see Box 5.2, page 165). The limb-like 'fins' were fleshy and able to support the weight of the animal. An ancestor of the coelacanth is thought to have led to the evolution of amphibian-like animals that crawled out of the water and onto land. The internal structure of the coelacanth limb shows the beginnings of the pentadactyl limb found in all terrestrial vertebrates.

The coelacanth has been around unchanged for over 400 million years – it is sometimes called a living fossil. It is a good example of stabilising selection with selection favouring the original form which has remained unchanged over time.

## Box 4.4  Sickle cell anaemia

Sickle cell anaemia is controlled by a single gene which has two alleles, **Hb$^A$** and **Hb$^S$**. The genetics of sickle cell anaemia is discussed in Chapter 3, page 68. Sickle cell anaemia is not an example of continuous variation, but it does illustrate how genes can confer a selective advantage.

The three possible genotypes are **Hb$^A$Hb$^A$**, **Hb$^A$Hb$^S$** and **Hb$^S$Hb$^S$**. People who are **Hb$^A$Hb$^A$** have normal haemoglobin, but are not resistant to malaria, and in malarial regions of the world they will have a selective disadvantage. People who are **Hb$^A$Hb$^S$** suffer from periodic bouts of sickling, causing severe anaemia as well as pain in the joints that affect mobility, but they have resistance to malaria and therefore have a selective advantage in malarial regions of the world.

It appears that the protoctistan parasite causing malaria, *Plasmodium* sp. (Figure 4.24), cannot live in sickle-shaped red blood cells. In parts of Africa and Asia, there are therefore large numbers of people with the **Hb$^S$** allele, far higher than elsewhere in the world. This is because people with the sickle cell trait, the heterozygotes, have resistance to malaria, which is endemic in these parts of the world.

The homozygous individuals, **Hb$^S$Hb$^S$**, will suffer from severe anaemia and usually die in infancy. Therefore in malarial regions of the world, the heterozygote is selected for, an example of **heterozygous advantage**. The two homozygotes are at a disadvantage, either through severe anaemia or through malaria.

This can be described as a type of stabilising selection favouring the heterozygote and reducing the homozygotes. This is only true in malarial areas and will not apply in areas where malaria is absent. In non-malarial areas, selection favours the homozygote **Hb$^A$Hb$^A$**, leading to directional selection in these areas. Figure 4.25 shows the distribution of the sickle cell allele **Hb$^S$** in two populations.

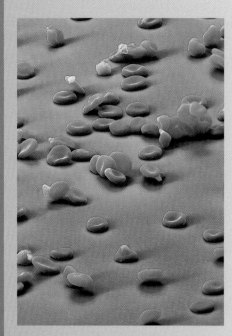

*Figure 4.24*  *The malarial parasite* Plasmodium falciparum *(seen here in yellow) amongst red blood cells*

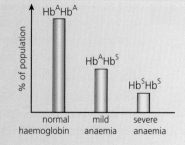

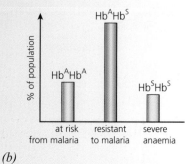

*Figure 4.25 Distribution of the sickle cell allele (a) in a non-malarial region, (b) in a malarial region*

(a)

(b)

Extension

### Box 4.4  Sickle cell anaemia continued

As the heterozygote is selected for, a large number of the population in malarial areas will carry the sickle cell allele, so increasing the chance of their having children with severe anaemia. Heterozygous advantage and stabilising selection therefore maintain a high frequency of the sickle cell allele in malarial areas. Figure 4.26 shows areas of the world with high levels of malaria in relation to the distribution of the sickle cell allele.

About 1 in 400 people of Afro-Caribbean origin in Britain has sickle cell anaemia and 1 in 10 carries the sickle cell allele.

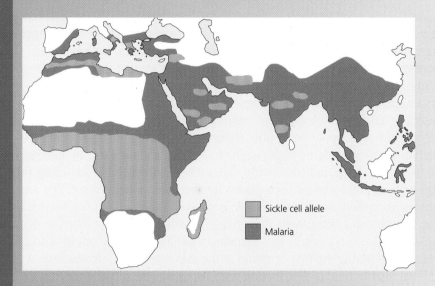

*Figure 4.26* Map showing the distribution of malaria in relation to the incidence of the sickle cell allele before 1930

4  Explain why, in Britain, sickle cell anaemia is more common in black Afro-Caribbeans than in the population as a whole.  *(2 marks)*

### Did You Know

Thalassaemia is a genetic disorder found in parts of the Mediterranean and Asia, causing severe anaemia in sufferers. This disorder also appears to give some protection to malaria and may explain the high rates of thalassaemia in areas where malaria was common.

## 4.8 Disruptive selection

This type of selection favours individuals at the phenotypic extremes. The intermediate phenotypes are selected against, so dividing or disrupting the phenotype into two distinct forms. Over time a population may be completely separated into two sub-populations. If the two groups do not interbreed, each population may give rise eventually to a new species. This type of selection is rare but important.

### Pacific salmon

A particularly good example is found in the Pacific salmon. These fish breed only once in their lifetime. The males fall into two categories – either large and aggressive fish, or small ones that hide amongst the rocks. Both types are successful at mating with the females. The large male fish fight between themselves for a position near a female and release their sperm once the female's eggs are released. The small male fish hide amongst the rocks and sneak out quickly to release their sperm at the appropriate time. Male fish of an intermediate size would stand little chance of fertilising the eggs, being too small to compete with the large, aggressive males and too large to hide between the rocks. The large and small male fish are both successful at fertilising the eggs and therefore pass on their alleles for fish size to their young. The less successful medium-sized male fish do not pass on their alleles for intermediate size, resulting in a decrease in male fish of middle size and an increase in those that are large or very small. This situation must have come about by disruptive selection favouring the two extremes at the expense of the intermediate. A bimodal distribution, or distribution with two peaks, results (Figure 4.27).

### Snails

The snail *Cepaea nemoralis* has been extensively studied and found to exist in a variety of colours and banding depending on the alleles present. When a variety of forms exists within a species, it is called **polymorphism** (many forms). Figure 4.28 shows the variety of forms that exists.

The snails are eaten by birds, in particular the song thrush, although a decrease in their number suggests that other selection pressures may now exist. Any snail that blends into the surroundings is less visible and has a greater chance of survival. In grassland areas the dominant snail type is the yellow, unbanded variety as it is well camouflaged and less visible to birds. Amongst leaf litter, however, these snails are at a disadvantage, and banded and brown shells predominate as they are better camouflaged.

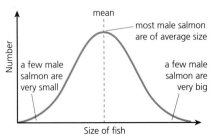

Original population – variation in fish size due to alleles (continuous variation, as no gap).

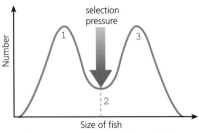

1  Small male salmon are successful at fertilising the eggs so passing on their alleles.
2  Intermediate size male salmon are unsuccessful at mating, so contributing few alleles to the next generation. They are selected against and decrease in numbers.
3  Large male salmon are able to fertilise the eggs and pass on the alleles for large size.

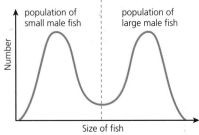

Continued selection pressure results in two groups of male fish, one small, the other large, with few intermediate size male fish.

This is an example of **discontinuous variation** – distinct groups.

***Figure 4.27*** *Disruptive selection in the Pacific salmon (male)*

Clearly, different phenotypes are selected for in different habitats. One population of *Cepaea* can contain both yellow and brown varieties, depending on the background colour of the habitat. Snails of other colours will not be selected for if they are not camouflaged, and their allele frequency will decrease. This may result in distinct colours only due to disruptive selection. Colour in snails is an example of discontinuous variation.

It is important to realise that there are other factors involved in selection other than predation. For example, it appears that yellow snails survive exposure to the sun and high temperatures better than other snails and this may contribute to their widespread distribution on sand dunes.

***Figure 4.28*** *Varieties of the snail* Cepaea nemoralis

**Summary of the three main types of natural selection**

**1 Directional selection**

● **Directional selection** favours one extreme in a range of phenotypes. Organisms with phenotypes at one end of a range have a selective advantage over others. This causes the mean to shift towards the phenotype providing the selective advantage. There is an increase in the allele frequency at one end of a range and a decrease in those at the other end. This causes the modal class to move one way (Figure 4.29a).

● **Examples of directional selection:**

| *What is selected* | *How it is an advantage* |
|---|---|
| long neck in giraffes | enables giraffes to reach leaves when food is scarce |
| ability to run fast | allows prey to escape from predators |
| long tail in peacocks | attracts a female for mating |
| resistance to antibiotics | enables bacteria to survive when antibiotics are used |
| large canine teeth in lions | assists in killing prey |

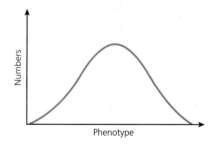
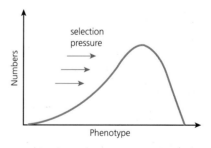

***Figure 4.29*** *(a) Directional selection*

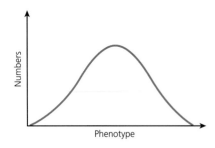

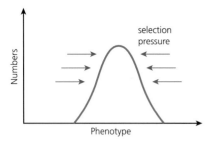

*Figure 4.29 (b) Stabilising selection*

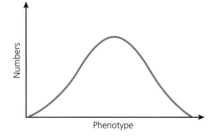

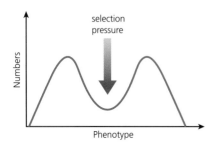

*Figure 4.29 (c) Disruptive selection*

## 2 Stabilising selection

● **Stabilising selection** favours the mean in a range of phenotypes. When the environment is stable and unchanging, the mean is selected for as it is best suited for the current conditions, and the extremes are selected against. The mean has come about by selection in particular conditions and the environment has not changed. There is therefore an increase in the allele frequency at the mean and a decrease in those at the extremities of the phenotype (Figure 4.29b).

● **Examples of stabilising selection**

| *What is selected* | *How it is an advantage* |
| --- | --- |
| average weight in human babies | increases their survival rate |
| sickle cell trait (heterozygote) | protects against malaria |
| flowering at a similar time within a species | ensures that pollination is successful |

## 3 Disruptive selection

● **Disruptive selection** favours the extremes in a range of phenotypes. Those organisms with intermediate phenotypes are selected against and more than one distinct phenotype is selected for, leading to more than one modal class (Figure 4.29c).

● **Examples of disruptive selection**

| *What is selected* | *How it is an advantage* |
| --- | --- |
| large and very small male Pacific salmon | able to fertilise the females' eggs |
| finches with short and long beaks | able to exploit different food sources |
| snails with yellow and brown shells | provides camouflage in habitat |

**?**

5 Farmers selected cattle to produce long horned and hornless varieties. Which type of selection does this represent? *(1 mark)*

6 Over time the colour of rose petals has become more vivid. Using your knowledge of natural selection, explain which type of selection is taking place. *(2 marks)*

## 4.9  Population genetics

A population is a group of the same species in a particular habitat. Members of the same species can interbreed and therefore their genes are free to mix within the **gene pool**. All the genes and alleles in a population at a particular time make up the gene pool. The proportion of different alleles in a population change over time due to mutations and selection, but remain constant if no selection or mutations are taking place.

Therefore, in the absence of change, the proportion of alleles of a particular gene in a population will remain constant from generation to generation. Under these circumstances the **Hardy–Weinberg equations** can be used to calculate the frequencies of alleles and genotypes in a population.

The Hardy–Weinberg equilibrium states that 'the frequency of dominant and recessive alleles in a population will remain constant from generation to generation provided certain conditions exist'.

Conditions in which the Hardy–Weinberg equations apply are:

● no selection is taking place; all alleles are equally advantageous
● no mutations are occurring
● mating is random
● the population is large
● no migration is taking place either into or out of the population.

These conditions can only really be met in an ideal (theoretical) population. If the above conditions are met then the allele proportions will not change as no selection is taking place, and the equations can be applied.

In reality, the stable situation rarely applies but it is a mathematical tool that can be of use in determining allele and genotype frequencies. If these are found to have changed over time, then evolution or change is taking place in a population and the Hardy–Weinberg equilibrium does not apply.

Two equations are involved. The first equation applies to **alleles**:

$$p + q = 1.0$$

The letter p represents the frequency of the dominant allele; q represents the frequency of the recessive allele; 1.0 represents the total frequency of all alleles of this gene in the population as a decimal (instead of using 100%). (Adding together the frequency of the dominant and recessive alleles would make up the whole population for this particular gene, i.e. 100% or 1.0.)

For example, let **A** represent the dominant allele and **a** the recessive allele. If the frequency of **A** in a population is 60%, then we can calculate the frequency of **a**.

First we change the % to a decimal:

     Frequency of A   = 60%
     Therefore p       = 60% or 0.6

Using the Hardy–Weinberg equation:

$$p + q = 1.0$$
$$\text{therefore } q = 1.0 - 0.6$$
$$= 0.4$$

Thus the frequency of the recessive allele, **a**, is 0.4 or 40%.

The second equation applies to **genotypes**:

$$\mathbf{p^2 + 2pq + q^2 = 1.0}$$

where  $p^2$ = frequency of the homozygous dominant individuals
$2pq$ = frequency of the heterozygous individuals
$q^2$ = frequency of the homozygous recessive individuals
$1.0$ = the total population.

Taking the example of the genotypes, **AA**, **Aa** and **aa**:

● the frequency of **AA** is represented by $p^2$
● the frequency of **Aa** is represented by $2pq$
● the frequency of **aa** is represented by $q^2$.

The two equations can be used to find out the frequency of both alleles and genotypes in a stable population.

**Using the Hardy–Weinberg equation**

**Example 1:**

In a population, some people can taste a chemical called phenylthiocarbamide; others can not. Tasters have the dominant allele; non-tasters do not. Nine per cent of the population are non-tasters.

**T** represents the allele for tasting and **t** the allele for not tasting. Therefore tasters must have the dominant allele and be either **TT** or **Tt** ($p^2$ or $2pq$). Non-tasters must have the genotype **tt** ($q^2$). Nine percent are non-tasters, so $q^2$ = 9% or 0.09.

To find q we find the square root of $q^2$ (0.09), which is 0.3.
Therefore the frequency of the recessive allele, q, is 0.3.

We can now use the equation to find p.

$$p + q = 1.0$$
$$\text{therefore } p = 1.0 - 0.3 = 0.7$$
$$\text{therefore the frequency of the dominant allele, p, is } 0.7$$

These allele frequencies can then be inserted into the genotype equation.

$$p^2 + 2pq + q^2 = 1.0$$
$$p^2 \text{ (\textbf{TT})} = (0.7)^2 = 0.49 = 49\%$$
$$2pq \text{ (\textbf{Tt})} = 2 \times 0.7 \times 0.3 = 0.42 = 42\%$$
$$q^2 \text{ (\textbf{tt})} = (0.3)^2 = 0.09 = 9\%$$

**Example 2:**

In ladybirds the two colours, red and black, are controlled by a single gene. The allele for black, **B**, is dominant to that for red colour, **b**. Figure 4.30 shows the red and black forms of ladybirds.

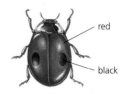

In a population of ladybirds the frequencies of the alleles **B** and **b** are the same. Using the Hardy–Weinberg equations it is possible to calculate the frequency of the genotypes and phenotypes in the population.

Let p represent the frequency of **B**, and q represent the frequency of **b**. If the frequency of the dominant and recessive alleles are the same, then p = q. As p + q = 1.0, then p = 0.5 and q = 0.5.

To find out the frequencies of the genotypes the equation $p^2 + 2pq + q^2 = 1.0$ is used:

$p^2$ (**BB**) = $(0.5)^2$ = 0.25 or 25%

2pq (**Bb**) = $2 \times 0.5 \times 0.5$ = 0.5 or 50%

$q^2$ (**bb**) = $(0.5)^2$ = 0.25 or 25%

As **B** is dominant, the genotypes **BB** and **Bb** produce black ladybirds. The frequency of black ladybirds is therefore 0.25 + 0.50 = 0.75 or 75%, and that of red ladybirds is 0.25 or 25%.

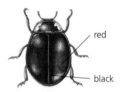

**Example 3:**

In a human population, one person in 2000 suffers from cystic fibrosis. The condition is controlled by a single gene with two alleles. People who have the dominant allele, **D**, do not suffer from this condition. People with the genotype **dd** suffer from cystic fibrosis. If one in 2000 is a sufferer, then one in 2000 must have the genotype **dd**.

**dd** or $q^2 = \dfrac{1}{2000}$ = 0.0005

therefore q = the square root of $q^2$

= 0.022

***Figure 4.30** The red and black forms of the two-spot ladybird*

The frequency of the recessive allele, q, is 0.022.

p + q = 1.0
therefore p = 1.0 − 0.022 = 0.978

The frequency of the dominant allele, p, is 0.978.

Using the formula $p^2 + 2pq + q^2 = 1.0$ we can calculate the frequency of the genotypes in the population:

$p^2$ (**DD**) = $(0.978)^2$ = 0.956 = 95.6%

2pq (**Dd**) = $2 \times 0.978 \times 0.022$ = 0.043 = 4.3% (these are carriers)

$q^2$ (**dd**) = $(0.022)^2$ = 0.0005 = 0.05%

**Remember**

- allele frequency:
  p + q = 1.0

- genotype frequency:
  $p^2 + 2pq + q^2 = 1.0$

This is the formation of two or more species from an existing one. As long as members of a population are able to interbreed, they are considered to be the same species sharing the same gene pool with free **gene flow** between organisms. Gene flow is the interchange of alleles within and between populations. If they become separated, two distinct populations result, each with their own gene pool. They are described as **reproductively isolated** if one part of the population is unable to reproduce with another. Mutations occur in each population producing new alleles, and natural selection takes place independently, selecting different alleles in each population depending on local conditions. For example, in one population a sandy soil may favour lizards that are a sandy colour and blend in. In another population grassland will favour green lizards, so leading to the selection of different alleles for colour in each population. Geographical isolation may prevent the alleles from one population spreading into the other, and as there is no gene flow between them, they have separate gene pools. Over time, with separate mutations and selection taking place, the two populations will become increasingly genetically different from each other. Each population will become adapted to the different local conditions. At this stage, mating between individuals from the different populations may no longer be successful as they have become genetically different. They have therefore become different species – **speciation** has occurred.

The fossil record shows some of these changes in a population preserved in rock. A variety of fossil types can be found:

- The entire body of an organism may be preserved but this is rare, e.g. insects trapped in amber resin, frozen woolly mammoths found in Siberia.
- The hard parts of an organism are preserved, such as bones, teeth or shells.
- A mould of an organism is found, embedded in rock.
- A copy of part of a plant or an animal is found in stone, formed when the original parts were replaced by mineral deposits, e.g. petrified wood.

Members of a species can interbreed successfully and produce fertile offspring. The problem here is how can we tell if organisms would be able to breed successfully if they are not around at the same time as each other.

**Types of speciation**

Two types of speciation are possible:

- **Sympatric** (same land) **speciation** between two groups in the same environment. There are several ways in which this type of speciation can occur.
- **Allopatric** (different land) **speciation** occurs due to geographical isolation. Geographical barriers include rivers, deserts and mountains, and lead to reproductive isolation and speciation in different places. Examples of this type of speciation are much more common than those of sympatric speciation.

**Definition**

A species is defined as a group of organisms which look alike and can interbreed successfully to produce fertile offspring.

## Box 4.5 Dog species

Biologists generally agree that the domestic dog, *Canis familiaris*, is descended from the wild grey wolf, *Canis lupus*. The grey wolf used to be common in Europe, North America and Asia. Despite its name, the grey wolf's coat comes in a variety of colours and the many different coat colours found in dogs today are due to this. The range of domestic dogs has spread in association with humans and they are able to interbreed with other *Canis* species successfully. Figure 4.31 shows successful interbreeding between members of the genus *Canis*.

### Did You Know

Wolf populations in Scandinavia may die out because they are mating with dogs. Wolves breed with dogs when there are too few wolves around and this could lead to the end of the wolf population and the development of a new hybrid.

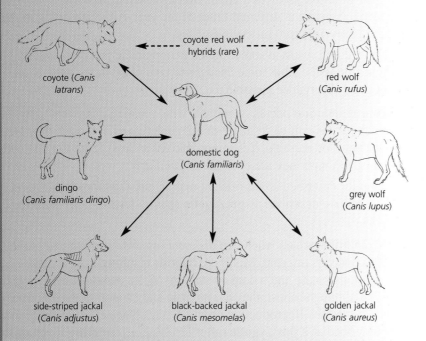

**Figure 4.31** *Interbreeding between dogs of the genus* Canis, *producing fertile hybrids*

It is interesting that the domestic dog acts as the 'common currency' linking the other members of the dog family. Although all the dogs can mate successfully with domestic dogs and produce fertile hybrids (this is not a common occurrence but is possible), most cannot mate successfully with the other *Canis* species. Only coyotes and red wolves can interbreed.

This indicates that members of the dog family share a similar gene pool. It is interesting to note that although three distinct species of jackals overlap in East Africa, the species do not interbreed which means they have separate gene pools. But they are all able to interbreed with the domestic dog which suggests that they and domestic dogs all share the same gene pool and should therefore be classified as a single species!

As you can see, defining a species is not easy and there are many exceptional circumstances. Problems with defining gull species is discussed fully in Chapter 5.

**Types of sympatric speciation**

- **Behavioural isolation** can prevent reproduction if, for example, one part of a population becomes active at night, (nocturnal) and the other is only active during the day (diurnal). They will be unable to reproduce even though they are in the same area. They are reproductively isolated and speciation may follow.

- **Ecological isolation** can cause reproductive isolation followed by speciation if different parts of a population occupy different niches, e.g. if one part lives in the tree tops whilst another is found on the forest floor. They occupy the same area but are reproductively isolated.

- **Temporal isolation** involves a population in which groups are reproductively active at different times of the year. Although living in the same area, they are reproductively isolated.

All of the above will result in some limited geographical isolation even in the same habitat, and will enable some changes to take place in the gene pool, which will make divergence possible.

- **Mechanical isolation** occurs in animals where mating is prevented by the non-matching position or structure of the genitalia, even though they are the same species. For example, it is impossible for a Great Dane to mate with a Chihuahua although they are both *Canis familiaris*.

All the examples described so far prevent mating and therefore fertilisation from taking place and are described as **prezygotic** (before fertilisation) **isolating mechanisms**.

- **Hybrid isolation** occurs when two different organisms interbreed but the hybrid produced is infertile. Mules are hybrids resulting from a cross between a horse mare and a donkey stallion. Mules demonstrate hybrid vigour but although robust, they are infertile and cannot produce normal gametes at meiosis. Reproductive isolation takes place after fertilisation and this is described as a **postzygotic** (after fertilisation) **isolating mechanism**.

**Did You Know**

The zebronkey is a hybrid produced from crossing a zebra with a donkey. Zebras have a 2n (diploid number) of 44 and donkeys a 2n of 62. Gametes from each have 22 and 31 chromosomes respectively, which join at fertilisation to produce a zebronkey with a 2n of 53. Clearly an odd number of chromosomes cannot pair up during meiosis to form gametes and so the hybrids are infertile.

*Figure 4.32* A zebronkey

## Examples of allopatric speciation

### ● Speciation in spiders

Starting with a single breeding population of one species of spider, species A. There is one gene pool and free gene flow within the whole population. There is no reproductive isolation.

Species A
• one gene pool
• free gene flow within whole population
• no reproductive isolation

Diversion of a river isolates part of the population. Spiders cannot swim across the river and so breeding is not possible between the two groups. They have become reproductively isolated with separate gene pools. As it is a geographical barrier this type of reproductive isolation is called geographical isolation.

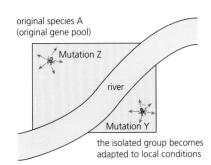

river

now reproductively isolated

The two groups have different mutations which cannot be introduced into the other group as there is no gene flow between them. Each isolated group becomes adapted to the different conditions as different features are selected in each habitat, and their gene pools become different, i.e. they become genetically different.

original species A (original gene pool)

Mutation Z

river

Mutation Y

the isolated group becomes adapted to local conditions

Another diversion of the river allows mixing of the two populations, but they are unable to reproduce successfully to produce fertile offspring. The gene pool of A is significantly different from that of B, and any offspring will be infertile as their chromosomes will no longer be homologous and will be unable to pair up during meiosis. Successful reproduction is therefore not possible between the two groups.

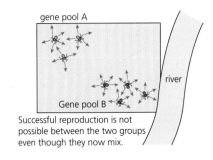

gene pool A

river

Gene pool B

Successful reproduction is not possible between the two groups even though they now mix.

Speciation has occurred – a new species B has evolved from the original species A. There are now two species, A and B, with separate gene pools, and although complete mixing is possible, successful reproduction cannot take place between them. The gene pool has changed in both groups – they have diverged.

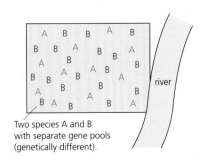

river

Two species A and B with separate gene pools (genetically different).

*Figure 4.33  Speciation in spiders*

## ● Darwin's finches

The classic study of speciation is shown by Darwin's finches. During his voyage on the *Beagle*, Darwin visited the Galapagos Islands and collected a variety of finches. He observed 13 different species of finch, each with a different beak type (Figure 4.34).

Darwin noticed that the finches on the Galapagos Islands were different to those on the mainland of South America 1000 km away. On the mainland the only niche available to the finches is that of a seed eater; the other niches being already occupied by other birds. Consequently there was only one species of finch with a beak adapted for eating seeds.

The Galapagos are isolated, volcanic islands and it is thought that the ancestral finches must have reached the islands from the mainland during freak weather conditions, blown by strong winds and resting on drifting vegetation and boats. When the ancestral finches reached the Galapagos islands there were very few other birds present due to their geographical isolation. Without competition, a wide range of ecological niches was available to the finches and there were few other selection pressures. The numbers of finches increased and distance from the mainland meant they were reproductively isolated from the original population. Mutations occurred and selection produced changes from the original ancestors. As a result the finches adapted to the local conditions and became different from those on the mainland.

Some finches flew to a second island where the environment was again different and lots of niches were available. This population of finches was isolated from those on the first island. Mutations took place followed by selection allowing adaptation to the local conditions. These finches became different to those on the first island and any finches that returned to the first island were unable to mate with finches there as speciation had taken place. This process was repeated as finches colonised more and more islands. In this way 13 species of finch evolved from one original species. The process was assisted by the remoteness of the islands. Lack of other birds reduced competition and opened up new feeding possibilities. Mutations in the finches which resulted in unusual beak shapes were often an advantage as they allowed the exploitation of yet another food source, whereas on the mainland they would be selected against as that niche was already occupied. The absence of other birds allowed the finches to diversify and adapt to a wide range of niches.

Many different species of finch therefore evolved from one original species, a process known as **adaptive radiation**. Also, within one island a variety of niches enabled isolation, specialisation and speciation to take place, resulting in many different species on some islands.

**Tree finches**

*Camarhynchus pallidus*
woodpecker finch
(insects)

cactus spine
used to probe
for insects

*C. heliobates*
mangrove finch
(insects)

*Certhidea olivacea*
warbler finch
(small flying insects)

*C. psittacula*
(large insects)

*C. crassirostris*
(buds, leaves,
fruit, seeds)

*C. pauper*
(medium insects)

*Pinaroloxias inornata*
Cocos Island finch
(insects)

*C. parvulus*
(small insects)

*G. scandens*
(cactus)

*Geospiza magnirostris*
(seeds)

*G. conirostris*
(cactus)

*G. fortis*
(seeds)

*G. difficilis*
(seeds)

*G. fulginosa*
(seeds)

**Ground finches**

***Figure 4.34*** *The variety of finch
species found on the Galapagos Islands*

### Remember

The mean is the average value and the mode is the most commonly occurring value. Modal class refers to the most commonly occurring group of values.

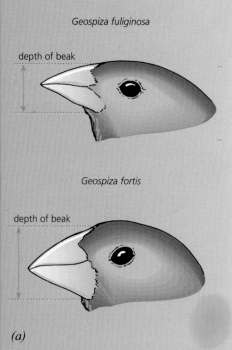

*Geospiza fuliginosa*

depth of beak

*Geospiza fortis*

depth of beak

*(a)*

**Figure 4.35** *(a) The heads of* Geospiza fuliginosa *and* Geospiza fortis, *(b) graphs showing range of beak size on three islands*

### Did You Know

Biologists have recently discovered another variety of finch on the Galapagos. A vampire finch has been observed that feeds on blood from open wounds on cattle and other large animals, again exploiting another available niche.

## Box 4.6 A case study of finches on the Galapagos Islands

A study of finches on three islands in the Galapagos has revealed some interesting findings regarding two species, *Geospiza fuliginosa* and *Geospiza fortis*, and variations in their beak size (Figure 4.35). Both species feed on plant seeds; the type of seed depending on the size of beak. Only *G. fortis* is found on Daphne Island. It has a mean beak depth and modal class of 10 to 10.5 mm, with a range on either side. In this example the mean and the modal class are the same.

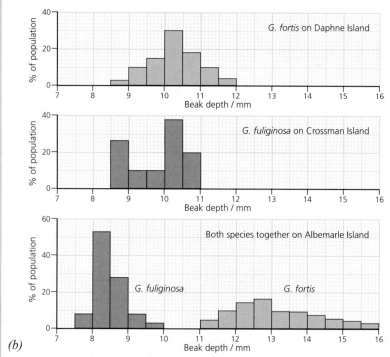

*(b)*

On Crossman Island, only *G. fuliginosa* is present, with the majority of birds, i.e. the modal class, also having a beak depth of 10 to 10.5 mm. The mean value, however, is smaller than the modal class.

When these two finch species are together on Albemarle Island the modal classes are quite different. *G. fuliginosa* has a modal class of 8 to 8.5 mm and *G. fortis* a modal class of 12.5 to 13 mm. They can therefore exploit different food sources and occupy different niches, so reducing the competition which would arise with similar-sized beaks and increasing their chances of survival. Directional selection has taken place favouring a smaller beak in *G. fuliginosa* and a larger beak in *G. fortis*. Selection has favoured those birds whose beak size will reduce competition and allow feeding to be more successful – they are the 'fittest' here.

**?**

**7** Why is reproductive isolation essential for speciation?　　*(2 marks)*

**Extension**

### Box 4.7  Mosquitoes and the London Underground

It is generally thought that a long period of time is necessary to allow speciation to occur, but a recent study has shown the emergence of new species of mosquitoes after only 100 years of isolation.

When the London Underground was being dug 100 years ago, mosquitoes invaded and colonised the tunnels. The original invaders were bird-biting mosquitoes. Over 100 years, without birds available, the mosquitoes in the tunnels evolved new feeding behaviours and began to feed on rats, mice and humans. Londoners sheltering in the underground during the Second World War were attacked by feeding mosquitoes (Figure 4.36).

Tube workers in particular are now plagued by mosquitoes appropriately named *Culex molestus*. The bird-biting species which entered the Underground 100 years ago was *Culex pipiens*. Scientists were amazed to discover that it was almost impossible to mate the mosquitoes living in the tunnels with those living above ground. This means that genetic differences are so great that they are almost separate species. Normally this takes place when species are isolated over thousands of years, not just a century. It appears that the original colonisers thrived in the tunnels with warm temperatures and water for breeding, and numbers increased rapidly. Once in the tunnels, the mosquitoes were reproductively isolated from those above ground. Separate mutations and selection took place leading to differences in the two populations and the emergence of a new species of mosquito. Amazingly, the differences are as great as if the species had been separated for thousands of years.

Further studies have shown genetic differences between mosquitoes along the different tube lines and this is thought to be due to draughts blowing mosquitoes along lines but not between them. Perhaps eventually a Northern line mosquito will be different from a Jubilee line one!

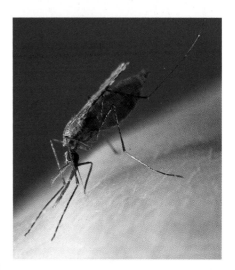

*Figure 4.36  A mosquito feeding*

### ● Speciation in *Tilapia*

Lake Turkana in Kenya contains fish of the genus *Tilapia*. Over the last 10 000 years the water level in the lake has fluctuated considerably. A volcanic island in the lake has three craters which filled with lake water and fish when the water level was high. Later they became isolated as the water level fell. Figure 4.37 shows a map of the island and three flooded craters.

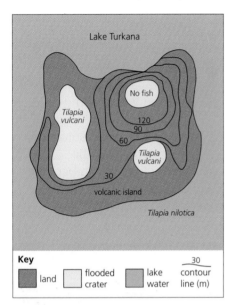

Lake Turkana

No fish

Tilapia
vulcani

120
90
60
Tilapia
vulcani
30

volcanic island

Tilapia nilotica

**Key**

☐ land  ☐ flooded crater  ☐ lake water  ─30─ contour line (m)

***Figure 4.37*** *Lake Turkana and the distribution of fish of the genus* Tilapia

*Tilapia nilotica* is found in the main lake but a different species, *Tilapia vulcani*, is found in the crater. The two species are very similar in appearance which suggests a common ancestor. When the water flooded into the crater it must have contained ancestors of *Tilapia nilotica* from the lake. Once in the crater, this population of *T. nilotica* was reproductively isolated from the rest of the population in the main lake. Mutations took place separately in each group and different features were selected for. Unable to interbreed, there was no gene flow between the two groups and the their gene pools became increasingly different. The two groups have now diverged to form different species, unable to interbreed.

Interestingly, there are differences in the gene pools of the populations of *T. vulcani* in the two crater lakes. This suggests that these populations are becoming increasingly different due to reproductive isolation. Their gene pools are separate and, if they remain apart, over time they will become increasingly genetically different leading to a new species.

**8** The Hawaiian islands in the Central Pacific Ocean are approximately 3000 km from the nearest mainland. They support 500 different species of the fruit fly, *Drosophila*, compared to only 32 in Britain. Suggest an evolutionary explanation for this difference. *(3 marks)*

Speciation can be summarised as follows:
● Originally there is one population of a particular species.
● Part of the population becomes reproductively isolated from the rest.
● Different mutations occur in the two groups.
● Different features are selected for in each population, in other words they become adapted to the local conditions.
● The two groups become increasingly genetically different.
● Eventually, successful reproduction is no longer possible between the two groups.
● A new species has formed; this is speciation.

**4.11 Artificial selection**

Artificial selection or selective breeding is selection by *deliberate human action* leading to a change in the allele frequency for a population. (Natural selection is selection by *environmental* factors.) It involves controlled breeding, with humans deciding which organisms can breed to produce offspring with desired characteristics.

Artificial selection is used in **crop improvement** to produce crops with:
● higher yield
● resistance to disease
● shorter stems, so less damage from wind
● tolerance to different climates.

Seeds with genes for these characteristics are economically more profitable and therefore in great demand.

There are *two* main techniques used in selective breeding:

● breeding within one variety or type of species
● cross breeding between different varieties of the same species.

## Breeding within one variety of species

Bread wheat is one of the most important food crops in the world. The grain or ear is a source of flour used to make bread (Figure 4.38). Bread wheat plants which produce large quantities of grain are selected and crossed with other similar, high-yielding plants. The offspring of this cross with the largest ears of grain are then crossed and this process is continued over many generations. Offspring with small ears are not used for breeding purposes.

This is deliberate directional selection, and over time leads to only bread wheat with 'heavy' grain. The alleles for large grain are selected; the alleles for small grain are not. There is an increase in the frequency of alleles producing heavy grain and a decrease in the frequency of alleles producing smaller grain. This change in allele frequency shifts the distribution curve to the heavier grain (to the right), so increasing the yield (Figure 4.39). Seeds produced this way are important economically as they have the potential to provide more food for increasing populations.

This method can be used to increase yields of crops, resistance to disease and tolerance to different climates. The same technique can be used in animals, for example to increase the milk yield in cows and the mass of poultry.

## Cross breeding between different varieties of the same species

There are many varieties of wheat with different characteristics. As they are the same species, cross breeding can take place between them. For example, variety A may have resistance to a disease but produce a low yield of grain; while variety B may have no resistance to disease but produce a high yield of grain.

It is possible to cross these two varieties to combine the useful features of both. Only offspring with the desired features from both parents are used for breeding purposes; the others are discarded. Offspring from unlike parents are called $F_1$ hybrids. The desired $F_1$ hybrid will be both resistant to disease and produce a high yield, combining the desirable features from both parents. This new variety will form the new breeding stock.

*Figure 4.38  Bread wheat*

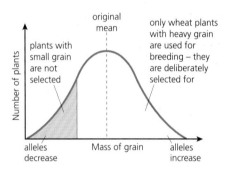

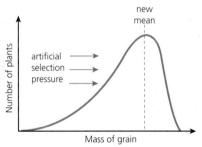

The yield of grain is increased, providing more food. Continued selective breeding will lead to the loss of the alleles producing small grain.

*Figure 4.39  Deliberate directional selection in bread wheat*

$$\begin{array}{ccc}
\text{variety A} & \times & \text{variety B} \\
\text{resistant to disease} & & \text{not resistant to disease} \\
\text{low yield} & & \text{high yield} \\
& \downarrow & \\
& F_1 \text{ hybrid} & \\
& \text{resistant to disease} & \\
& \text{high yield} &
\end{array}$$

Any $F_1$ hybrids that produce a low yield or are not resistant are not used.

Cross breeding to combine features from two unlike parents is also carried out in animals. Farm animals like pigs are bred in similar ways to produce varieties that are more profitable. For example, Chinese pigs are fast growing but small and these can be crossed with slower growing but larger varieties. The $F_1$ hybrid which combines fast growing with larger size is economically more desirable and can form the basis of a new variety. Other $F_1$ hybrids are not used for breeding.

### The dangers of artificial selection

As a result of artificial selection in pigs, there are now only five breeds of commercial importance remaining in Britain. Farmers only keep the most profitable varieties and other varieties have been lost. The diversity has therefore been reduced and this reduces the gene pool. Alleles may be lost which might give resistance to future diseases or allow adaptation to changing environmental conditions.

Artificial selection in crops has led to the same reduction in diversity as farmers only want profitable varieties. From 1903 to 1994 the number of varieties of maize plants has decreased from 790 to 30. Similarly, bean varieties have fallen from 690 to 21 in the same time period (Figure 4.40).

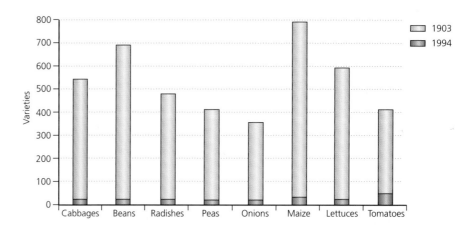

*Figure 4.40* *Loss of crop diversity 1903–1994 (USA)*

Selective breeding has therefore led to a significant reduction in alleles and varieties of crops. As only selected offspring are used for future breeding, all future generations have this narrow range of alleles. A wide range of alleles is necessary to provide the ability to adapt to changing conditions. If many alleles have been lost, this capability is severely reduced. In effect, artificial selection reduces the range of alleles and so reduces the phenotypes possible.

## Preventing loss of diversity

Genetic diversity is essential to maintain a wide range of alleles and phenotypes in a population. The greater the variation, the greater the ability to adapt to new conditions. To prevent the permanent loss of varieties of animals and plants, the following methods are available:

- **Protection of species in the field** – Valuable ecosystems are recognised and protected in the natural state. Areas of tropical rain forest, game reserves and national parks are examples.
- **Protection of species in an artificial environment** – Rare breeds of cattle, pigs and sheep are conserved and allowed to breed in rare breeds parks. Zoos play an important role in protecting endangered species and encouraging their breeding.
- **Germ banks** – Sex cells from plants and animals, and seeds from rare plants are stored in a dormant state. Thousands of varieties of plants have been conserved in this way for possible use in the future. A new seed bank opened in the year 2000 at Wakehurst Place in Sussex.

Many varieties of tomato plants have been lost by artificial selection, but germ banks have prevented their total loss. Some of the conserved varieties have features that are being re-introduced into modern varieties. For example, there are many wild varieties of tomato that have the following useful features:

- they can absorb water from sea water
- they can tolerate high temperature and humidity
- they are resistant to pests and viruses
- they have a high vitamin A and C content.

> **Remember**
> Biodiversity is the diversity of living things. There are two main causes of a reduction in biodiversity – artificial selection and habitat destruction. Artificial selection leads to the reduction of varieties of plants and animals through controlled breeding. Deforestation, flooding for reservoirs, desertification and large-scale mining all reduce species variety through destruction of their habitats.

*Figure 4.41  Many wild varieties of tomatoes have been conserved in germ banks, enabling their useful features to be re-introduced into modern varieties*

## Summary – (4) Evolution

- Differences between organisms of the same species are called intra-specific variation.

- Continuous variation exists when a characteristic shows a complete gradation from one extreme to the other, producing a normal distribution curve.

- Discontinuous variation is when a characteristic falls into distinct categories, with no intermediates.

- Variation can be caused by meiosis and fertilisation (which produces new combinations of existing alleles) and mutations (which give rise to new alleles).

- The theory of natural selection explains evolutionary change by the process of selection favouring offspring with favourable alleles.

- Evidence for natural selection is found in the fossil record, as well as in the results of studies of comparative anatomy, embryology and biochemistry.

- The distribution of the peppered moth and the incidence of the sickle cell allele demonstrate selection in action.

- Selection causes a change in the allele frequency leading to a change in the gene pool.

- The three types of selection are directional, stabilising and disruptive selection.

- The Hardy–Weinberg equilibrium applies in populations where no change is taking place. The equations are:
  $p + q = 1.0$ (allele frequencies)
  $p^2 + 2pq + q^2 = 1.0$ (genotype frequencies).

- Reproductive isolation of part of a population can lead to speciation – the formation of a new species from an existing one.

- Allopatric speciation takes place when a population is isolated geographically, in different lands.

- Sympatric speciation occurs when reproductive isolation takes place in the same land.

- Humans select animals or plants for breeding to produce organisms with desired features. This is artificial selection and leads to a change in the allele frequency and therefore characteristics by deliberate human action.

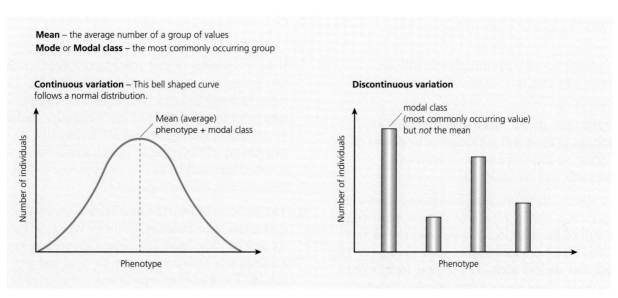

**Mean** – the average number of a group of values
**Mode** or **Modal class** – the most commonly occurring group

**Continuous variation** – This bell shaped curve follows a normal distribution.

Mean (average) phenotype + modal class

*Number of individuals*

*Phenotype*

**Discontinuous variation**

modal class (most commonly occurring value) but *not* the mean

*Number of individuals*

*Phenotype*

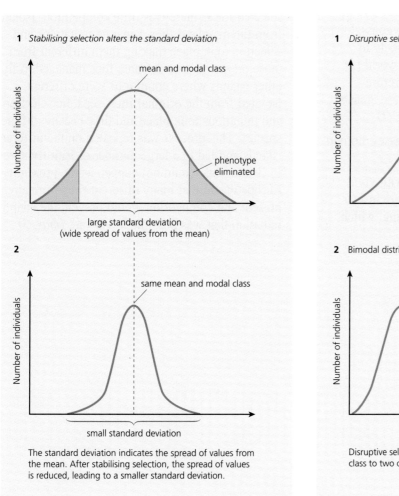

**1** *Stabilising selection alters the standard deviation*

mean and modal class

*Number of individuals*

phenotype eliminated

large standard deviation (wide spread of values from the mean)

**2**

same mean and modal class

*Number of individuals*

small standard deviation

The standard deviation indicates the spread of values from the mean. After stabilising selection, the spread of values is reduced, leading to a smaller standard deviation.

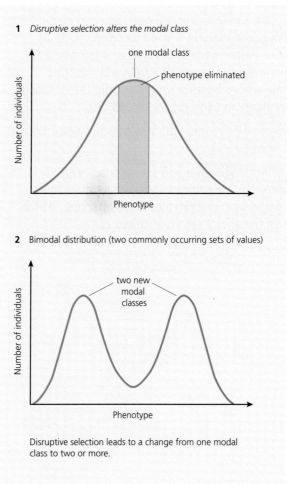

**1** *Disruptive selection alters the modal class*

one modal class

phenotype eliminated

*Number of individuals*

*Phenotype*

**2** Bimodal distribution (two commonly occurring sets of values)

two new modal classes

*Number of individuals*

*Phenotype*

Disruptive selection leads to a change from one modal class to two or more.

**Figure 4.42** *Summary of terms used to describe patterns of variation*

# 4 Evolution

## ? Answers

**1** As cats fall into two distinct groups, with and without tails, this is an example of discontinuous variation *(1)*.

**2** Somatic cells are not found in gametes and any mutations in them will not therefore be passed on to offspring, so they have no effect on future generations and evolution *(1)*.

**3** Stick insects blend into their background, protecting them from predators *(1)*. Those that are most stick-like will be the most likely to survive *(1)* and pass on their alleles producing more stick-like insects *(1)*. Those that are less stick-like are more likely to be eaten and will not pass on their unhelpful alleles to the next generation *(1)*.

**4** People with the sickle cell trait (heterozygous), have resistance to malaria and are therefore likely to survive better in malarial regions of the world *(1)*. Malaria is rife in parts of Africa and therefore people whose families originate from these areas are more likely to carry the sickle cell allele *(1)*.

**5** Disruptive selection, with two modal classes, horned and hornless *(1)*.

**6** Directional selection will favour roses with vivid colours *(1)*. They are likely to attract more insects for pollination and produce more offspring, which inherit the alleles for vivid colour *(1)*.

**7** If a population is split so reproduction between the two groups is not possible, they are said to be reproductively isolated. Mutations and selection will take place independently in the two groups leading to changes in each gene pool *(1)*. Eventually, the two groups will become so genetically different that if mixed they will be unable to reproduce as they have become different species *(1)*.

**8** The isolation of the Hawaiian islands means that only a few species can reach them. When the fruit fly arrived, there were few other flies present and consequently little competition for food and a habitat. The fruit flies reproduced and adapted to the local conditions. A wide variety of features was selected for as there was little competition. Isolated from the mainland flies, separate mutations and selection took place making them different from those on the mainland. Some flies managed to fly to other islands where conditions were different. Isolated from the original flies, separate selection and mutations took place and they became different species. This process was repeated on the different islands and led to a large number of fruit fly species. In Britain this would not happen as the proximity to Europe meant that many different flies were already present and competition would prevent the adaptive radiation that was possible in Hawaii. *(max. 3)*

## End of Chapter Questions

1   The histogram shows the heights of wheat plants in an experimental plot.

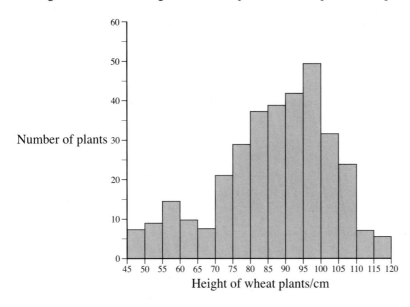

(a)  What evidence from the data suggests that there were two strains of wheat growing in the experimental plot?                  *(1 mark)*        N2.1

(b)  (i)  Which type of variation is shown by the height of each of the strains of wheat plants? Give the reason for your answer.          *(2 marks)*

(ii)  Explain why the height of the wheat plants varies between 45 cm and 120 cm.                                                   *(1 mark)*

*NEAB 1997*                                                   *(Total 4 marks)*

2   An investigation into the heights amongst two groups of school students was carried out. The results are shown in the table below.          N3.2
                                                                          N3.3

| Height range/cm | Number of students in that range | |
|---|---|---|
|  | Group 1 | Group 2 |
| 146–150 | 1 | 0 |
| 151–155 | 3 | 1 |
| 156–160 | 9 | 4 |
| 161–165 | 11 | 7 |
| 166–170 | 4 | 14 |
| 171–175 | 2 | 3 |
| 176–180 | 2 | 1 |
| 181–185 | 0 | 2 |
| 186–190 | 0 | 0 |
| 191–195 | 0 | 1 |

(a) Using the mid-point values for each height range (that is 148, 153, 163, etc.), calculate the mean height (to 2 decimal places) of students in group 1. Show your working. *(2 marks)*

(b) On graph paper, and using one pair of axes **only**, plot these data in the form of an appropriate graph. *(5 marks)*

(c) What is the name given to the type of variation shown by the data? *(1 mark)*

(d) Suggest **two** reasons for the differences shown between the two groups *(2 marks)*

*OCR 1999*                         *(Total 10 marks)*

**3** The diagram below shows the chromosomes from a human white blood cell grown in culture. The individual from whom the culture cells were obtained had Down's syndrome.

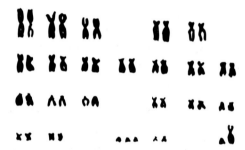

(a) How does the chromosome set shown differ from that of a person without Down's syndrome? *(1 mark)*

(b) Suggest how this difference may have come about. *(1 mark)*

(c) How many chromosomes would you expect to find in: (i) sperm cells, and (ii) liver cells of the individual with Down's syndrome? *(2 marks)*

(d) Describe how Down's syndrome can be diagnosed pre-natally. *(2 marks)*

(e) Suggest another characteristic which may be determined by the method you have described in (d). *(1 mark)*

(f) Down's syndrome is an example of a chromosomal mutation. Name **two** other types of chromosomal mutation you have studied. *(2 marks)*

*Oxford 1998*                         *(Total 9 marks)*

**4** There are 64 chromosomes in each body cell of a horse and 62 chromosomes in each body cell of a donkey.

(a) Complete the table to show the number of chromosomes in the nuclei of these animals at the end of the various stages in cell division.

| Animal | Number of chromosomes in one of the nuclei formed at the end of | | |
| | mitosis | first division of meiosis | second division of meiosis |
| Horse | | | |
| Donkey | | | |

*(2 marks)*

**(b)** A mule is the offspring of a cross between a horse and a donkey.

(i) How many chromosomes would there be in a body cell from a mule? Give a reason for your answer. *(1 mark)*

(ii) Suggest why a mule is unable to produce fertile sex cells. *(2 marks)*

(iii) Explain why a horse and a donkey are regarded as different species. *(1 mark)*

*NEAB 1998*      *(Total 6 marks)*

**5 (a)** Ladybirds are small beetles. In one species of ladybird, individuals vary in the amounts of yellow and black colouring on their wing-cases. They range in colour from all yellow to all black but most individuals are between 30% and 60% black.

(i) Name the type of variation shown by this example. *(1 mark)*

(ii) Studies have shown that the variation in colour of the wing cases of this ladybird is mainly genetic. What does the pattern of variation suggest about the genes controlling the colour of the wing-cases? *(1 mark)*

**(b)** The colour of the wing-cases in the 2-spot ladybird is determined by a single gene. The allele for red wing-cases is dominant to that for black wing-cases. In a series of mating experiments, equal numbers of homozygous males and females of the two colour forms were put together and observed. The number of matings between the different colour forms is shown in the table.

N2.2
N2.3

| | | Male | |
| | | red | black |
| Female | red | 47 | 26 |
| | black | 39 | 35 |

(i) Give two general conclusions that might be drawn from the results of this investigation about the pattern of mating of the male ladybirds. *(2 marks)*

(ii) Explain how this pattern of mating could affect the frequency of these alleles in the next generation of ladybirds. *(2 marks)*

*NEAB 1998*      *(Total 6 marks)*

**6** The beetles belonging to the genus *Colophon* are unable to fly and are found on hilltops in South Africa. The dotted lines on the map show the distribution of three species.

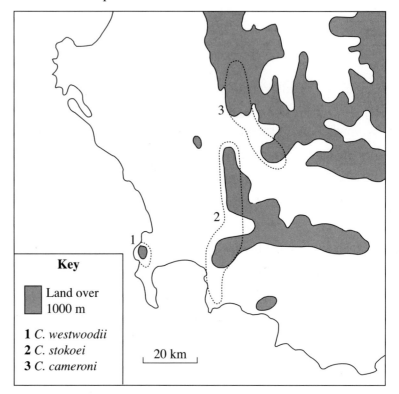

Suggest an evolutionary explanation for each of the following statements.

**(a)** All of these beetles are of very similar general appearance. *(1 mark)*

**(b)** There are slight differences between the species of *Colophon* found in the three areas. *(2 marks)*

**(c)** The fact that beetles of the genus *Colophon* are unable to fly has been important in the evolution of twelve different species of the genus in a small area of South Africa. *(2 marks)*

*AEB 1995*

*(Total 5 marks)*

# 5 Classifying living organisms

Over about fifteen hundred million years of the Earth's history, many millions of different species have evolved. Of these, more than ten million species are known to exist today. Many others must exist which have not yet been discovered. In order to make sense of the living world, it is essential to understand what a species is and to organise these species into some sort of order which reveals their relationships with each other.

**This chapter includes the following:**
- the definitions and significance of a 'species'
- the classification of species into groups
- the five kingdoms of living organisms
- summary of important features of each kingdom
- outline survey of each kingdom.

## 5.1 What is a species?

Species have been regarded in the past as different types of organism such as dogs or cats, but this idea is far too vague to stand up to any serious study.

A generally accepted, brief definition of a species is as follows:

*A species is a group of closely related organisms which are capable of interbreeding with each other and producing viable and fertile offspring. Members of a species are similar to each other anatomically and biochemically.*

This definition depends upon the fact that, as members of the same species have arisen recently from one or a few common ancestors, they have many genes in common. This fact is responsible for similarity in appearance and chemical make-up of members of the group. As they have so many similar genes, members of the same species usually have similar **karyotypes** (see *AS Biology*, Box 6.1, page 157) with similar types and numbers of chromosomes. This similarity allows an offspring of two members of the same species to inherit the usual karyotype of that species with homologous pairs of chromosomes.

If the offspring is fertile, it must be able to produce gametes (see *AS Biology*, Section 7.2, page 183). The production of gametes involves the division of certain cells by **meiosis**. Meiosis involves the pairing and organised separation of homologous chromosomes (see *AS Biology*, Figure 6.14, page 169). So, normally, offspring are fertile only if they inherit homologous pairs of chromosomes.

Members of different species may interbreed occasionally and produce viable **hybrid** offspring. However, these offspring are infertile as illustrated in Figure 5.1.

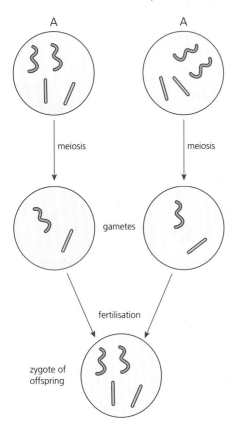

Cross between two members of species A (2n = 4)

A    A

meiosis    meiosis

gametes

fertilisation

zygote of offspring

The offspring has the same karyotype as the parents. It has homologous pairs of chromosomes and so meiosis can take place. It can, therefore, produce gametes. It is **fertile**.

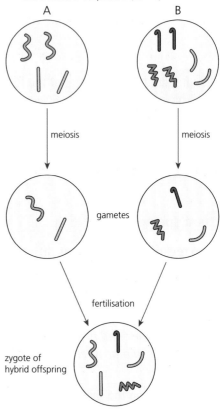

Cross between a member of species A (2n = 4) and a member of species B (2n = 6)

A    B

meiosis    meiosis

gametes

fertilisation

zygote of hybrid offspring

In the unlikely event that such a zygote can live and produce an adult, meiosis cannot take place as there are no homologous chromosomes to pair. So no gametes can be produced. It is **sterile**.

**Figure 5.1** *Why hybrids are sterile (infertile)*

**! Did You Know**

In the well-known case of the female horse (2n = 64) and male donkey (2n = 62) producing a sterile mule, the mule inherits 63 chromosomes and most of these are poorly matched into pairs. Meiosis cannot take place and the mule theoretically cannot produce gametes.

However, *very* occasionally, a female mule has been reported to be fertile. This is such an irregular occurrence in recent times that it makes headlines. In Roman times, there was a saying, in Latin, meaning 'When a mule foals'. A modern equivalent would be 'Once in a blue moon'.

There are other ways of regarding a species. These do not conflict with the definition given on page 155. For example:

● *Members of the same species are potentially members of the same gene pool.* This means that the genes of one member of a species can be inherited along with the genes of any other member, if sexual reproduction takes place.

● *A species is the fundamental unit upon which living organisms are classified.* This means that, just as polymers are built from monomers or houses are built from bricks, all systems of grouping living organisms are built from species.

### 5.2  Determining whether or not organisms belong to the same species

If a biologist has the task of determining whether or not two or more specimens belong to the same species, a range of techniques is available to help. The biologist uses as many of the following as possible.

● Observation of similarities in structural features (both large and microscopic) in all specimens, whether fossil, recently dead or living. Close similarity can be a guide, but does not provide proof that the specimens belong to the same species.

● Chemical analysis of dead or living specimens and some fossils. Again some chemical similarities act as a guide. For example, members of the same species would usually have identical (or very similar) amino acid sequences in the protein parts of their cytochrome molecules.

● Study of karyotypes and DNA analysis on dead or living specimens and some fossils. This can give very strong clues about whether the specimens could potentially interbreed and produce fertile offspring.

● Study of courtship and mating behaviour together with observation of the fertility of offspring. This can give conclusive answers, but is only possible if the specimens are living!

?

1 List the following procedures in a likely descending order of significance when determining if individuals belong to the same species.

A  comparison of leg structure

B  comparison of chemical composition of cytochromes (respiratory coenzymes)

C  comparison of karyotypes

D  observation of reproductive behaviour

E  comparison of composition of DNA.            *(2 marks)*

**Extension**

**Box 5.1  Difficulties when applying the definition of a species**

The definition of a species is easy to understand and it is useful to apply when considering whether or not individuals are members of the same species, as long as the following limitations are considered:

*   Despite having many genes in common, sometimes members of the same species can look different from each other. These structural differences can occur because of different environmental conditions in which different members of the same species live. The different conditions can cause different genes to be expressed. For example, if one of two identical hydrangea plants is grown in acid soil (pH 5.5 or lower) and the other in more neutral or alkaline soil, the one in acid soil will produce blue flowers and the other will produce pink flowers (Figure 5.2).

*   Sometimes, members of the same species have distinctly different structural features from each other. This is known as **polymorphism**. For example, in many species there are distinct differences between the sexes (**sexual dimorphism**). Honey bees (*Apis mellifica*) have three distinct forms – workers, queens and drones (Figure 5.3).

*   **Directional selection** (see Section 4.6 page 123) may have taken place. Some members of a species may differ in appearance from each other quite considerably as a result of differences in genetic make-up resulting from selection. However, they may still have enough genes in common to allow them to interbreed and produce fertile offspring. This selection may be natural or artificial (Figure 5.4).

*   Some species usually reproduce asexually, for example some bacteria. This makes it difficult to be sure whether members are potentially capable of interbreeding to produce fertile offspring and are therefore members of the same species. A related problem arises when studying fossils. It is not possible to be sure whether two, apparently similar fossil organisms had been able to interbreed or not.

*(a)*

*(b)*

***Figure 5.2*** Hydrangea macrophylla *growing in* (a) *acid soil (pH 5.5 or less),* (b) *soil with a pH greater than 5.5*

***Figure 5.3*** *(a) A peacock and peahen – an example of sexual dimorphism,* *(b) Polymorphic forms of the honey bee* (Apis mellifica)

*(a)*

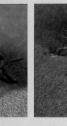

*(b) Worker*

*Drone*

*Queen*

Extension

*Red setter*

*Labrador*

*German Shepherd*

**Figure 5.4** *Artificial selection by dog breeders has led to members of the same species having very different appearances!*

**Figure 5.5** *Gulls around the northern hemisphere. Do they all belong to the same species? (Just five of the 10 races are shown.)*

## Box 5.1  Difficulties when applying the definition of a species continued

• Some individuals behave as if members of the same species in one part of the world, but as different species elsewhere. For example, in the UK, the lesser black-backed gull looks and behaves as a different species from the herring gull. The two species do not interbreed although living alongside each other. However, the lesser black-backed gull can interbreed with the Scandinavian black-backed gull although there are some differences between them, as directional selection has occurred. The Scandinavian black-backed gull can interbreed with the Siberian vega gull, which can, in turn, interbreed with the American herring gull. The interbreeding group encircles the northern hemisphere, as shown in Figure 5.5. The accumulated differences result in the American herring gull being able to interbreed with the herring gull in the UK, but not the lesser black-backed gull.

North America

American herring gull

Siberian Vega gull

Asia

Scandinavian black-backed gull

Lesser black-backed gull (*Larus fuscus*)

Europe

Herring gull (*Larus argentatus*)

**Key** interbreed and produce fertile offspring

**Systematics** is the study of biological diversity. It includes:

- **Taxonomy** – the study of the principles and methods used in biological classification.

- **Nomenclature** – the naming of biological groups.

## 5.3 The reasons for classifying species into groups

As there are thought to be about ten million living (**extant**) species on the Earth today, any study of biology requires that some order and method of grouping is established. (Imagine going into a library in order to study the work of one particular author or to compare crime novels, only to find ten million books in one big disorganised heap!)

Systems of **biological classification** (arranging species into groups) have been known since the times of the ancient Greeks. They fall into two main sorts of systems, **artificial classification** and **natural classification**.

### Artificial classification

This involves separating organisms into groups according to their differences. These differences may be chosen for convenience rather than for any biological significance. Such a system may have uses but it gives no guide to true relationships or evolutionary patterns. For example, some of the **dichotomous keys** used in the field which help biologists to identify organisms in a particular habitat may be based on artificial principles.

In such dichotomous keys:

- at each step, organisms with an exclusive feature are selected and placed in a sub-group

- each separation is made on the basis of a single character. Unrelated forms may be included in the same group.

For example, the animals in Figure 5.6a may be separated into two groups by asking the question: Do they have legs? This produces:

- group A with legs – crayfish and frog

- group B with no legs – snake and earthworm.

These groups contain totally unrelated forms. Examination of more features (including the possession of a vertebral column) shows that the snake is more closely related to the frog than either of the others.

*Figure 5.6 (a) Artificial classification*

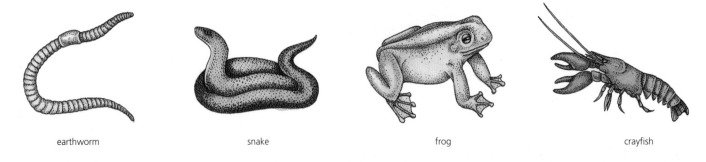

earthworm          snake          frog          crayfish

In another example, a gardener may collect six weeds from the garden (A to F in Figure 5.6b) and identify them using an artificial key such as the one below. Such a key may help the gardener, but, as it uses easily observable characteristics with no biological significance, it provides no information about evolutionary relationships between the plants.

*Figure 5.6 (b) Artificial classification – a dichotomous key*

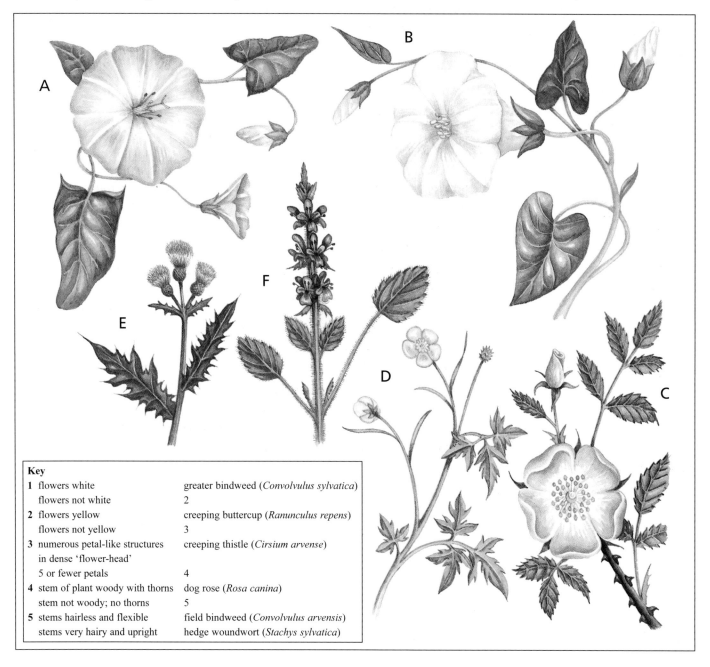

**Key**

| 1 | flowers white | greater bindweed (*Convolvulus sylvatica*) |
| | flowers not white | 2 |
| 2 | flowers yellow | creeping buttercup (*Ranunculus repens*) |
| | flowers not yellow | 3 |
| 3 | numerous petal-like structures in dense 'flower-head' | creeping thistle (*Cirsium arvense*) |
| | 5 or fewer petals | 4 |
| 4 | stem of plant woody with thorns | dog rose (*Rosa canina*) |
| | stem not woody; no thorns | 5 |
| 5 | stems hairless and flexible | field bindweed (*Convolvulus arvensis*) |
| | stems very hairy and upright | hedge woundwort (*Stachys sylvatica*) |

**2** Using the key, give the correct common names for plants A to F in Figure 5.6b. *(6 marks)*

**Figure 5.7** *A hierarchy of groups. Grouping according to number of significant features in common results in grouping more closely related organisms together*

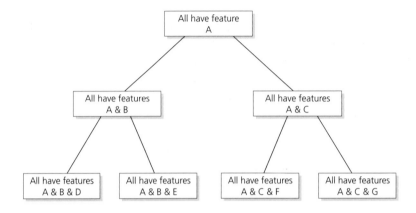

---

💡 **Definition**

* **Homologous** characters in two or more species have a similar basic nature rather than function and they have the same evolutionary origin.

* **Analogous** characters look superficially similar because they have similar functions, but their basic nature can be quite different, having evolved separately. They can occur in unrelated organisms.

---

### Natural classification

In a natural classification, there is a **hierarchy** of groups (a system where groups can be sub-divided into smaller and smaller groups). Figure 5.7 shows how, in general, organisms are grouped together into smaller groups, the more features they have in common. However, adjustments are made for particular cases where evolutionary studies suggest that members belong in a particular related group even though they may now have some unusual features. So the system collects more closely related organisms together in smaller groups. The grouping reflects natural relationships and evolutionary descent (**phylogeny**). Usually, the more features in common two species have, the more recently they have evolved from a common ancestor.

These features may include biochemical, chromosomal and DNA similarities as well as structural similarities. Great care must be taken when grouping organisms according to structural (and some chemical) similarities. Only **homologous** characters must be considered and **analogous** characters should be ignored.

The wing of a bird, a whale's flipper and the front leg of a horse, although superficially dissimilar, can all be seen to have similar bone structure and on further study can be seen to have evolved from a common form of forelimb, known as the **pentadactyl limb** (Figure 5.8). They are **homologous** characters.

**Figure 5.8** *The forelimbs of birds, whales and horses are homologous characters. Their bone structure has evolved from the same pattern*

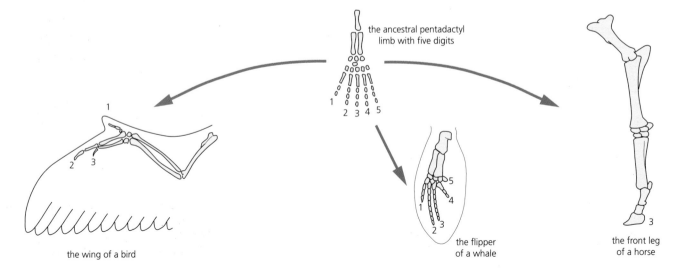

the wing of a bird

the ancestral pentadactyl limb with five digits

the flipper of a whale

the front leg of a horse

On the other hand, the wings of a butterfly and humming bird may be flapping extensions but they are totally different in basic structure. They are **analogous** characters. The eyes of vertebrates and the eyes of molluscs, such as squids, are remarkably similar superficially as they both allow image perception. Close examination shows differences in their development and a study of possible evolutionary origins shows that these features are analogous and have evolved quite separately in the two groups.

In devising a system of natural classification, every effort is made to group organisms using characters with evolutionary significance.

## 5.4  The groups of the natural classification system

The basic unit of the system is the species. Related species are grouped together into a **genus**. Related **genera** (plural of genus) are placed into larger groups and so on. This produces a hierarchy of groups as seen in Figure 5.9 and Table 5.1.

The groups form a series of subdivisions or **taxonomic ranks** and each group is referred to as a **taxon** (plural **taxa**).

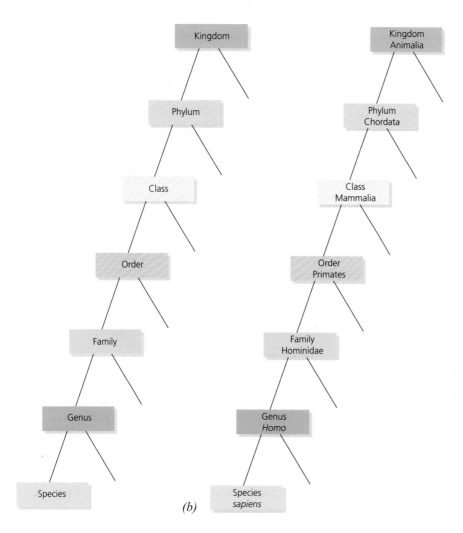

**Remember**

The system of natural classification is important as:

- it creates order out of chaos and helps the study and referencing of organisms

- it reveals natural evolutionary relationships and is a guide to the study of evolutionary pathways

- it helps with the identification of newly discovered organisms

- it forms a basis for an international system of **nomenclature** (the naming of organisms) (see Section 5.5, page 164).

**Remember**

The taxa in descending order of rank are listed as:

**k**ingdom, **p**hylum, **c**lass, **o**rder, **f**amily, **g**enus and **s**pecies.

Remember – **k**eep **p**ond **c**lean **o**r **f**rog **g**ets **s**ick!

**Figure 5.9** *The taxanomic ranks used in classification (a) the names of the taxa, (b) the classification of humans*

# 5 Classifying living organisms

| Taxa | Bacterium (*E. coli*) | Seaweed (Serrated wrack) | Mushroom | Stinging nettle | Tiger |
|---|---|---|---|---|---|
| Kingdom | Prokaryotae | Protoctista | Fungi | Plantae | Animalia |
| Phylum | Gracilicutes | Phaeophyta | Basidiomycota | Angio-spermophyta | Chordata |
| Class | Proteobacteria | Phyophyceae | Hymeno-mycetes | Dicotyledoneae | Mammalia |
| Order | (gamma subdivision) | Fucales | Agaricales | Urticales | Carnivora |
| Family | Entero-bacteriaceae | Fucaceae | Agaricaceae | Urticaceae | Felidae |
| Genus | *Escherichia* | *Fucus* | *Agaricus* | *Urtica* | *Felis* |
| Species | *coli* | *serratus* | *campestris* | *dioica* | *tigris* |
| | x 1000 | | | | |

**Table 5.1** *Classification of five organisms showing a representative of each Kingdom*

**Figure 5.10** *'The father of taxonomy' – Carl Linnaeus (1707–1778) was a Swedish botanist who devised the system of classification and nomenclature which forms the basis of that in use today (Here portrayed in Lapp costume!)*

## 5.5 The naming of living organisms (biological nomenclature)

Carl Linnaeus, a Swedish botanist who lived in the 18th century, devised the system of natural classification and used the same groups as are used today. While the basic taxonomic ranks have not changed, the placing of many organisms within these groups has undergone radical changes in the light of more recent ideas and discoveries. He grouped organisms simply according to similar structural features he could observe at the time.

Linnaeus also devised a system of naming organisms, based on his classification. This is still used today as an international naming system. Such a universal system is essential if there is to be communication between scientists who speak different languages. Without this, confusion could also arise because many organisms have different common names in different parts of the same country.

Some of the present rules based on Linnaeus' system are as follows:

- Latin (or Latinised) names are used.
- Each organism is referred to by two names, the name of its genus followed by the name of its species. Therefore, it is known as a **binomial system**.
- The genus takes a capital letter and the species name does not.
- The names are written in italics when in print or underlined when in handwriting.
- The first time an organism is referred to, its full name should be given, e.g. *Homo sapiens*. Later in the same passage, it can be abbreviated to *H. sapiens*.

- If the species is not certain but the genus is known, it is possible to refer to an organism by the name of the genus followed by 'sp.' or, if plural, 'spp.' (short for species), e.g. *Paramecium* sp.

**?**

3 Write the international names of the organisms classified in Table 5.1 using the correct style. *(6 marks)*

4 The large white butterfly (*Pieris brassicae*) is a common pest of cabbages and similar plants. It lays its eggs on the leaves of the plant. The caterpillars which hatch from these eggs feed voraciously on the leaves. The table below classifies this butterfly.

Which terms do the letters A–E represent?

| | |
|---|---|
| Kingdom | Animalia |
| Phylum | Arthropoda |
| A | Insecta |
| B | Lepidoptera |
| C | Pieridae |
| Genus | D |
| Species | E |

*(5 marks)*

If a new species is discovered, the discoverer may be allowed to choose the species name as long as it is Latinised and acceptable. Some biologists immortalise themselves or give credit to someone else by modifying a chosen name. Often the name describes some feature of the organism in Latin or it may refer to where the specimen was found.

**Did You Know**

Imagine the inaccuracies which could arise if a scientist were trying to study the countrywide distribution of stinging nettles and received local reports of Hokey-pokeys, Jinny nettles or Devil's playthings. These are just three of the local names for the same plant.

**Extension**

### Box 5.2   What's in a name?

Coelacanths are fish which were only known by their fossil forms. They lived at the time of the dinosaurs and were thought to have died out 80 million years ago. However, fishermen's tales of seeing similar fish were proved to be correct when, in 1938, the first living coelacanth was caught by accident at the mouth of the Chalumna river off the east coast of South Africa. The catch was reported to the local museum in the nearby small town of East London. The director of the museum, Miss Courtney-Latimer, realised the importance of the find and alerted the South African expert, Dr J.L.B. Smith, who examined the specimen. The discovery was soon world news. The specimen was considered to belong to a new genus as well as being a new species, so there were two new names to be given. Dr Smith unselfishly named the fish ***Latimeria chalumnae*** (after the curator and the river).

Since then, many more specimens of this fish have been found. You can see a coelacanth in Figure 4.23 on page 127.

Extension

*Figure 5.11  Giant sequoia or Wellingtonia*

**Box 5.3  Size matters**

What is meant by the largest organism on Earth? Does it mean the tallest, the heaviest or the one occupying the greatest volume? Does it mean the one which extends over the greatest area of Earth or the one which forms the largest, solid undivided mass? There is no satisfactory answer to these questions.

For many years, the giant sequoia (sometimes called Wellingtonia) (*Sequoiadendron giganteum*), while not the world's tallest tree, has been regarded as the largest and heaviest living species (Figure 5.11). Particular specimens in the western USA reach up to 83 m in height, having enormous trunks with girths of up to 33 m at the base. The weight of the timber in such trunks can only be estimated, but is thought to be about the same as that of 15 blue whales!

However, DNA analysis has revealed that some organisms, previously believed to be separate individuals, have identical DNA. Further examination has shown that these forms are not just clones produced by asexual reproduction but are still connected to each other underground. This means that they are technically part of one individual (although sometimes regarded as colonies).

Quaking aspen trees (*Populus tremuloides*) in the Utah mountains, which form a wood extending over 40 hectares, have reproduced by suckers and remain interconnected by an underground root system. An estimated total weight of over 6000 tonnes makes the total mass much greater than the largest giant sequoia. (However, it cannot be certain that some parts have not become separated.)

An even stranger contender for largest organism is the fungus, *Armillaria bulbosa*. A specimen which DNA analysis suggests must have grown from one spore now has a mycelium which extends over 12 hectares, closely associated with tree roots. The mushroom-like fruiting bodies which appear above ground are connected by a network of underground hyphae.

In a similar way the Cnidarians, whose interconnected polyps form coral reefs, may prove to be some of the world's largest animals.

## 5.6 Classifying organisms into five kingdoms

All organisms are at this time usually grouped into five kingdoms, although new discoveries lead to new ideas and this grouping may change in the future. Viruses are not included in this system as many biologists do not consider them to be living organisms. These five kingdoms are:

- Kingdom Prokaryotae
- Kingdom Protoctista
- Kingdom Fungi
- Kingdom Plantae
- Kingdom Animalia.

### Important features – a summary

Members of each kingdom have features in common. However, some of these characteristics are possessed by members of other kingdoms too. For example, members of all the kingdoms except the Prokaryotae have eukaryotic cells. Such features cannot be regarded as **distinguishing characteristics** of a kingdom. Distinguishing characteristics of a kingdom (or any taxon) are possessed by members of that kingdom (or taxon) only.

However, a *combination* of characteristics may be unique to one group. For example, only the Fungi have eukaryotic cells with no undulipodia (true flagella or cilia). The Prokaryotae have no undulipodia because prokaryotic cells do not have these (see *AS Biology*, Section 1.20, page 26).

Table 5.2 on pages 168 and 169 presents a list of the important features of each kingdom which need to be memorised for examinations. Important **distinguishing characteristics** are in **bold type** and marked with asterisks (*). (Remember, there are usually a few atypical forms in most kingdoms.)

This list may seem a little daunting when presented without any background material. For this reason Extension boxes 5.4 to 5.8 contain brief surveys of each kingdom. The material shown in these extension boxes does not need to be memorised in detail but should help you to relate the features listed in the table to real organisms.

| Kingdom | Important features |
|---|---|
| Prokaryotae | • **\*bodies made of prokaryotic cells (cells without membrane-bound organelles such as nuclei or mitochondria – see *AS Biology*, Section 1.20, page 25 for details)**<br>• **\*cell walls are always present but not made of chitin or cellulose** (This is a feature of prokaryotic cells.)<br>• bodies unicellular or made of short chains or clusters of cells<br>• some are heterotrophic, some are autotrophic<br>• usually store carbohydrates as glycogen and lipids as oil |
| Protoctista | • have eukaryotic cells (some have cells with no cell walls; others have cell walls, some of which are made of cellulose)<br>• bodies unicellular; made of chains, or have simple branching bodies with very little differentiation into different types of cells<br>• **\*no clear distinguishing characteristics – it is the range of simple body forms with eukaryotic cells which distinguishes this group**<br>• some are photosynthetic, some are heterotrophic<br>• usually store lipid as oil |

**Table 5.2** *Important features of each kingdom*

**?**

5 List four differences between prokaryotic and eukaryotic cells.(*4 marks*)

6 Name the kingdom(s) to which organisms possessing the cells shown in Figure 5.12 may belong. Ignore the sizes and shapes of the cells.

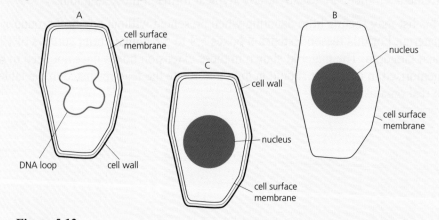

**Figure 5.12**                                        *(6 marks)*

7 List one characteristic which is *only* found in each of the following kingdoms:

(a) Fungi

(b) Animalia

(c) Prokaryotae.            *(3 marks)*

| Fungi | • have eukaryotic cells |
|---|---|
| | • **\*cells have cell walls of chitin** |
| | • **\*never have undulipodia, although cells are eukaryotic** |
| | • bodies usually multicellular |
| | • **\*bodies are made up of thread-like hyphae which may make a loose mesh or be compacted into a mass. The whole body is referred to as a mycelium** |
| | • do not possess chlorophyll; they are all heterotrophic, feeding as saprobionts or parasites (see *AS Biology*, Section 13.7, page 400 and Section 13.8, page 403) |
| | • store carbohydrates as glycogen and lipids as oil |
| | • reproduce sexually and asexually by spores |
| Plantae | • have eukaryotic cells |
| | • **\*all cells have cell walls composed largely of cellulose and they have large permanent cell vacuoles** |
| | • bodies are multicellular |
| | • **\*have chlorophyll *a* and *b* and other photosynthetic pigments contained in chloroplasts – they are photosynthetic** |
| | • **\*store carbohydrates as true starch** |
| | • store lipids as oil |
| | • **\*life cycles involve alternation of generations**. This means that a haploid gamete-producing generation alternates with a diploid spore-producing generation during a complete life cycle. Spores are involved in sexual reproduction but not asexual reproduction |
| | • body form usually branched with growth from special groups of dividing cells (meristems) |
| Animalia | • have eukaryotic cells |
| | • **\*cells never have cell walls** |
| | • bodies are multicellular |
| | • do not possess chlorophyll, all are heterotrophic |
| | • store carbohydrates as glycogen |
| | • **\*usually store lipids as fats** |
| | • **\*have nervous systems and so have nervous coordination** |
| | • growth not restricted to meristems |

*Table 5.2* continued

**?**

**8** State *two* features which may be present in both members of each of the following pairs:

(a) Fungi and Animalia

(b) prokaryotic and eukaryotic cells

(c) Protoctista and Plantae.  *(6 marks)*

Extension

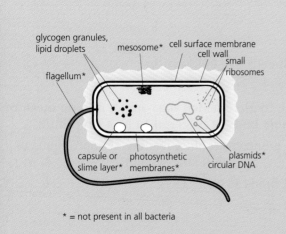

glycogen granules, lipid droplets
mesosome*
cell surface membrane
cell wall
small ribosomes
flagellum*
capsule or slime layer*
photosynthetic membranes*
circular DNA
plasmids*

\* = not present in all bacteria

*Figure 5.13 (a) A generalised bacterial cell*

*Figure 5.13 (b) Students sampling Archaea from a hot spring in Yellowstone National Park*

*Figure 5.13 (c) Blue-green bacteria*

## Box 5.4  Kingdom Prokaryotae

This kingdom is made up of archaea and bacteria of many types and contains some of the earliest organisms to evolve. All these organisms are **microscopic** with bodies consisting of one or a few **prokaryotic** cells. Figure 5.13a shows the features of a generalised bacterial cell.

The shapes of the cells may vary and sometimes the cells are grouped to form chains or clumps. Despite this basic similarity in structure, there are many thousands of species and they inhabit a wide range of habitats – from around vents of hot water on the sea bed to inside the human colon. Different forms also undergo all the different methods of nutrition.

The Prokaryotae are difficult to classify into groups as mutations, high selective pressures and short life cycles mean that their forms change rapidly. Also the variations between the different types are sometimes difficult to observe.

### Archaea

This group has recently been recognised as being distinct from other bacteria. Archaea look like most true bacteria but differ from them significantly, genetically and biochemically. They usually live in extreme environments with low levels of oxygen. Some can only survive in very hot springs (Figure 5.13b) or near hot rift vents in deep sea beds where pressures as well as temperatures are high. These can only reproduce at temperatures over 100 °C. Some can live in hot sulphuric acid while others are only found in very salty conditions such as those around the Dead Sea.

### Eubacteria (true bacteria)

All other living bacteria belong to this group.

### Blue-green bacteria (Cyanobacteria)

These are found in fresh or sea water and are photosynthetic. They may be single celled or consist of chains of cells as shown in Figure 5.13c. Some of them can fix nitrogen.

## Other true bacteria

The range of these bacteria is enormous. Some examples are:

- parasitic forms, of which some are harmless while others cause disease (**pathogenic**)

- those which live in symbiotic relationships with other organisms, for example *Rhizobium* sp. in leguminous plants, and bacteria which break down cellulose in ruminants such as cows

- nitrifying bacteria which are free living in soil and water. Their actions are important in the recycling of nitrogen.

**spherical**

Diplococcus sp.
some cause pneumonia

Staphylococcus sp.
some cause abscesses

Streptococcus sp.
some cause sore throats

**rod-shaped**

**vibrio**

**spiral-shaped**

Bacillus tuberculosis
can cause TB

Vibrio cholerae
can cause cholera

Spirillum sp.
free living

*Figure 5.13* (d)  *Bacteria can be of varied shapes*

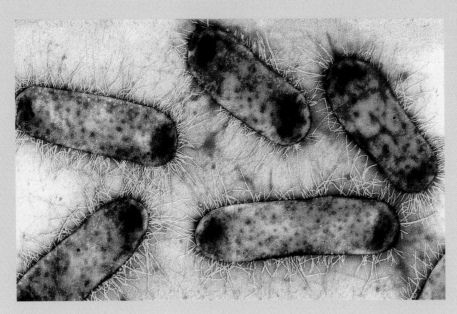

*Figure 5.13* (e)  Escherichia coli – *these bacteria are common in human intestines; only a few strains are harmful (× 12 500)*

Extension

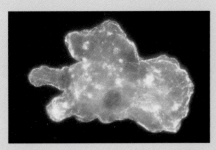

Amoeba – *a free-living species* ( × 85)

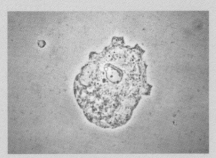

Entamoeba histolytica

***Figure 5.14*** *(a)  Two examples of amoebas*

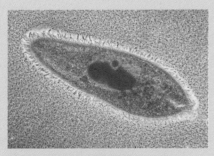

Paramecium *sp.* ( × 200)

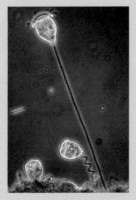

Vorticella *sp.*
(× 35)

***Figure 5.14*** *(b)  Ciliated protoctistans*

### Box 5.5  Kingdom Protoctista

This is a varied group which may be reorganised in the future. At present it consists of simple organisms, with **eukaryotic** cells, which are not included elsewhere! They include organisms previously known as '**protozoa**', whose bodies consist of one cell or a cluster of a few cells. Some of these have cell walls, some do not, and a wide range of nutrition methods is found. The group also includes forms previously classified as '**algae**'. These can have unicellular or very simple multicellular bodies. Their cells have cellulose cell walls and they are all photosynthetic.

There are many phyla, and changes in groupings are constantly being made. Representative examples of five phyla are shown and give some idea of the range of organisms in this kingdom.

**Amoebas** (Phylum Rhizopoda)

These are usually unicellular with no cell walls. They move and capture food using temporary extensions, **pseudopodia**, which are made by changing shape. They are found in fresh water, soil water, the seas or as internal parasites (e.g. *Entamoeba histolytica* which can cause dysentery).

**Ciliated forms** (Phylum Ciliophora)

These are unicellular and their outer surfaces have cilia which can beat in an organised way. The cilia can be used for movement and/or for capturing food. Some, such as *Paramecium*, swim freely in fresh water using their cilia. Others, such as *Vorticella*, live in ponds or seas and are anchored by stalks which can suddenly contract. They use their cilia to create whirlpool-like currents (vortices) which draw in bacteria on which they feed.

**Flagellated forms** (Phylum Discomitochondria)

This is a wide-ranging group. Some, like *Euglena viridis*, are photosynthetic and live in fresh water where they can be so abundant that they form a green scum on the surface. It is essential that they remain in the light near the surface and they each possess a light-sensitive organelle and pigmented red spot which enable them to detect the direction of light. They swim using a flagellum and squirming movements.

In Africa, *Trypanosoma gambiense* lives in the blood plasma of humans and other mammals. It wriggles about by means of a flagellum attached to a membrane. It is a parasite transmitted into humans by tsetse flies. It can invade the fluid surrounding the brain and spinal cord and eventually cause sleeping sickness.

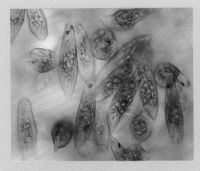
Euglena *sp.* (× 90)

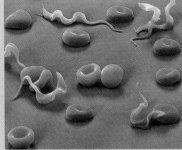
Trypanosoma gambiense (× 2000)

Fucus vesiculosus

***Figure 5.14*** *(c) Flagellated forms*

## Brown algae (Phylum Phaeophyta)

These are the large, familiar seaweeds of the seashore and shallow seas. They have large multicellular bodies. They are photosynthetic and appear brown because, as well as having green chlorophylls *a* and *c*, they have brown fucoxanthin pigments which are efficient at absorbing the blue light which penetrates water further than red light. Their bodies are usually differentiated into gripping **holdfasts** (root-like structures), a short stem-like **stipe** and long photosynthetic **fronds** which, in some forms such as the kelps (*Laminaria* spp.), can be up to 100 m long. Others have shorter fronds which in *Fucus vesiculosus* have air-filled bladders. These help the fronds to float in the surface water which is penetrated by light.

## Green algae (Phylum Chlorophyta)

These can be unicellular, filamentous or consist of small, spherical colonies of cells. A few are large but simple in form, such as the sea lettuce (*Ulva* sp.) which is found in the higher zones of sea shores. Green algae are usually found in fresh water or in damp situations such as the north side of tree trunks in a damp wood, where numerous algae give the bark a greenish colour. *Spirogyra* is a filamentous alga with spiral chloroplasts and is found floating on the surface of ponds. Members of this phylum have similar photosynthetic pigments to plants and are probably closely related to the ancestors of plants.

*Laminaria sp.*

***Figure 5.14*** *(d) Two examples of brown algae*

Spirogyra *sp.* (× 400)

***Figure 5.14*** *(e) Two examples of green algae*    Ulva *sp.*

Extension

## Box 5.6  Kingdom Fungi

This group contains the familiar mushrooms and toadstools as well as moulds and the single celled yeasts. Fungi were previously grouped with the plants simply because, like the plants, they cannot move from place to place in order to feed. However, they were placed into a separate kingdom when the significance of the differences between the groups was recognised. For example, plants are photosynthetic autotrophs whereas fungi never have chlorophyll and feed heterotrophically. The group contains important decomposers as well as parasitic pests.

### Pin moulds (Phylum Zygomycota)

Pin moulds such as *Mucor* spp. are white moulds which usually grow and feed saprobiontically on moist organic matter (including forgotten food). The hyphae make a loose white mesh and bear spore cases which look like black pin-heads at the ends of upright hyphae.

*Figure 5.15 (a) Scanning electronmicrograph of* Mucor mucedo *(× 80)*

### Mushrooms, toadstools and bracket fungi (Phylum Basidiomycota)

These are the most well known fungi as the hyphae can grow into compact masses called 'fruiting bodies', which form the visible mushrooms, toadstools and puff balls. However, many more hyphae are present, spread out underground or in an associated organism.

*Figure 5.15 (b) Examples of basidiomycetes*

Some forms live alone, feeding saprobiontically on dead organic matter in the soil or dead trees. However, many have a mutualistic relationship with plant roots. These relationships are called **mycorrhiza**. For example, some fungi have hyphae which surround and penetrate air spaces in the roots of trees such as conifers or beech. The fungi receive organic compounds such as carbohydrates from the tree, while absorbing and making minerals such as nitrates and phosphates available to that tree. The fruiting bodies can be seen around the bases of the trees, especially in autumn.

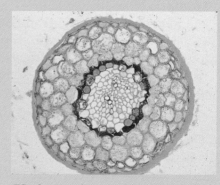

*TS of a root showing the presence of mycorrhizal fungal hyphae (stained blue)*

*fly agaric toadstool (*Amanita muscaria)

Polypora *sp. growing on birch*

*field mushroom (*Agaricus campestris)

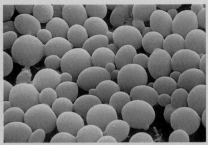
yeast (Saccharomyces cerevisiae) (× 350)

hyphae of Penicillium mould (× 600)

Penicillium mould growing on a peach

**Figure 5.15** (c) *Yeasts and moulds*

elm trees suffering from Dutch Elm disease

rose leaves with 'black spot'

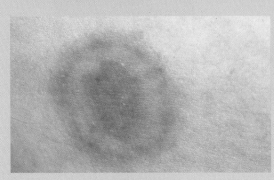
ringworm infection on human skin

**Yeasts and penicillium moulds** (Phylum Ascomycota)

This is the largest group of fungi which contains many economically important forms. Moulds such as *Penicillium* spp. have hyphae, while others like yeast (*Saccharomyces* spp.) are unicellular. Some *Penicillium* spp. form the bluish moulds which grow saprobiontically on food and other organic matter. Many secrete chemicals which prevent the growth of competing saprobionts (leaving more food for *Penicillium*!). The discovery and extraction of these chemicals led to the development of antibiotics.

Yeasts live on the surface of fruit such as grapes, using exuding sugars for respiration. Since ancient times, humans have taken advantage of the fermentation reactions of yeast to produce alcoholic drinks and to leaven bread.

**Fungi and disease**

Some parasitic forms can be pathogenic, causing diseases such as Dutch elm disease which has made a healthy elm tree a rare sight in the UK. The very common but less harmful 'black spot' on roses discolours and destroys rose leaves. Fungal diseases in humans have tended to be minor and include skin complaints such as athlete's foot and ringworm. However, diseases such as histoplasmosis and some fungal lung diseases can be very serious in AIDS patients owing to their poor immune responses.

**Figure 5.15** (d) *Examples of pathogenic fungi*

### Box 5.7 Kingdom Plantae (plants)

The members of this kingdom are prominent in all but the most urban landscapes. They range in size from tiny floating duckweeds less than 1 mm long to trees which can reach heights of over 120 m.

**Liverworts and mosses** (Phylum Bryophyta)

Although liverworts and mosses are thought to have evolved separately they have many features in common. They are small, simple land plants which can only live actively and reproduce in damp places. They require superficial water as they have no xylem to conduct water through the plant, no cuticle to restrict water loss, and during sexual reproduction the flagellated male gametes swim to the female gametes.

Liverworts such as *Pellia* spp. appear as flat, undifferentiated, branching green 'mats' in damp, undisturbed spots. Mosses such as *Funaria* spp. appear more complex, with leaf-like blades. However, their simple bodies are not differentiated into true roots, stems and leaves although they have small, thread-like extensions (**rhizoids**) for attachment. Sexual reproduction in both groups involves swimming male gametes which fertilise female gametes producing zygotes. These grow into spore-producing generations, each of which consists of a stalk and spore capsule. At certain times of year, these spore-producing capsules can be seen clearly.

Mosses are very abundant. *Sphagnum* mosses which grow in acid bogs make up the bulk of **peat** which has formed by partial decomposition over many years.

**Ferns** (Phylum Filicinophyta)

Ferns are easily recognised by large, often feathery, leaf-like fronds which are arranged spirally around a short stem. They have vascular tissue but are still restricted to habitats which are damp for part of the year as their flagellate male gametes swim to the female gametes. The fern fronds bear clusters of spore cases (sori) on their undersides. Many hillsides are covered with bracken. This is a tall fern whose spores should be avoided as they contain cancer-causing chemicals!

*Funaria sp. – a moss*

*Pellia sp. – a liverwort*

Sphagnum *moss in a bog*

*Figure 5.16* (a) *Examples of mosses and liverworts*

*Figure 5.16* (b) *Examples of ferns and detail of spore cases*

*bracken (*Pteridium *sp.)*

*spore cases on the underside of fronds of* Polypodium *sp.*

*royal fern (*Osmunda *sp.)*

## Plants which produce seeds

### Conifers (Phylum Coniferophyta)

*coast redwood (*Sequoia sempervirens*)*

These are trees or shrubs which usually have needle-shaped leaves and produce seeds which are borne in cones. They are not restricted to damp habitats as they do not require superficial water for reproduction. The male gamete is transported to the female gamete inside a pollen tube (see *AS Biology*, Section 8.6, page 223). Some of the world's tallest plants, the Coast Redwoods, are to be found in this group (see Box 5.3, page 166). Conifers native to the UK are the yew, juniper and Scots pine but many other species have been imported for timber production and garden decoration.

*yew (*Taxus baccata*)*

*juniper (*Juniperus communis*)*

*Scots pine (*Pinus sylvestris*)*

**Figure 5.16** *(c)  Examples of conifers*

### Flowering plants (Phylum Angiospermophyta)

This is the largest group of plants with at least 260 000 known species. There is a wide range of form, size and lifespan. Their reproductive structures are contained within flowers and they produce seeds which are protected inside fruits. Like the conifers, they are not restricted to damp habitats as their male gametes are transported to the female gametes inside pollen tubes.

**Figure 5.16** *(d)  Examples of flowering plants*

*duckweed (*Lemna sp.*)*

*dog rose (*Rosa canina*)*

*oak tree (*Quercus robur*)*

## Box 5.8  Kingdom Animalia (animals)

This is by far the most diverse kingdom. Although the range of size is far more restricted than in plants, the variety of form is immense and only a few phyla are represented below.

### Corals, jelly-fish and sea anemones (Phylum Cnidaria)

This group contains animals whose body organisation is simple, based on two layers of cells (**diploblastic**) around a central cavity. They are **radially symmetrical**. This means that their body parts are arranged evenly around a line through the centre. They have tentacles arranged in a circle around the mouth and they have no anus. This is the only group of animals to have special cells called stinging cells (**cnidoblasts**) which can shoot out stinging threads when triggered. Their tissues do not work together in organs and they are said to be at the **tissue level of organisation**.

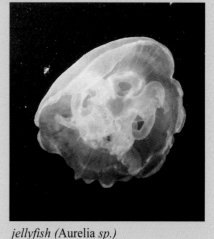

*sea anemone (Actinia sp.)*    *'skeleton' of brain coral (Zoantheria sp.)*

*jellyfish (Aurelia sp.)*

**Figure 5.17 (a)  Examples of cnidarians**

### Flatworms (Phylum Platyhelminthes)

The body is simple but based upon three layers of cells (**triploblastic**) arranged around a gut cavity (although some forms have lost this cavity). There is no separate body cavity (**coelom**) – they are **acoelomate**. The group is characterised by its flat body shape. The bodies are **bilaterally symmetrical**. This means that the shape of the body can be cut through in one plane only to produce two mirror image halves. Although some, such as *Planaria* spp., are free living in streams, many are parasitic. Tapeworms and flukes of many kinds are parasites of humans.

**Figure 5.17 (b) Examples of flatworms**

*free-living flatworm (Euplanaria sp.)*

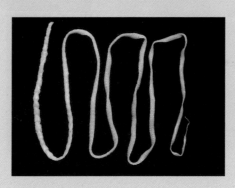

*tapeworm (Taenia sp.)*

*earthworm*

*leech (*Hirudo medicinalis*) which can take in three times its own weight in blood in one meal!*

*ragworm (*Nereis *sp.)*

## Segmented worms (Phylum Annelida)

These are triploblastic animals having a fluid-filled body cavity (coelom) as well as the central gut cavity. This is called **coelomate** organisation. They are bilaterally symmetrical. Their cylindrically shaped bodies show clearly marked sections along their length, known as segments. Many but not all of the organs of the body are repeated in each segment. This type of body structure is known as **metameric segmentation**. Most groups of worms have prominent bristles called **chaetae** which help them to move or grip. Many segmented worms live in water and swim or burrow in the mud. Earthworms are an exception but are restricted to damp habitats as their body surfaces must remain moist as they are respiratory surfaces (see *AS Biology*, Section 9.3, page 233).

## Crustacea, insects, millipedes and spiders (Phylum Arthropoda)

This phylum contains over three quarters of the species of the animal kingdom. This means that there are many varied forms but they have such clear features characteristic of the group that they are usually easily recognisable as arthropods. They are all triploblastic, bilaterally symmetrical, coelomate animals (although the coeloms are reduced and replaced by a blood-filled cavity). Their bodies are metamerically segmented. A characteristic feature of the group is the outer cuticle based on chitin with hardening materials. This forms an external rigid but jointed skeleton to which muscles are attached internally. This exoskeleton contributes to the success of the group as it is highly protective. When covered with a waxy layer preventing dehydration, it allows animals to inhabit extremely dry habitats. On the other hand, its weight restricts size and moulting is necessary during growth. Arthropods have paired jointed limbs which have various functions.

*Figure 5.17 (c) Examples of segmented worms*

*crab (*Cancer *sp.)*

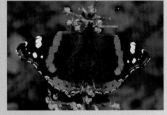

*red admiral butterfly (*Vanessa atalanta*)*

*desert hairy scorpion (*Hadrurus arizonensis*)*

*millipede*

*Figure 5.17 (d) Examples of arthropods*

### Box 5.8  Kingdom Animalia (animals) continued

**Vertebrates** (Phylum Chordata, sub-phylum Craniata)

This is probably the most prominent group on Earth but, in fact, contains less than 4% of animal species. While also being triploblastic, coelomate, bilaterally symmetrical and metamerically segmented like most animals, vertebrates also have several distinguishing characteristics. The column of vertebrae which supports the body and protects the spinal cord gives the group its common name. The front end of the spinal cord is expanded to form a brain which is protected by the skull. Although humans are an exception, most vertebrates have tails.

*shark – a cartilaginous fish*

*frog – an amphibian*

*flamingo – a bird*

*giraffe – a mammal*

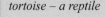
*tortoise – a reptile*

**Figure 5.17** *(e)  Examples of vertebrates*

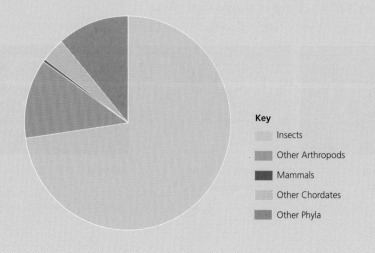

**Key**

- Insects
- Other Arthropods
- Mammals
- Other Chordates
- Other Phyla

### 💡 Did You Know

Almost three quarters of the animal species are insects! There are about 72 times as many species of beetles as there are species of mammals! The pie chart gives some idea of the relative numbers of species in the Phylum Arthropoda and the Phylum Chordata.

**Figure 5.18** *Number of species as a percentage of the total number of animal species*

## Summary – ⑤ Classifying living organisms

● A **species** is a group of similar, closely related organisms which are capable of interbreeding with each other and producing viable and fertile offspring.

● Members of the same species have many genes in common and similar karyotypes.

● Offspring of two members of the same species inherit the same karyotype. They are fertile as they have homologous pairs of chromosomes. This allows meiosis and gamete formation to occur.

● It is difficult to determine if individuals belong to the same species if:
  – their appearance differs because of environmental factors
  – there are polymorphic forms of the species
  – directional selection has led to differences between individuals
  – the members of the species mainly reproduce asexually
  – only fossil forms are known.

● The ways of determining whether two individuals belong to the same species include:
  – study of similarities of structural features
  – study of similarities of chemical features
  – study of similarities of karyotypes
  – study of similarities of DNA by DNA analysis
  – direct observation of mating behaviour and fertility of offspring.

● Species are classified into groups. Such **classification** may be **artificial** where grouping is based on convenient similarities and differences.

● The **natural** classification system groups species into a hierarchy of groups which reflect relationships and evolutionary descent. Species with the most similar chromosomal and DNA structure as well as similar **homologous** structural and chemical features are grouped together in the smallest groups.

● The hierarchy of groups in descending size order is – **Kingdom, Phylum, Class, Order, Family, Genus, Species**.

● The international system of naming species is known as the **binomial system of nomenclature**. Each species is known by its genus name followed by its species name.

● Living organisms are grouped into five kingdoms. The features of the five kingdoms are summarised in Table 5.2 on pages 168–169.

## ❓ Answers

1  D, E, C, B, A *(2)*.

2  A = field bindweed; B = greater bindweed; C = dog rose; D = creeping buttercup; E = creeping thistle; F = hedge woundwort *(6)*.

3  *Escherichia coli, Fucus serratus, Agaricus campestris, Urtica dioica, Felis tigris (1 mark for each if capitals used correctly, additional mark if handwriting underlined – 6)*.

4  A = Class, B = Order, C = Family, D = *Pieris*, E = *brassicae* *(5)*.

5

| Prokaryotic cells | Eukaryotic cells |
| --- | --- |
| no membrane bound organelles | have membrane bound organelles (e.g. nuclei, mitochondria, endoplasmic reticulum, plastids) |
| circular DNA not bound to histones | linear DNA bound to histones |
| cell walls contain peptidoglycan | if cell walls present, do not contain peptidoglycan |
| small 70S ribosomes present | 80S and 70S ribosomes present |
| no undulipodia | undulipodia may be present |
| smaller, usually less than 0.5 μm diameter | larger, usually 10–100 μm diameter |

*(4 max.)*

6  A – Prokaryotae, B – Protoctista, Animalia, C – Protoctista, Fungi, Plantae *(6)*.

7  (a) Chitin cell walls/eukaryotic cells without undulipodia/hyphae.

(b) Cells never have cell walls/nervous system present/store lipids as fats.

(c) Any of characteristics of prokaryotic cells (see above) *(max. 3)*.

8  (a) eukaryotic cells; multicellular; all heterotrophic; no chlorophyll; store carbohydrates as glycogen *(any 2)*.

(b) cell surface membrane; DNA; cytoplasm; ribosomes *(any 2)*.

(c) eukaryotic cells; cellulose cell walls are/sometimes are present; chlorophyll is/sometimes is present *(any 2)*.

## End of Chapter Questions

**1** For each of the following statements, list the kingdom(s) for which it is true.

**(a)** Some or all are heterotrophic.

**(b)** The cells all have or may have cell walls.

**(c)** Lipid is mainly stored as oil.

**(d)** Carbohydrate is stored as starch.

**(e)** Mitochondria are present in the cells.

**(f)** Some members may be unicellular.

**(g)** Some or all members may photosynthesise.

**(h)** Hyphae are present.

**(i)** All members are heterotrophic.

**(j)** Undulipodia may be present. *(Total 10 marks)*

**2** The diagram shows how four species of pig are classified.

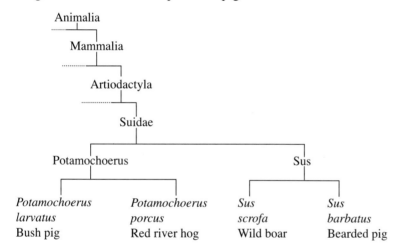

**(a)** (i) To which family does the red river hog belong? *(1 mark)*

(ii) To which genus does the bearded pig belong? *(1 mark)*

**(b)** Some biologists think bush pigs and red river hogs belong to the same species. The list below summarises some features of the biology of bush pigs and red river hogs.

N2.3
C2.2

• The bush pig has a body length of 100–175 cm and a mass of 45–150 kg.
The red river hog has a body length of 100–145 cm and a mass of 45–115 kg.

• The red river hog is found in West Africa. The bush pig is found in East Africa.

• Both animals are omnivorous but feed mainly on a variety of underground roots and tubers.

- The ranges of these animals overlap in Uganda. In this area populations of animals which have all characteristics intermediate between those of bush pigs and red river hogs have existed for many years.

Do you think that bush pigs and red river hogs belong to the same or to different species? Explain how the information above supports your answer. *(3 marks)*

*AQA/A 2000*

*(Total 5 marks)*

**3** Hares are small mammals similar to rabbits. The diagram shows how some of the hares found in southern Africa are classified.

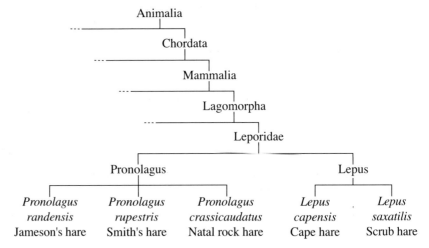

**(a)** (i) Name the genus to which the scrub hare belongs. *(1 mark)*

(ii) Name the order to which the Natal rock hare belongs. *(1 mark)*

**(b)** The map shows the distribution of three of these species of hare.

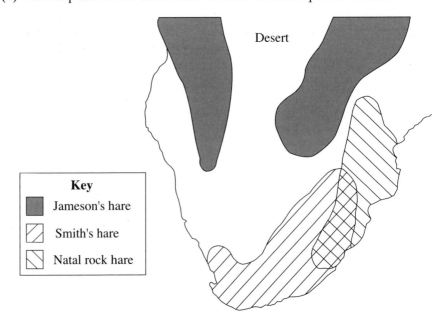

(i) What is the evidence from the map that suggests that Smith's hare and the Natal rock hare are different species? *(1 mark)*

(ii) Suggest what has caused the gene pools of the two populations of Jameson's hare to differ. *(2 marks)*

*AEB 1999* *(Total 5 marks)*

**4** The diagram shows one way of representing the classification of living organisms into five kingdoms. Box **A** has been drawn overlapping the other boxes since its members share characteristics with the other kingdoms.

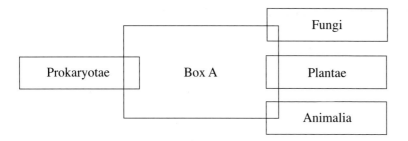

**(a)** Which kingdom is represented by the box labelled **A**? *(1 mark)*

**(b)** Give **one** structural characteristic that the members of the kingdom represented by box **A** may share with:

(i) Fungi *(1 mark)*

(ii) Prokaryotae *(1 mark)*

**(c)** Give **two** reasons why the Fungi are placed in a separate kingdom from the Plantae. *(2 marks)*

*AEB 1995* *(Total 5 marks)*

**5** A cladogram is a simple branching diagram which shows relationships between different groups of organisms. The cladogram below suggests, for example, that Fungi and Plantae are more closely related to each other than they are to Animalia.

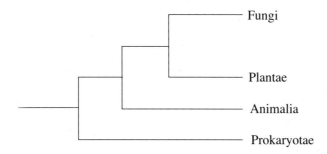

Give:

**(a)** **one** feature shared by Fungi and Plantae which suggests that they are more closely related to each other than they are to Animalia; *(1 mark)*

**(b)** **two** features which suggest that this view is not correct and Fungi are more closely related to Animalia than they are to Plantae; *(2 marks)*

**(c)** **two** features of Prokaryotae which suggest that they are not closely related to any of the other three kingdoms. *(2 marks)*

*AEB 1994* *(Total 5 marks)*

**6** The diagram shows how living organisms may be classified into five kingdoms.

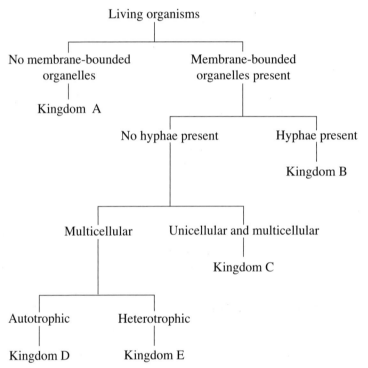

**(a)** Name:

(i) Kingdom **B**;

(ii) Kingdom **C**. *(2 marks)*

**(b)** Which of Kingdoms **A**, **B**, **C**, **D** or **E**, include organisms that possess:

(i) undulipodia;

(ii) a cell wall? *(2 marks)*

*AEB 1992* *(Total 4 marks)*

7 The table below gives features of four different kingdoms. Copy and complete the table by writing the name of the kingdom in the boxes.

| Features | Kingdom |
|---|---|
| Organisms usually consist of a mass of hyphae with cell walls containing chitin. | |
| Heterotrophic, multicellular organisms with nervous coordination. | |
| Multicellular photosynthetic organisms with cellulose cell walls. | |
| Eukaryotic organisms which are often single-celled or consist of groups of similar cells. | |

*Edexcel 2000*                                                (Total 4 marks)

8 The diagram below shows the way in which four species of monkey are classified.

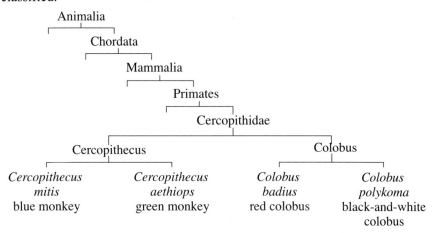

(a) This system of classification is described as hierarchical. Explain what is meant by hierarchical. *(1 mark)*

(b) (i) To which genus does the green monkey belong? *(1 mark)*

   (ii) To which family does the red colobus belong? *(1 mark)*

(c) What does the information in the diagram suggest about the similarities and differences in the genes of these four species of monkey? *(2 marks)*

*AQA 2000*                                                (Total 5 marks)

# 6 Neurones and the nerve impulse

In mammals, the functions of communication within the body and the coordination of the body's activities are carried out by the **nervous** and **endocrine (hormonal) systems**. This chapter will consider how information is transmitted around the body by the **neurones** (nerve cells) of the nervous system. Neurones are highly adapted for this function in various ways, e.g. they have long extensions called **axons** which act like wiring.

Information is carried by neurones as **nerve impulses**. These are signals called **action potentials** which move as a wave along the neurone. The nerve impulse is transmitted from one neurone to another at junctions called **synapses**. **Neurotransmitters** – chemical messengers – carry the impulse across the synapse. Many different chemicals can affect synapses, including medicines, illegal and legal drugs.

Other aspects of the nervous system, including the brain and **receptors** such as the eye, are described in Chapter 7. The endocrine system is referred to in Chapters 9 and 10.

**This chapter includes:**
- the structure and function of the different types of neurone
- the myelin sheath and the role of Schwann cells
- the nature and formation of the resting potential
- the nature and formation of the action potential
- the nature of the nerve impulse and its conduction along neurones
- the structure and function of synapses
- the significance of synapses in terms of the overall nervous system
- the effects of drugs and other chemicals on the synapse.

## 6.1 The structure of neurones

The nervous system is made up of two main groups of cells. **Neuroglia**, e.g. **Schwann cells**, are packing cells providing structural and metabolic support to the other type of nerve cell, the **neurones**. Neurones are highly specialised cells adapted to conduct nervous impulses. They can be classified according to their function (Figures 6.1 and 6.2).

- **Motor** or effector neurones conduct impulses to an **effector**, e.g. a muscle.
- **Sensory** neurones conduct impulses away from a **receptor**, e.g. from a pressure receptor in the skin.
- **Relay** or connector neurones conduct impulses between a sensory and a motor neurone, e.g. in the spinal cord.

How these three types work together in a **reflex action** is described in Section 7.9, page 246.

**?**

1 Most neurones make connections (synapses) with many other neurones. In which part of the body would you expect neurones to have the most connections with other neurones? *(1 mark)*

*Figure 6.1* *A myelinated motor neurone*

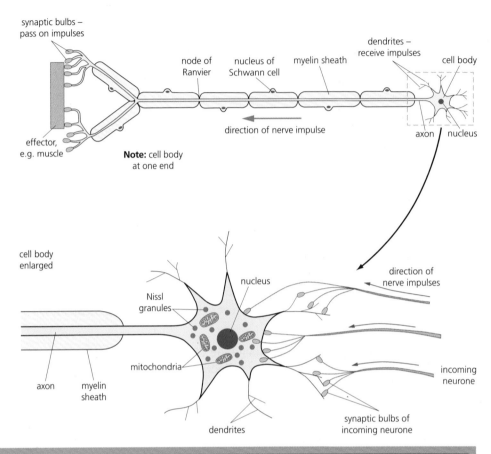

synaptic bulbs –
pass on impulses

node of
Ranvier

nucleus of
Schwann cell

myelin sheath

dendrites –
receive impulses

cell body

← direction of nerve impulse

effector,
e.g. muscle

**Note:** cell body
at one end

axon    nucleus

cell body
enlarged

nucleus

direction of
nerve impulses

Nissl
granules

mitochondria

axon    myelin
sheath

dendrites

synaptic bulbs of
incoming neurone

incoming
neurone

| Motor neurone: how structure is adapted for function | |
|---|---|
| Highly branched dendrites | Increase the surface area so that synapses (junctions) can be made with many other neurones. Receive incoming impulses. |
| Many mitochondria in cell body and in axon | Produce ATP for the sodium/potassium pumps – needed to maintain resting potential. |
| Long thin axon | Provides the pathways for the nervous system. Able to conduct impulses quickly over long distances with a single cell. Using many shorter cells would be slower because of the synapses between neurones. |
| Myelin sheath | Acts as an electrical insulator preventing the formation of action potentials. This speeds up conduction because the nerve impulse has to jump from node to node along the axon (see Section 6.8, page 202). |
| Nodes of Ranvier | These are gaps in the myelin sheath. The absence of the sheath is essential to allow action potentials to form. Action potentials are the basis of the nerve impulse. |
| Branched terminal ends of axon | Increase surface area so that the axon can form many synapses with effector. They pass on the impulses. |
| Synaptic bulb | Adapted to transmit impulses to effector, e.g. contains vesicles with neurotransmitter. |

**sensory neurone**

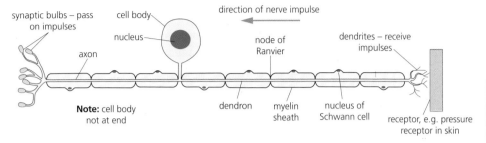

*Figure 6.2 Sensory and relay neurones*

**relay (connector) neurone**

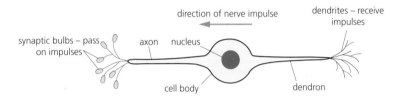

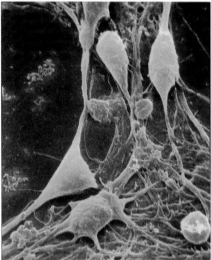

*Figure 6.3 False colour scanning electronmicrograph of relay neurones in the brain (× 400)*

Neurones vary in function, and in size and shape, but many of their structural features are the same (see Figures 6.1 and 6.2). These features include the following:

● A large cell body containing the nucleus and other cell organelles, including mitochondria and **Nissl granules**. Nissl granules are groups of rough endoplasmic reticulum and ribosomes, and are associated with the synthesis of proteins, including the neurotransmitters.

● Extending from the cell body are thin branches or extensions which are often referred to collectively as **fibres**. Fibres which conduct impulses towards the cell body are called **dendrons**. Smaller fibres called **dendrites** converge to form dendrons. **Axons** are long fibres which conduct impulses *away* from the cell body. They may be over 1 m in length. The terminal end of the axon is divided into a number of branches with swollen endings called **synaptic bulbs** (or knobs). In mammals many axons and dendrons are **myelinated** – covered with a **myelin sheath**.

**multipolar neurone**, e.g. motor neurone

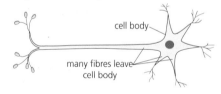

**unipolar neurone** (or pseudo unipolar) e.g. sensory neurone

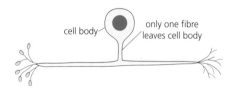

**bipolar neurone**

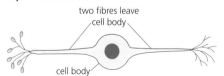

*Figure 6.4 Types of neurone based on structure*

**Box 6.1 Classification of neurones**

Neurones can also be classified according to their structure – by how many extensions or fibres they have leaving the cell body. These fibres include the dendrites and the axon. Multipolar neurones have many fibres, bipolar have two and unipolar (or pseudo-unipolar) have one (Figure 6.4). Pseudo-unipolar neurones are so called because they have more than one fibre when first formed and only become unipolar later in development. Motor neurones are multipolar whereas sensory neurones are unipolar. Examples of bipolar neurones are the ganglion neurones in the retina and in the brain (Figure 6.3).

## 6.2 Myelination and Schwann cells

The myelin sheath is formed by a type of neuroglia cell called a **Schwann cell** which wraps itself round and round the nerve fibre like the rolling up of a mat. This process occurs during early development (Figures 6.5 and 6.6).

*Figure 6.5 Schwann cells and the myelin sheath*

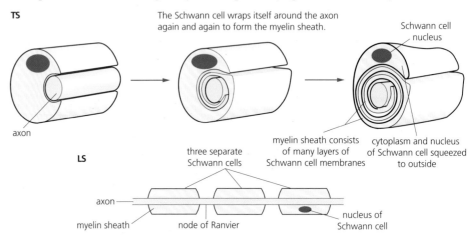

Each Schwann cell is able to form a sheath around approximately 1 mm length of nerve fibre. There are small spaces between Schwann cells giving rise to gaps in the myelin sheath called **nodes of Ranvier**. At these nodes the fibre is exposed. Because the myelin sheath consists mainly of layer upon layer of cell membrane there is a high proportion of phospholipid present and the sheath acts as an electrical insulator.

Action potentials cannot be formed where the sheath is present, they can only be formed at the nodes of Ranvier. Therefore the nervous impulse has to jump from node to node – a process called **saltatory conduction**. This speeds up the rate of the conduction of the nervous impulse. For further details see Section 6.8, page 202.

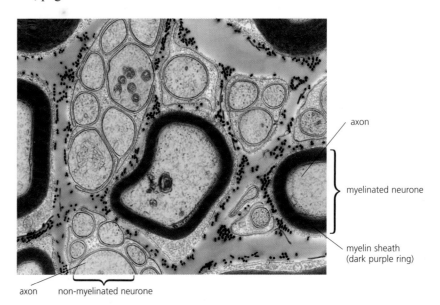

*Figure 6.6 Electronmicrograph of transverse sections of myelinated and non-myelinated neurones*

## 6.3  Potential differences

**Potential difference** is an important feature in this topic. It is caused by a difference in the electrical charge on the inside of the plasma membrane compared to the outside. In a resting neurone both sides of the membrane have a positive charge but the inside is less positive than the outside, i.e. the inside is negative *compared to* the outside. Potential differences always refer to the inside of the membrane and they refer to a specific numerical value, e.g. –70 millivolts (mV). In this case the inside is 70 mV *less* positive than the outside, hence –70 mV.

All animal cells have a potential difference across their plasma membranes – the membranes are said to be **polarised**. In most cells this is fixed. Neurones, muscle cells and receptors are **excitable**. The potential difference can be reversed – a process called **depolarisation**. It is called depolarisation because the potential difference is decreased and then cancelled out before being reversed. Potential differences are set up due to the unequal distribution of ions (charged particles) on either side of the membrane (Figure 6.7).

There are various types of potential difference. These are summarised in Table 6.2 on page 215.

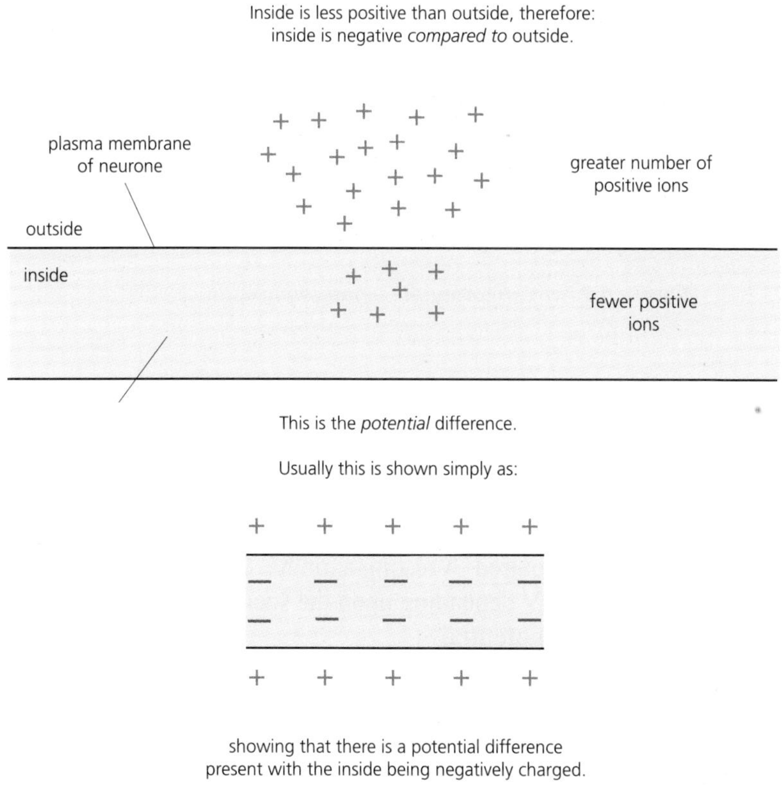

### Definitions

A membrane is said to be **polarised** if a potential difference is maintained across it with the inside being negative with respect to the outside.

**Depolarisation** is a decrease and then temporary reversal of the potential difference so that the inside of the membrane becomes positive with respect to the outside. (The membrane is actually still polarised but the polarity is reversed.)

Inside is less positive than outside, therefore:
inside is negative *compared to* outside.

plasma membrane of neurone

greater number of positive ions

outside

inside

fewer positive ions

This is the *potential* difference.

Usually this is shown simply as:

showing that there is a potential difference present with the inside being negatively charged.

*Figure 6.7 Potential differences are caused by the unequal distribution of positively charged ions on either side of cell membranes*

**2** In terms of their function, suggest why muscle cells need to be excitable.
*(1 mark)*

**Extension**

*Figure 6.8  Andrew Huxley (left) and Alan Hodgkin (right)*

membrane of giant axon
electrical stimulus applied here
internal electrode
interior of axon
impulse
external electrode
cathode ray oscilloscope

### Box 6.2  Investigating the nerve impulse

Because the events of a nervous impulse are electrical they can be investigated using apparatus sensitive to small electrical changes, in particular by the cathode ray oscilloscope. These studies were pioneered by the British scientists Alan Hodgkin and Andrew Huxley at Cambridge University and they were awarded the 1963 Nobel Prize for work on the nerve impulse.

Very tiny microelectrodes are used, with one placed inside an axon and another on the outer surface of the axon membrane. The axon is stimulated and the resulting electrical changes are detected by the microelectrodes. Signals from the electrodes are amplified and displayed on a cathode ray oscilloscope. This has a screen like a TV screen and the signals are displayed as a trace (Figure 6.9).

Hodgkin and Huxley used the very large 'giant' axons found in squid. These have diameters of up to 1 mm and were big enough to allow the microelectrode to be inserted inside the axon. Giant axons do not have myelin sheaths – they are unmyelinated. These giant axons allow very rapid movement of the squid when escaping from danger. The nerve impulses can be conducted very rapidly to the muscles when danger is detected.

Since these early days, further refinements of the methods and the materials mean that nerve fibres of much smaller sizes can be used.

*Figure 6.9  Investigating the nerve impulse*

### Remember

In discussing potential differences, including resting and action potentials, it is usually more convenient to just refer to the axon and the axon membrane. It should be remembered, however, that these processes can occur in *all parts* of the neurone plasma membrane and not just the axon membrane.

### 6.4  The resting potential

An axon is 'at rest' when it is not conducting an impulse. Although it is 'at rest' it is not inactive. The maintenance of the resting state requires the use of ATP for active transport. The axon membrane is being held in a state of readiness to conduct an impulse when required. A resting axon has a potential difference of between –50 mV and –90 mV depending upon the species. A typical value is –70 mV. This is the **resting potential**.

The resting potential results from the unequal distribution of charged ions, particularly sodium ($Na^+$) and potassium ($K^+$) on either side of the membrane. (Other ions are involved – see Box 6.4 on page 197 – but an understanding of these is not essential at A level.) The total number of positive ions ($Na^+$ and $K^+$ together) on the outside of the membrane is greater than the total number on the inside. The ions inside the membrane are in the cytoplasm of the cell body and the **axoplasm** (cytoplasm) of the axon. The ions outside the membrane are in the **tissue fluid** which surrounds all cells (see Figure 6.7 and Table 6.1).

| Ion | Concentration/mmol dm$^{-3}$ | |
|-----|------------------|------------------|
| | inside the membrane | outside the membrane |
| Na$^+$ | 50 | 460 |
| K$^+$ | 400 | 20 |
| total | 450 | 480 |

*Table 6.1* *The distribution of ions on either side of the plasma membrane in a resting squid axon*

Note: There is a higher total concentration of + ions outside, therefore the inside of the membrane is negative compared to the outside. This is the resting potential.

This unequal distribution is caused by two things – the differential permeability of the neurone membrane to Na$^+$ and K$^+$ ions and the action of **sodium/potassium (cation) pumps**. In the resting state, concentration gradients for both Na$^+$ and K$^+$ ions are maintained. There is a higher concentration of Na$^+$ outside the axon membrane and a higher concentration of K$^+$ inside the membrane (Figure 6.10).

> ### Definition
> The resting potential is the potential difference which exists across the axon membrane when the neurone is not conducting an impulse. The inside of the membrane is negative compared to the outside.

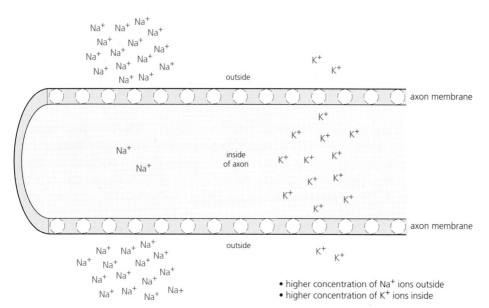

*Figure 6.10* *Concentration gradients are maintained for Na$^+$ and K$^+$ ions*

• higher concentration of Na$^+$ ions outside
• higher concentration of K$^+$ ions inside

Because of these concentration gradients, Na$^+$ ions tend to diffuse into the axon and K$^+$ ions tend to diffuse out. Diffusion occurs due to both electrical and chemical gradients – **electrochemical gradients** (see Box 6.3).

The neurone membrane is differentially permeable – it is more permeable to K$^+$ than Na$^+$. The permeability of the membrane to ions depends upon proteins in the membrane which act as **ion channels**. These channels are specific. Some allow Na$^+$ ions to diffuse into the axon through the membrane and others allow the diffusion of K$^+$ ions out of the axon. The channels are also **gated**; they have gates which can be open or closed. When closed, ions cannot diffuse through the channels. The permeability of the membrane to these ions depends upon the proportion of gates open or closed.

In a resting axon, most of the sodium gates are closed, but proportionally more of the potassium gates are open. As a result the membrane is approximately 20 times

> ### Box 6.3  Electrochemical gradients
> **Extension**
>
> The movement of ions by diffusion into and out of the axon is caused by two gradients – electrical and chemical. Ions have an electrical charge and will tend to move towards the opposite charge. Ions will also diffuse in the usual way from a higher concentration of the ion to a lower concentration down a chemical gradient. Therefore when Na$^+$ ions diffuse into the axon this is due to two gradients – hence the term *electrochemical* gradient.

more permeable to K⁺ than to Na⁺. Therefore the rate of K⁺ ions diffusing out is greater than the rate of Na⁺ ions diffusing in. This helps produce a build-up of a greater combined total of positive ions (both Na⁺ and K⁺) outside the membrane (Figure 6.11).

*Figure 6.11 The membrane is more permeable to K⁺ ions than to Na⁺ ions – it is differentially permeable*

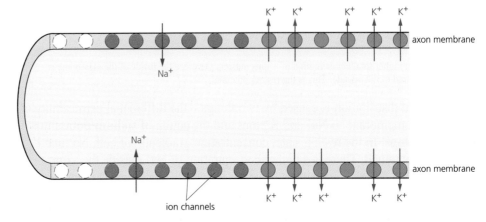

- most Na⁺ channels are closed
- membrane has low permeability to Na⁺ ions
- slow rate of diffusion of Na⁺ ions *into* axon

- most K⁺ channels are open
- membrane has high permeability to K⁺ ions
- fast rate of diffusion of K⁺ ions *out of* axon

Also present in the membrane are other proteins which act as **sodium/potassium pumps** (Na⁺/K⁺ pumps), actively transporting ions against their concentration gradients. Three Na⁺ ions are pumped out for every two K⁺ ions pumped into the axon. These pumps require energy from the hydrolysis of ATP. The action of these pumps maintains the relative concentration gradients for Na⁺ and K⁺ and prevents diffusion eventually leading to equilibrium (Figure 6.12).

*Figure 6.12 The action of the Na⁺/K⁺ pumps helps to maintain the concentration gradients for the ions*

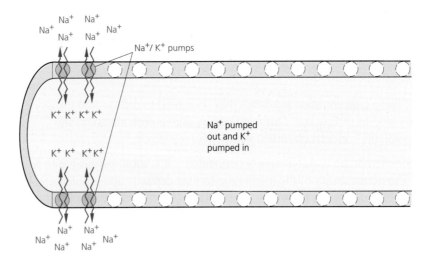

It is the *combined* effect of the differential permeability of the membrane to Na⁺ and K⁺ ions and the action of the Na⁺/K⁺ pumps which create the resting potential. These two processes are summarised in Figure 6.13.

*Figure 6.13* *The resting potential*

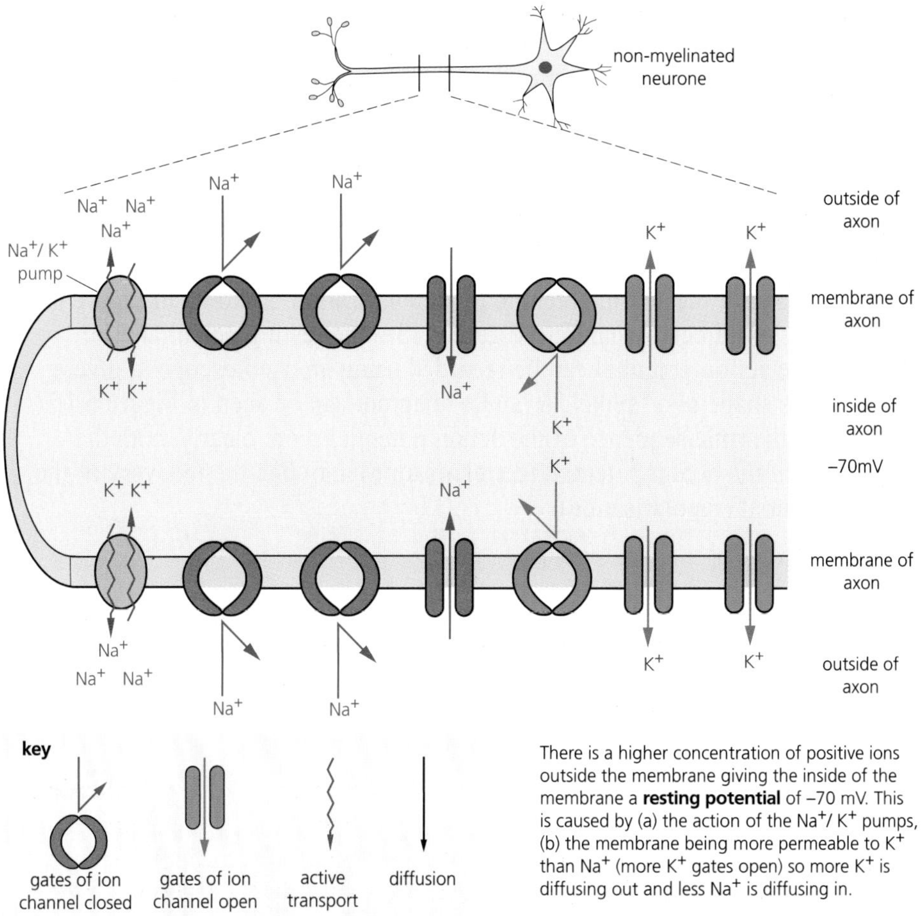

**key**

gates of ion channel closed | gates of ion channel open | active transport | diffusion

There is a higher concentration of positive ions outside the membrane giving the inside of the membrane a **resting potential** of −70 mV. This is caused by (a) the action of the Na⁺/ K⁺ pumps, (b) the membrane being more permeable to K⁺ than Na⁺ (more K⁺ gates open) so more K⁺ is diffusing out and less Na⁺ is diffusing in.

**Extension**

### Box 6.4 Other ions

In addition to Na⁺ and K⁺, other ions, including chloride ions (Cl⁻) and charged proteins, are present on either side of the neurone membrane. These are involved in nerve impulses. For example, the organic protein ions are negatively charged and are only found inside the membrane. The membrane is impermeable to these ions which therefore remain inside and contribute to the resting potential difference. A detailed understanding of these is not essential at A level.

### ! Did You Know

There are between 100–200 Na⁺/K⁺ pumps per square micrometre (µm²) of neurone membrane. (1 µm = 1/1000 mm.)

**?**

3  Cyanide is a poison which stops the production of ATP. Nerve fibres treated with cyanide soon lose the ability to conduct impulses. Why?

*(1 mark)*

### Definition

An action potential is the change in the potential difference across an axon membrane which occurs during the passage of a nerve impulse. There is a temporary reversal of the resting potential so that the inside is positive compared to the outside.

## 6.5 The action potential

Nerve impulses depend upon a stimulus causing a sudden and brief reversal of the potential difference. For a very short time (about 1 millisecond [ms]) the inside of the axon membrane is positive with respect to the outside – this is the **action potential**. It only occurs in one small section of the membrane at a time. A nerve impulse consists of the movement or propagation of an action potential along a nerve axon.

The production of the action potential by depolarisation is shown in Figure 6.14. This figure should be compared to Figure 6.13 – the resting potential. The events of the action potential can be recorded using an oscilloscope to give a record in the shape of a 'spike'. A 'spike' diagram can be seen in Figure 6.15. This shows the time sequence of the action potential more clearly, both the production of the action potential (depolarisation) and then the recovery of the resting potential (repolarisation).

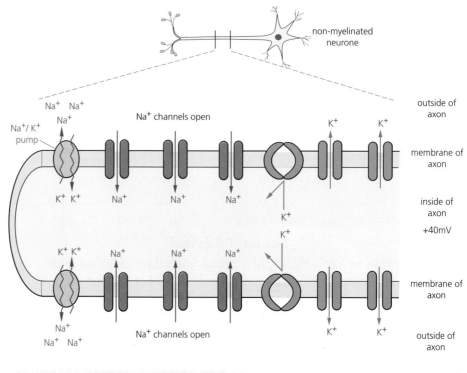

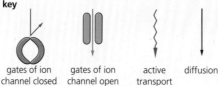

*Figure 6.14 The production of the action potential*

A stimulus causes the gates of the Na$^+$ channels to open and Na$^+$ ions to diffuse rapidly into the axon. The moving in of these positive ions reverses the potential difference. The inside of the membrane is now positive. This is the **action potential** with a value of +40 mV.

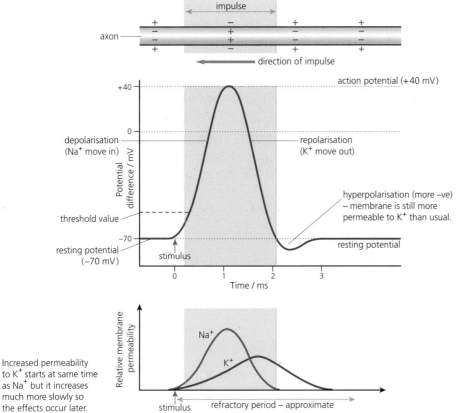

*Figure 6.15* *The action potential 'spike'*

## The events of the action potential

The following sequence of events occurs during an action potential. Both Figures 6.14 and 6.15 should be referred to.

1   A stimulus causes a sudden increase in permeability of the axon membrane to $Na^+$ ions by opening the $Na^+$ gates (the gates of the $Na^+$ channel proteins).

2   $Na^+$ ions diffuse rapidly into the axon down an electrochemical gradient. As these are positively charged they begin to make the inside of the axon positive, i.e. to depolarise the membrane. This depolarisation causes even more $Na^+$ gates to open and even more $Na^+$ ions to enter.

3   Eventually the potential difference of the inside of the membrane is reversed from $-70$ mV to approximately $+40$ mV. This is the action potential.

(From now on refer to Figure 6.15 only.)

4   The membrane now becomes less permeable to $Na^+$ ions as the $Na^+$ gates close. At the same time permeability to $K^+$ ions increases as $K^+$ gates open. This can be seen in the lower graphs in Figure 6.15. A large increase in permeability to $Na^+$ is followed by a decrease in permeability to $Na^+$ and an increase in permeability to $K^+$.

5   Positively charged $K^+$ ions diffuse rapidly out of the axon, making the inside of the membrane negative again.

6   The membrane has been **repolarised** and the resting potential of approximately $-70$ mV has been restored.

> **Remember**
>
> **Exam hint:** For examination answers it is important to be clear about the *precise* sequence of the events causing the action potential. Study the text carefully.

## Did You Know

More than one third of the ATP (energy) used by a resting body is used for the operation of all the $Na^+/K^+$ pumps, including those found in neurones.

**Extension**

### Box 6.5  Absolute and relative refractory periods

The refractory period is divided into two parts. For the first 1 ms no new action potential can be formed, however large the stimulus, because the $Na^+$ channels cannot be stimulated to open. This is the absolute refractory period. For a further period a new action potential can be formed but only if the stimulus is much larger than usual. This is the relative refractory period.

7   For a brief time the membrane becomes **hyperpolarised** – it is *more* negative than the usual resting potential. This is because the membrane still has an increased permeability to $K^+$ ions.

8   Finally the $K^+$ gates close and resting levels of permeability are re-established for both $K^+$ and $Na^+$ ions. The action of the $Na^+/K^+$ pumps also restores the normal concentrations for $Na^+$ and $K^+$ ions on the inside and outside of the axon membrane.

Note that ATP is *not* required to directly cause the action potential but is required for the $Na^+/K^+$ pumps that maintain the resting potential. The pumps maintain the ion concentration gradients so that $Na^+$ can diffuse in passively when required to produce the action potential.

**?**

4   What would be the effect on the action potential if the stimulus caused the $Na^+$ and $K^+$ gates to open at the same time and to the same extent?

*(2 marks)*

## 6.6   The refractory period

When an action potential has just occurred at a section of the axon there is a short period of time when a second action potential cannot be generated (fired) at that same point. This is called the **refractory period**. It occurs because the $Na^+$ ion channels are closed and cannot be stimulated to reopen immediately. The membrane cannot be depolarised and a new action potential cannot be formed. The refractory period has two important consequences:

● It imposes an upper limit on the frequency of nerve impulses. Because the refractory period lasts for approximately 4–8 ms, it limits the number of nerve impulses that can pass along an axon in a given time (see Figure 6.20 on page 205).

● It ensures that nerve impulses can only pass in one direction along an axon. They can only pass from an active region to a resting region. They cannot 'double back' because the membrane immediately behind the active region is in the refractory period and cannot be depolarised again.

 **Did You Know**

Fugu is a fish dish which is a great delicacy in Japan. Unfortunately it is made from *Fugu rubripes,* a species of puffer fish, parts of whose bodies contain a lethal nerve toxin (poison) called tetrodotoxin. If eaten, this tetrodotoxin blocks sodium channels, stopping the formation of action potentials and the transmission of nerve impulses. This is fatal. Highly trained chefs carefully remove the skin and other parts containing the toxin so as to avoid serving a deadly meal.

*Figure 6.16  The puffer fish* Fugu rubripes

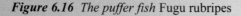

## 6.7  Conduction of nerve impulses

Information is transmitted or conducted by nerves as **nerve impulses**. In simple terms a nerve impulse is the movement of an action potential along the axon of a neurone.

The formation of the action potential at a particular section of axon membrane involves temporary depolarisation – reversal of the resting potential difference. This section of membrane now has the opposite electrical charge to the adjacent resting membrane. This sets up local electrical circuits as positive ions are attracted by the negative zone and flow from the positive region to the negative region. These electrical circuits act as a stimulus to the resting membrane. Figure 6.17 illustrates conduction in a non-myelinated axon.

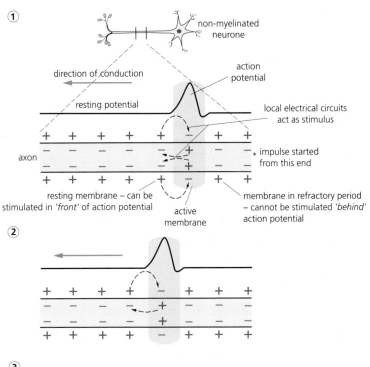

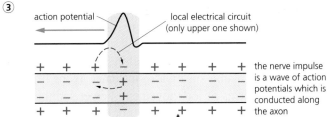

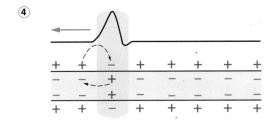

*Figure 6.17  Conduction of the nerve impulse in a non-myelinated neurone*

This stimulus triggers depolarisation and the formation of an action potential in the adjacent resting section of axon membrane. This process is repeated at each small section of membrane along the whole length of the axon. In effect a wave of action potentials or wave of depolarisation travels along the axon.

Sometimes the term **propagation** is used instead of conduction, i.e. propagation of the impulse. It refers to the fact that the impulse does not lose strength as it travels along. A new action potential is generated at each point so the size of the action potential at the end of a long axon is the same as at the beginning. Also, conduction implies a flow of electrons through a conductor rather than a movement of ions as in the action potential.

A nerve axon could theoretically conduct an impulse in either direction but in a living animal neurones are always stimulated at one end, meaning that the impulse only travels in one direction along an axon.

### 6.8 Myelinated axons and saltatory conduction

Some neurones are myelinated – their axons have a myelin sheath formed by Schwann cells (see page 192). Mammals have both myelinated and non-myelinated neurones. The relay neurones in the spinal cord (see Section 7.9, page 248) and the neurones of the autonomic nervous system (see Section 7.11, page 259) are non-myelinated. The sensory and motor neurones involved in the typical spinal reflex arc (see Section 7.9, page 248) are myelinated.

The myelin sheath acts as an electrical insulator to the axon membrane. Action potentials cannot form where the myelin is present because the membrane cannot be stimulated by the local electrical circuit and because $Na^+$ ions cannot diffuse into the axon. However, action potentials can form at the nodes of Ranvier where the myelin sheath is absent. The nerve impulse therefore 'jumps' from node to node (Figure 6.18).

The action potential which forms at one node causes the stimulation of the axon membrane at the next resting node. The node 'behind' is not stimulated because it is still recovering and is in the refractory period. This type of conduction is called **saltatory** from the Latin *saltus* meaning 'to leap'.

The main effect of myelination (the presence of the myelin sheath) is to greatly speed up nervous conduction. Typical speeds are $3\,m\,s^{-1}$ for non-myelinated neurones and $100\,m\,s^{-1}$ for myelinated axons. By jumping from node to node action potentials do not have to be formed at all the sections of axon membrane in between the nodes. The myelin sheath also insulates axons from each other and it is more efficient in terms of ion movement. Fewer sodium and potassium ions are involved because action potentials are not being formed at every point along the axon.

**Definition**

Saltatory conduction occurs in myelinated neurones. The nerve impulse jumps from one node of Ranvier to the next because action potentials are prevented by the myelin sheath – they can only occur at the nodes. The nodes of Ranvier are *gaps* in the myelin sheath which occur at intervals along the axon.

5 Suggest a disadvantage of the nodes of Ranvier being (a) closer together and (b) further apart. *(2 marks)*

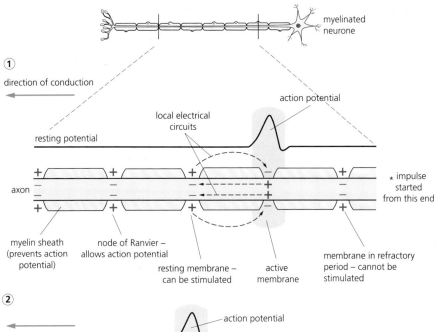

*Figure 6.18 Saltatory conduction in a myelinated neurone*

① 

direction of conduction

resting potential

local electrical circuits

action potential

axon

+ − + − + − − +
− − − + −
+ − + − + − − +

* impulse started from this end

myelin sheath (prevents action potential)

node of Ranvier – allows action potential

resting membrane – can be stimulated

active membrane

membrane in refractory period – cannot be stimulated

② 

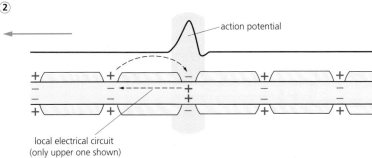

action potential

local electrical circuit (only upper one shown)

③ 

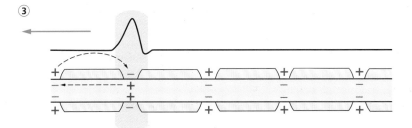

The nerve impulse is conducted by jumping from node to node along the axon – this is saltatory conduction.

## 6.9  Features of nerve impulses

### 1  Thresholds and the 'all or nothing' response

A stimulus must be at or above a certain minimum strength to cause the firing of an action potential. This minimum strength is the **threshold value**. Enough $Na^+$ gates must be opened and enough $Na^+$ ions must enter the axon to cause a large enough depolarisation (change in polarity). If the depolarisation is large enough the full action potential is then fired. Action potentials are always the same size – they have the same value (usually +40 mV). A depolarisation larger than the threshold does not produce an action potential with a larger value.

***Figure 6.19*** *'All or nothing' – the production of the action potential and the nerve impulse*

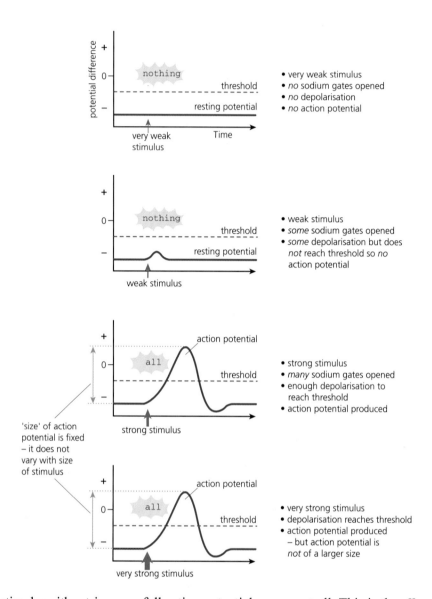

A stimulus either triggers a full action potential or none at all. This is the **all or nothing response** (Figure 6.19).

### 2  The impulse is non-decremental

Because action potentials are generated separately at each section of membrane as the nerve impulse travels, and because action potentials are 'all or nothing', the impulse stays the same strength. The nerve impulse is **non-decremental** – it does not decrease in strength from one end of the axon to the other.

### 3  The refractory period and the frequency of the impulse

As seen earlier, the maximum frequency depends upon the length of the refractory period. The refractory period is the minimum period between consecutive impulses (Figure 6.20). This allows the body to discriminate between separate stimuli, so making coordination more accurate. For example, in delicate movements of the fingers, such as playing the piano, the muscles of individual fingers need to contract, relax and then contract again very quickly.

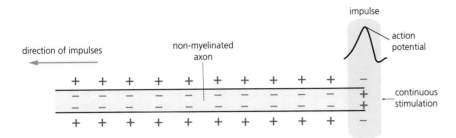

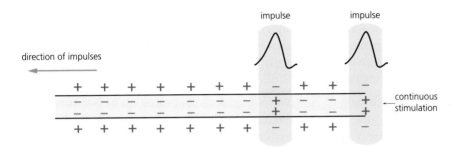

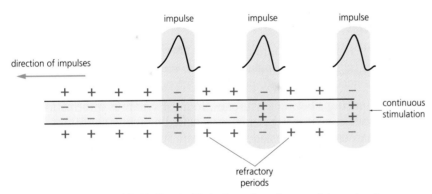

**Figure 6.20** *The frequency of the nerve impulse*

**The refractory period** (behind impulses) limits the maximum frequency for nerve impulses. There must be a minimum 'gap' between separate impulses.

Frequency is also used to indicate the relative strength of a stimulus. As described above, action potentials and hence nerve impulses, are always the same size. Relative size of impulse cannot be used to distinguish between a smaller and a larger stimulus – frequency of impulse is used instead. Larger stimuli fire off more impulses per second than smaller stimuli. The body can distinguish the size of the original stimuli from the relative frequency of the impulses. For more detail see Box 7.4, page 239.

6 For most neurones the number of nerve impulses which can pass along an axon is rarely more than 200 per second. Suggest why. *(2 marks)*

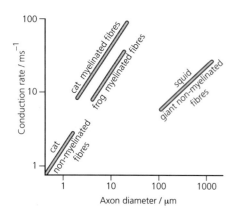

**Figure 6.21** *Axon diameter and myelination as factors in the speed of conduction of impulses along nerve fibres*

## 4 The speed of conduction

A number of factors affect the speed of conduction of the nerve impulse:

- The diameter of the axon – the greater the diameter of the neurone axon the greater the speed of transmission. A larger diameter axon reduces resistance to the nerve impulse. For example, as mentioned earlier, squids have some 'giant axons' with a much larger diameter than the others in their body. Speed of transmission in these giant axons is $45 \text{ m s}^{-1}$ compared to a speed of $0.5 \text{ m s}^{-1}$ for their other axons.

- The myelin sheath – myelinated fibres conduct impulses faster than non-myelinated fibres (see saltatory conduction, page 202). The effects of axon diameter and myelination on conduction speed are shown in Figure 6.21.

- Temperature – an increase in temperature can speed up transmission. A higher temperature means ions in solution have a greater kinetic energy and therefore there is a faster rate of diffusion for $Na^+$ ions into the axon to cause the action potential.

- The concentration of $Na^+$ ions – a reduced concentration of $Na^+$ ions surrounding the axon (e.g. because of illness) can slow down the diffusion of $Na^+$, slowing down the formation of the action potential and therefore slowing down conduction.

### 6.10 Synapses and synaptic transmission

To achieve their function of coordination and control, neurones need to connect with each other (e.g. as in the spinal reflex arc, see Section 7.9, page 248) and nerve impulses need to be able to pass from one neurone to another.

**Synapses** are junctions between neurones. In most synapses information is transmitted between neurones by chemicals called **neurotransmitters**. These 'chemical' synapses will be considered in detail. A small number of synapses are electrical rather than chemical. These are considered briefly in Box 6.6.

### The structure and function of a synapse

There are different types of chemical synapse, using different neurotransmitters, but here we will consider a typical **excitatory** synapse which uses **acetylcholine** (ACh) as the neurotransmitter. This is one of the most common neurotransmitters for excitatory synapses. Because acetylcholine is used such synapses are called **cholinergic**.

Excitatory synapses increase the chance of the transmission of the nerve impulse to the postsynaptic neurone. Other synapses are **inhibitory** and tend to block transmission. These are considered on page 213.

Excluding synapses in the brain, the main neurotransmitter used in addition to acetylcholine is **noradrenaline**. Synapses using this are known as **adrenergic**. Other transmitters are found and some of these are discussed on page 215. The structure of a synapse is shown in Figures 6.22 and 6.23.

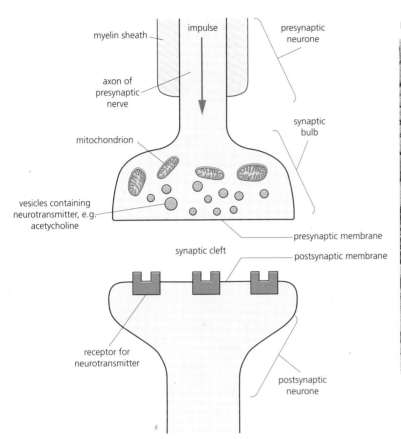

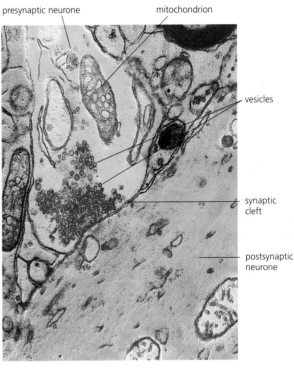

*Figure 6.23* *Electronmicrograph showing the structure of a synapse*

*Figure 6.22* *The structure of a typical synapse*

### Box 6.6  Electrical synapses

In electrical synapses the neurones are almost touching and are connected by hollow protein channels. This allows the wave of action potentials to pass directly from one neurone to the next. Transmission at the synapse is electrical and chemical transmitters are not used. Transmission is much quicker than for chemical synapses because the time taken for the neurotransmitters to diffuse across the synapse is not needed. Because neurotransmitters are not used, these synapses cannot fatigue due to the neurotransmitters running out (see page 211). They are also unaffected by chemicals and drugs. Electrical synapses are rare but they are found in the ganglion cells in the human retina (see Section 7.4, page 235).

The main disadvantage of electrical synapses is that chemical neurotransmitters are much more effective in changing the potential difference of membranes than direct electrical stimulation. Chemical synapses can therefore be much smaller and still be effective. Because they are so small, one neurone can receive synaptic endings from many other neurones allowing more interconnections between neurones and more complex coordination (see Nerve networks on page 211).

### Did You Know

Electric eels (genus *Electrophorus*) and electric rays (genus *Torpedo*) live in water and are able to stun their prey by producing a large 500 V jolt of electricity. Their electric organs consist of muscle cells modified so as to consist almost entirely of neuromuscular junctions (a type of synapse). Box 6.7 on page 210 gives details of neuromuscular junctions. The changes in potential difference of these synapses are added together to produce the electric shock.

*Figure 6.24  The mechanism of transmission at a synapse*

The mechanism of transmission at a cholinergic excitatory synapse is shown in Figure 6.24. (See also Figures 6.22 and 6.23.)

1   The axon of the presynaptic neurone ends in one or more swellings called **synaptic bulbs** (or synaptic knobs). A nerve impulse travels down the axon of the presynaptic neurone and arrives at the synaptic bulb.

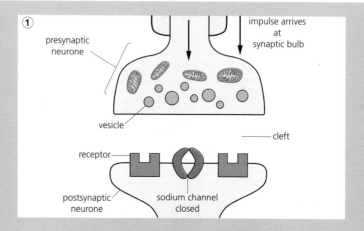

2   The impulse causes calcium channels in the presynaptic membrane to open so that the membrane is suddenly more permeable to **calcium ions**. Calcium ions then diffuse rapidly into the bulb from the surrounding tissue fluid.

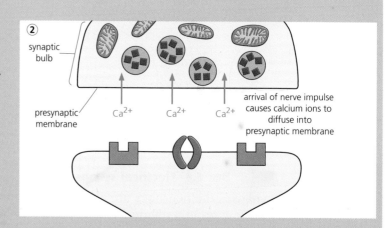

3   Entry of calcium ions causes **vesicles** containing the transmitter **acetylcholine** to move to the presynaptic membrane. These vesicles fuse with the membrane and release the acetylcholine by exocytosis into the **synaptic cleft**.

● The molecules of acetylcholine diffuse across the cleft down a concentration gradient. The cleft is only about 20 nanometres (nm) wide but this is wide enough to prevent direct electrical stimulation. (1 nm = one millionth of a mm)

● Diffusion across the synapse takes about 1 ms and this short time is known as the synaptic delay as it slows down nervous transmission.

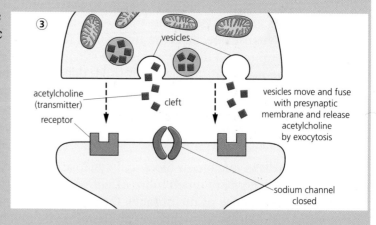

4   The acetylcholine fits into (or binds to) specific **receptor** molecules on the membrane of the postsynaptic neurone. These receptors are made of protein. The acetylcholine fits into the receptors because it has a molecular shape which is complementary to the shape of the receptors. The binding of the acetylcholine to the receptors causes **sodium channel proteins** in the postsynaptic membrane to open. The postsynaptic membrane is suddenly more permeable to $Na^+$ ions which diffuse rapidly into the postsynaptic neurone.

5   The entry of the positive $Na^+$ ions makes the inside of the membrane less negative. This change in potential difference is called the **excitatory postsynaptic potential (EPSP)**. Further details of the EPSP are given on page 212 and Table 6.2 on page 215 summarises the various potential differences. If the EPSP is large enough to reach threshold, this causes an action potential in the postsynaptic membrane and a nerve impulse is started. The nerve impulse is conducted along the postsynaptic neurone away from the synapse.

6   Once the acetylcholine transmitter has acted on the postsynaptic membrane it is immediately broken down into its components (acetyl and choline) by the enzyme **cholinesterase**. The components are a different shape and they leave the receptors. This ends transmission and prevents the continued stimulation and firing of impulses in the postsynaptic neurone. This process also allows the 'recycling' of the transmitter so that supplies are maintained on the presynaptic side of the synapse. These 'inactive' components diffuse back to the presynaptic membrane.

7   The components enter the synaptic bulb of the presynaptic neurone and are used to reform acetylcholine. This requires energy from ATP which is provided by the many mitochondria present in the bulb. The acetylcholine is stored in vesicles ready for the next transmission.

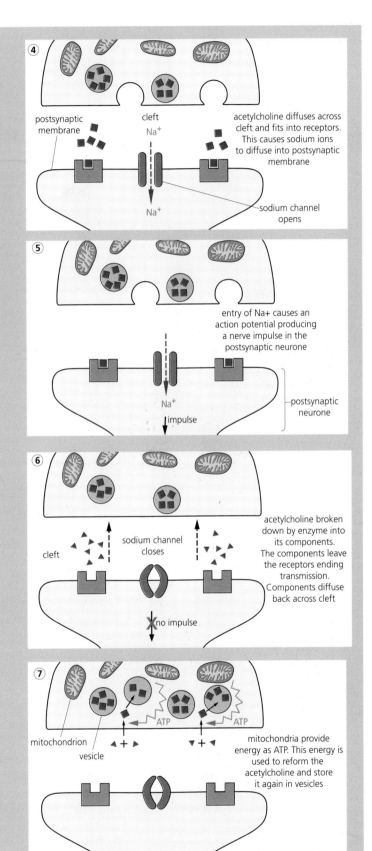

④ postsynaptic membrane / cleft $Na^+$ / $Na^+$ / sodium channel opens — acetylcholine diffuses across cleft and fits into receptors. This causes sodium ions to diffuse into postsynaptic membrane

⑤ $Na^+$ / impulse / postsynaptic neurone — entry of Na+ causes an action potential producing a nerve impulse in the postsynaptic neurone

⑥ cleft / sodium channel closes / no impulse — acetylcholine broken down by enzyme into its components. The components leave the receptors ending transmission. Components diffuse back across cleft

⑦ mitochondrion / vesicle / ATP / ATP — mitochondria provide energy as ATP. This energy is used to reform the acetylcholine and store it again in vesicles

**7** After continued stimulation for some time a synapse ceases to work. Suggest why this happens. *(2 marks)*

**Figure 6.25** *The neuromuscular junction – a synapse between a motor neurone and skeletal muscle*

**Box 6.7  The neuromuscular junction**

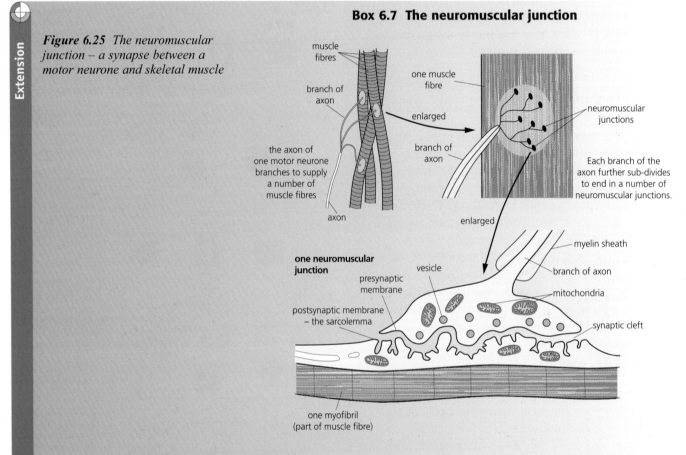

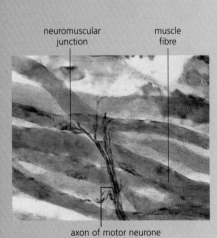

**Figure 6.26** *Photomicrograph showing neuromuscular junctions*

The neuromuscular junction (Figures 6.25 and 6.26) is a specialised form of synapse found between a motor neurone and skeletal muscle fibres. It allows motor neurones to stimulate muscle contraction. Both the membrane at the end of the motor axon (presynaptic membrane) and the sarcolemma membrane of the muscle (postsynaptic membrane) are highly folded. Other than this, both the structure and the way it functions are very similar to a normal excitatory synapse. The end of the axon contains many vesicles containing the neurotransmitter acetylcholine, and many mitochondria. The arrival of a nerve impulse at the end of the axon causes the same sequence of events as described earlier for a synapse. Eventually the entry of Na$^+$ ions into the sarcolemma causes a change in the potential difference called an end plate potential. If this reaches threshold an action potential is formed which spreads to the muscle fibres causing muscle contraction. For further details see Section 8.6, page 277.

## 6.11   The functional significance of synapses

Although their basic function of the transmission of nerve impulses between neurones seems fairly straightforward, synapses in fact play a complex role in the overall nervous system. Some of these aspects will now be considered.

### Unidirectionality

Synapses are 'one way' – they can only transmit impulses in one direction. This is because:

- the neurotransmitter is only produced on the presynaptic side
- only the postsynaptic membrane has the receptors for the neurotransmitter
- the neurotransmitter crosses the cleft by diffusion and must diffuse down a concentration gradient from high to low.

The unidirectional nature of synapses imposes one way transmission on the whole of the nervous system. This means nervous transmission is more controlled. It avoids impulses spreading in all directions which would be inefficient and potentially dangerous (Figure 6.27).

### Fatigue

Synapses protect nerve networks and effectors from over-stimulation. The continuous use of a synapse to transmit an impulse means that eventually supplies of neurotransmitter are exhausted. The neurotransmitter cannot be recycled quickly enough and not enough is present in the synaptic bulb to carry the impulse across the cleft. The synapse is said to be **fatigued**.

### The filtering out of low-level stimuli

Many low-level stimuli, e.g. the background noise of traffic 1 mile away, are filtered out. Only small amounts of neurotransmitter are released and these are not enough to cause an impulse in the postsynaptic neurone so that the information does not travel further than the synapse. This helps to prevent the nervous system from being overloaded, allowing it to concentrate on more important stimuli (see also Figure 6.30 on page 212).

### Learning and memory

The formation of millions of synapses between the huge numbers of neurones in the brain produces complex networks which change over time. These changes may form the physical basis of learning and memory (see Section 7.10, page 251).

### Nerve networks

Synapses are involved in the assembly of complex networks and pathways. Most neurones form synapses with many other neurones, e.g. those in the brain as mentioned above (see also Figure 6.28 and Figure 6.1 on page 190). Two types of pathway are **divergence** and **convergence**. These are shown in Figure 6.29 on page 212. Synapses are the crucial junctions for these pathways. An important example of convergence is retinal convergence in the retina of the eye (see Section 7.6, page 241).

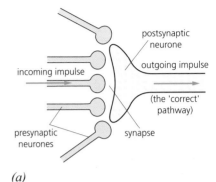

*(a)*

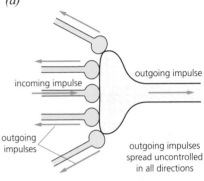

*(b)*

***Figure 6.27*** *How one way transmission is controlled at synapses (a) with synapses, (b) without synapses*

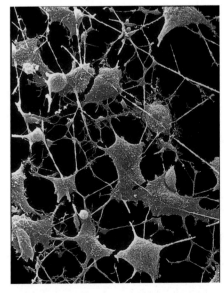

***Figure 6.28*** *Nerve networks: false colour scanning electronmicrograph of the many synaptic junctions with each neurone (× 3000)*

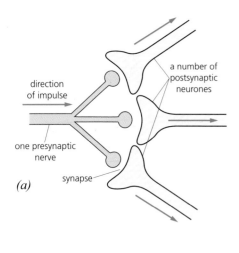

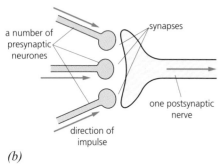

*(a)*

*(b)*

**Figure 6.29** *(a) Divergence and (b) convergence*

### Summation

A single nerve impulse arriving at a synapse may cause the release of only a small amount of neurotransmitter. The excitatory post synaptic potential (EPSP) this produces at the postsynaptic membrane may not be large enough to reach threshold and no action potential will be formed. A nerve impulse can only be fired in the postsynaptic neurone if sufficient neurotransmitter has been released and fitted into the receptors on the postsynaptic membrane.

One way to achieve sufficient neurotransmitter is by **summation**. There are two types (Figure 6.30).

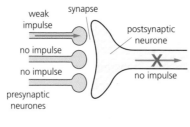

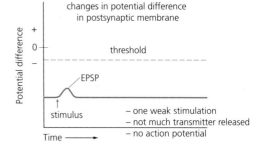

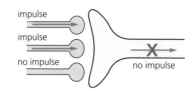

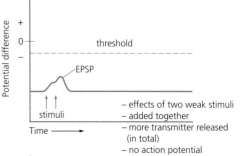

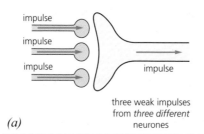

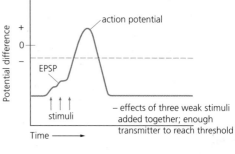

*(a)*

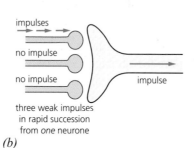

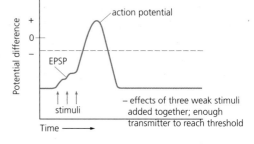

*(b)*

**Figure 6.30** *(a) Spatial and (b) temporal summation*

In **spatial summation** a number of presynaptic neurones converge onto and form synapses with a single postsynaptic neurone. If just one presynaptic bulb releases neurotransmitter this may not be sufficient to fire an impulse in the postsynaptic neurone, but if neurotransmitter is released from a number of bulbs there may be enough to reach threshold EPSP and to fire an impulse in the postsynaptic neurone.

In **temporal summation** a single presynaptic bulb fires a number of times in rapid succession with neurotransmitter released each time. More neurotransmitter is released before the first amount has been broken down so that the neurotransmitter builds up.

### Inhibitory synapses and inhibition

The synapses considered so far are excitatory. They tend to excite the postsynaptic neurone and make transmission of the nerve impulse more likely. Other synapses are **inhibitory** and block transmission, making the generation of a nerve impulse in the postsynaptic neurone less likely.

Whether a synapse is excitatory or inhibitory depends upon either the type of receptors present on the postsynaptic membrane, or on the type of neurotransmitter. In some cases the same neurotransmitters are used in both excitatory and inhibitory synapses but the receptors are different. In inhibitory synapses when the neurotransmitter fits into the receptors this opens different ion channels. Instead of opening $Na^+$ channels, potassium and chloride channel proteins are opened. Potassium ions diffuse out and negative chloride ions diffuse in. Together these make the inside of the postsynaptic membrane more negative than usual – a process of **hyperpolarisation**. The more negative potential difference which is produced is called the **inhibitory postsynaptic potential (IPSP)** and it can be as low as –90 mV (Figure 6.31 on page 214). A summary of the various types of potential difference is given in Table 6.2 on page 215.

A postsynaptic neurone may possess both excitatory and inhibitory synapses. If the inhibitory synapses fire they make the membrane more negative. It will therefore be less likely that threshold can be reached when the excitatory synapses fire because they will need to release even more neurotransmitter in order to transmit the impulse. An impulse will not be generated in the postsynaptic neurone unless the combined effects of the excitatory impulses and inhibitory impulses exceed the threshold value.

One specific use of inhibitory synapses is in antagonistic pairs of muscles, e.g the biceps and triceps in the arm. When one of the pair contracts it is vital that the other stays relaxed otherwise damage will be caused. Inhibitory synapses are fired, preventing stimulation and contraction of the relaxed muscle. In general this arrangement allows greater variability in control and coordination. A given neurone can produce a variable response depending upon the relative firing of its excitatory and inhibitory synapses.

**Definition**

**Summation** involves adding together the effects of a number of weak transmissions at a synapse so as to reach threshold and produce an impulse in the postsynaptic neurone.

**Figure 6.31** *The role of inhibitory synapses*

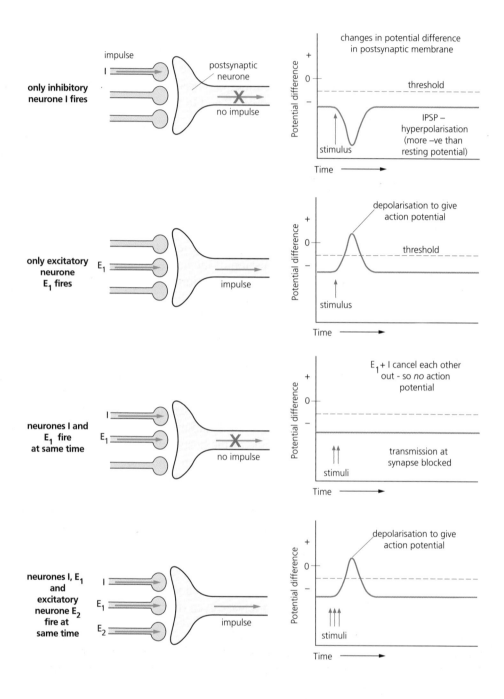

❓

8 When you accidentally touch something very hot, a reflex causes you to move your hand away quickly. It is possible to choose to 'override' this reflex and not move your hand. Use your knowledge of synapses to suggest a mechanism for how this is achieved. *(3 marks)*

| Type | Definition | Charge | Location | Further information |
|---|---|---|---|---|
| resting potential | the potential difference which exists across the membrane when the neurone is not conducting an impulse | inside of membrane is negative (compared to outside) | found in all cells but here mainly related to neurones | Section 6.4 |
| action potential | a temporary reversal of the potential difference across a neurone membrane during the passage of a nerve impulse | inside of membrane is positive | neurones, particularly the axons | Section 6.5 |
| excitatory post-synaptic potential (EPSP) | a temporary reversal of the potential difference across the membrane of the receiving neurone at an excitatory synapse | inside of membrane is positive if threshold is reached | the neurone receiving the transmission at an excitatory synapse | Section 6.11 and Figure 6.30 |
| inhibitory post-synaptic potential (IPSP) | an increase in the potential difference so that it is more negative than usual. This occurs across the membrane of the receiving neurone of an inhibitory synapse | inside of membrane is more negative than usual. This is hyperpolarisation | the neurone receiving the transmission at an inhibitory synapse | Section 6.11 and Figure 6.31 |
| generator potential | the change in the potential difference across the membrane of a receptor cell caused by a stimulus | inside of membrane can be either positive or negative depending upon the type of receptor | receptor cells, e.g. rods and cones in the eye | Figure 7.5, page 232, and Figure 7.13, page 238 |

*Table 6.2 Types of potential difference*

## 6.12   Chemicals and synapses

Many chemicals affect synapses and the influence of chemicals on synapses is an ongoing and complex area of research, particularly in relation to brain function. These chemicals include natural neurotransmitters, poisons and both legal and illegal drugs.

### Types of natural neurotransmitter

Over 50 naturally occurring neurotransmitters have been found in humans. The two main ones found outside the brain are acetylcholine, found in many parts of the nervous system and at neuromuscular junctions, and noradrenaline, found in the sympathetic part of the autonomic nervous system (see Section 7.11, page 259, for details of the autonomic nervous system). Acetylcholine can be excitatory, e.g. at junctions with skeletal muscle, or inhibitory, e.g. at junctions with heart muscle. Noradrenaline is excitatory.

*Figure 6.32* *B-endorphins produced during exercise give many people a 'high'*

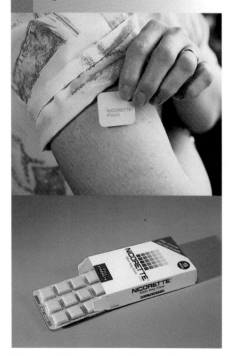

*Figure 6.33* *Anti-smoking aids – nicotine patch and nicotine gum*

Another group of neurotransmitters are amino acids. One example is glycine which is an inhibitory neurotransmitter found in the spinal cord. It inhibits the sending of nerve impulses to muscles, keeping them relaxed and helping prevent inappropriate contraction. Endorphins are a group of neuropeptide neurotransmitters found in the brain. They are natural painkillers, blocking pain pathways. They can also produce pleasant sensations – B-endorphin levels increase during exercise helping to explain the 'high' that some people get from running or working out in the gym (Figure 6.32).

### Drugs

'Drug' is an imprecise term including any chemical not normally found in the body which has an effect on the body. It includes substances which are currently illegal, such as cocaine, and legal medicines prescribed by doctors, e.g. antidepressants. Other legal 'drugs' include substances such as nicotine, caffeine and alcohol.

All of these chemicals affect synapses in two main ways. **Agonistic** drugs amplify (increase) synaptic transmission – they increase the effects of the synapse. **Antagonistic** drugs decrease synaptic transmission – they decrease the effects of the synapse. Agonistic drugs work in various ways, including:

- They mimic a natural neurotransmitter, fitting into the same receptors (e.g. nicotine mimics acetylcholine). Their molecules are a similar shape to those of the neurotransmitter.

- They interfere with the normal enzyme breakdown of a neurotransmitter – the neurotransmitter stays in the receptors and continues to stimulate the postsynaptic membrane, e.g organophosphate insecticides. One effect of this can be the continuous stimulation and contraction of muscles – convulsions.

## Box 6.8 Nicotine

Nicotine, found in tobacco, mimics the neurotransmitter acetylcholine by fitting into the same receptors and stimulating the postsynaptic cell or effector. It is excitatory (agonistic) and acts as a stimulant. Nicotine affects many neurones including those of the parasympathetic nervous system (see Section 7.11, page 259) that are involved in the constriction of blood vessels. Increased constriction can cause raised blood pressure and this can contribute to heart disease and circulatory problems. However, nicotine does not cause cancer – that is due to chemicals in tobacco tar.

Nicotine is addictive, which is one reason why smokers find it difficult to give up. The reasons for the addiction are not fully understood but nicotine does seem to stimulate the release of the neurotransmitter dopamine in the brain. One of the effects of dopamine is to stimulate 'pleasure' pathways, increasing a sense of well-being. (The stimulant caffeine, found for example in tea and coffee, probably also causes dopamine release.)

The ways in which antagonistic drugs work include:

- They prevent the release of a neurotransmitter, e.g. botulin is a poisonous toxin produced by the bacterium *Clostridium botulinum*. It causes botulism food poisoning and is the most lethal natural toxin known. Fifty per cent of cases are fatal. Muscles cannot be contracted – they are paralysed, and as a result breathing is not possible.

- They block the action of a neurotransmitter at the receptors on the postsynaptic membrane. For example, curare is a natural poison used in South America for hunting. It blocks the action of acetylcholine at neuromuscular junctions, preventing muscle contraction. It kills by stopping the muscle contraction needed for breathing. It is now also used in a controlled way as a muscle relaxant during surgery. Breathing is maintained artificially.

### Did You Know

One teaspoon of botulin toxin, e.g. in the water supply, would be enough to kill the entire population of a large city.

## Chemicals and inhibitory transmitters

All the examples given so far involve excitatory synapses. It is important to remember that some neurotransmitters and some synapses are inhibitory and not excitatory. An antagonistic chemical will block an excitatory synapse, reducing the visible effects, e.g. curare will cause paralysis – stopping muscle contraction.

Strychnine is a poison which also has an antagonistic effect. However, it affects the inhibitory neurotransmitter glycine. Glycine normally prevents inappropriate muscle contraction. Strychnine blocks the glycine receptors so that glycine is not effective. Inappropriate muscle contraction is *not* prevented and the person suffers from extensive muscular convulsions (spasms) which can be fatal. The antagonistic (inhibitory) chemical has actually *increased* the visible effects – there is too much muscular contraction.

# ⑥ Neurones and the nerve impulse

💡 **Summary – ⑥ Neurones and the nerve impulse**

● There are three types of neurones. Sensory neurones carry impulses towards the cell body from receptors. Relay or connector neurones connect motor and sensory neurones. Motor neurones carry nerve impulses away from the cell body towards effectors.

● Neurones have long extensions or fibres called dendrons and axons. These act as the 'wiring' of the nervous system.

● Some neurones are myelinated. The fibres are enclosed in a myelin sheath which acts as an electrical insulator and speeds up the conduction of nerve impulses. The myelin sheaths are formed by Schwann cells.

● When a neurone is not conducting an impulse it has a potential difference across its membrane so that the inside is negative compared to the outside. This is called the resting potential. An unequal distribution of ions is maintained on either side of the membrane.

● When the neurone is stimulated $Na^+$ ions diffuse in through the membrane causing depolarisation (a temporary reversal of the potential difference). The inside is now positive. This is the action potential.

● A nerve impulse consists of a wave of action potentials moving along an axon.

● The impulse travels faster in myelinated neurones as it moves by saltatory conduction – jumping between gaps in the myelin sheath called nodes of Ranvier.

● Impulses also travel faster along axons with a larger diameter.

● Junctions between neurones are called synapses and the impulse is carried across the cleft of the synapse by chemicals called neurotransmitters.

● Neurotransmitters fit into specific receptors on the membrane of the postsynaptic neurone generating an impulse which travels away from the synapse.

● Because neurotransmitters are only produced on the presynaptic side, synapses are 'one way' – transmitting impulses only in one direction. Synapses impose unidirectionality on the nervous system.

● Synaptic transmission can be affected by a variety of chemicals including illegal drugs and legal drugs such as nicotine.

● These chemicals have different effects. Some are agonistic, increasing transmission at the synapse. At excitatory synapses they act as stimulants, e.g. nicotine. Others are antagonistic, tending to block transmission and acting as depressants.

## ? Answers

1  In the brain *(1)*.

2  So they can contract when stimulated *(1)*.

3  It stops the action of the $Na^+/K^+$ pumps which need ATP *(1)*.

4  There would be no action potential formed *(1)*. The change in potential difference caused by the $Na^+$ entering would be cancelled out by the $K^+$ leaving *(1)*.

5  (a)  The impulse would be slower as it would have to make more jumps *(1)*.

   (b)  No impulse because the nodes are too far apart to form a circuit/too far for impulses to jump *(1)*.

6  Due to the refractory period; a second impulse cannot pass until the refractory period has finished; an action potential cannot be formed during the refractory period *(2 max.)*.

7  The bulb runs out of neurotransmitter to carry the impulse *(1)*. It cannot be recycled quickly enough *(1)*.

8  Moving the hand involves excitatory synapses; stopping the hand moving involves inhibitory synapses; the effects of firing the inhibitory synapses are sufficient to cancel out the effects of the excitatory synapses *(3)*.

### End of Chapter Questions

**1** The diagram below shows a neurone and some of the structures associated with it.

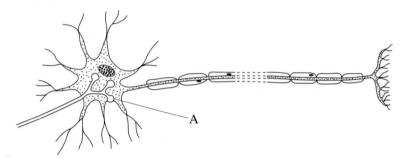

(a) (i) Name the type of neurone shown in the diagram. *(1 mark)*

   (ii) Give *one* reason for your answer. *(1 mark)*

(b) (i) Name the part labelled A. *(1 mark)*

   (ii) Describe how A is involved in the passage of nerve impulses to the cell body of the neurone. *(4 marks)*

*Edexcel 1999* *(Total 7 marks)*

**2** The graph shows changes in the potential difference across the cell surface membrane during an action potential.

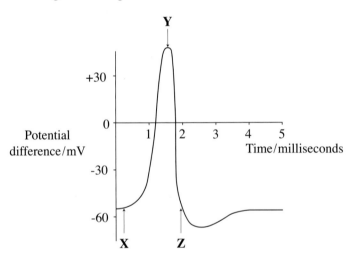

(a) Describe the events which lead to the change in potential difference between:

   (i) times **X** and **Y**; *(2 marks)*

   (ii) times **Y** and **Z**. *(2 marks)*

**(b)** The table shows the rate of conduction of nerve impulses along three different axons.

| Axon | Diameter/μm | Rate of conduction/m s$^{-1}$ |
|------|-------------|-------------------------------|
| **A** | 7 | 1.2 |
| **B** | 15 | 90 |
| **C** | 500 | 33 |

Suggest an explanation for the rate of conduction along axon B.

*(2 marks)*

*AQA A 1997*

*(Total 6 marks)*

**3** Read through the following passage about the mammalian nervous system, then write on the dotted lines the most appropriate word or words to complete the account.

The nervous system consists of several types of neurones. Of these, ..................... neurones carry impulses to muscles and glands while ................... neurones carry impulses from receptor cells to the central nervous system. The interior of a nerve fibre has a lower concentration of ......................... ions than its surroundings, as a result of the action of a ......................... in its membrane. This imbalance of ions creates an ............................. potential in the fibre, which is reversed during the passage of an impulse. When this happens, ............................. ions flood into the fibre, after which there is a compensating outward movement of ......................... ions.

*Edexcel 1996*

*(Total 7 marks)*

4 The diagram shows the permeability of the cell-surface membrane of a resting nerve cell to various ions.

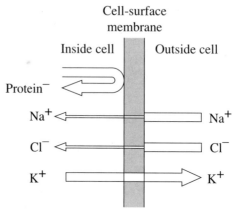

(a) Explain how the difference in permeability to particular ions leads to a resting potential of –70 mV. *(3 marks)*

(b) Describe the changes in permeability which take place when an action potential is initiated. *(2 marks)*

*AQA/A 1992* *(Total 5 marks)*

5 The diagram below shows part of two nerve fibres and a synapse. The figures indicate the value in mV of the potential across the membrane between the cytoplasm of the fibres and the extracellular fluid at intervals along each fibre.

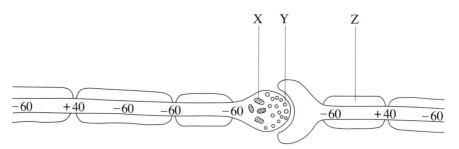

(a) (i) Draw a circle around one region of the diagram where an action potential exists. Explain your choice. *(2 marks)*

(ii) By means of an arrow on the diagram, indicate the direction in which action potentials would normally travel along these fibres. Explain your choice of direction. *(2 marks)*

(b) Identify structures X and Y, and state how each is involved in the transmission of nerve impulses. *(4 marks)*

(c) (i) What is the major chemical constituent of structure Z? *(1 mark)*

(ii) State two effects of structure Z on the transmission of action potentials. *(2 marks)*

*Edexcel 1989* *(Total 11 marks)*

**6** **(a)** Describe the sequence of events that takes place when a nerve impulse arrives at a synapse. *(4 marks)*

**(b)** The diagram below shows the changes in membrane potential in a presynaptic neurone and postsynaptic neurone when an impulse passes across a synapse.

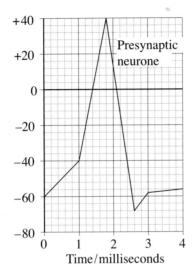

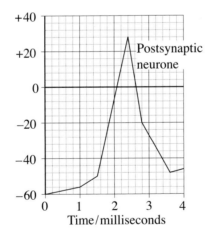

(i) Explain how depolarisation occurs in the presynaptic neurone. *(3 marks)*

(ii) The maximum depolarisation in the presynaptic neurone is +40 mV. What is the maximum depolarisation in the postsynaptic neurone?  N2.1 *(1 mark)*

(iii) How long is the delay between the maximum depolarisation in the presynaptic and postsynaptic neurones? *(1 mark)*

(iv) What is the cause of this delay? *(1 mark)*

**(c)** Describe how nicotine affects synaptic transmission. *(2 marks)*

*Edexcel 1996* *(Total 12 marks)*

**7** The diagram shows some of the events which occur in a synapse after the arrival of an impulse at the presynaptic membrane.

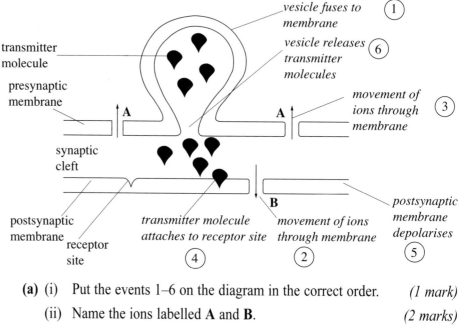

**(a)** (i)   Put the events 1–6 on the diagram in the correct order.   *(1 mark)*

(ii)  Name the ions labelled **A** and **B**.   *(2 marks)*

(iii) By what process do transmitter molecules move across the synaptic cleft?   *(1 mark)*

(iv)  Name **one** transmitter molecule released by synaptic vesicles.

*(1 mark)*

**(b)** One impulse arriving at the presynaptic membrane does not produce an action potential in the postsynaptic neurone, but several impulses arriving in close succession do.

(i) Explain this observation.   *(2 marks)*

(ii) What name is given to the process described?   *(1 mark)*

*AQA/B 1997*   *(Total 8 marks)*

**8** The diagram shows a normal nerve–muscle junction.

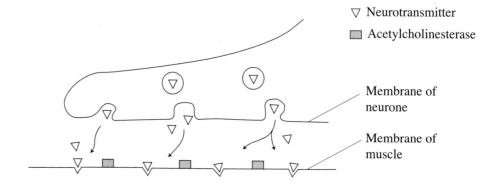

Using the evidence from the following diagrams, describe and explain the effect on muscle activity of

**(a)** curare, *(2 marks)*

**(b)** organophosphates, *(2 marks)*

**(c)** botulin toxin. *(2 marks)*

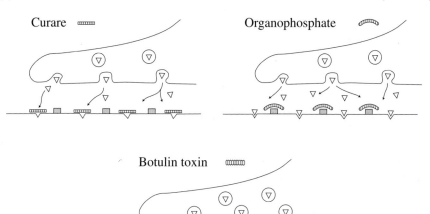

*AQA/B 1998* *(Total 6 marks)*

**9** The transmission of information at synapses may be modified in four ways: unidirectionality; convergence; summation; inhibition. Use the information in the diagram and table below to explain what is meant by these four ways.

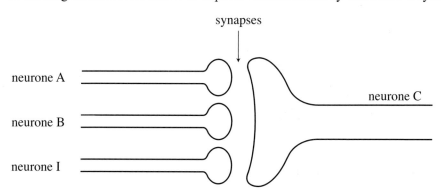

|  | Impulse in neurone C |
|---|---|
| Impulses in neurone A | ✗ |
| Impulses in neurones A + B | ✔ |
| Impulses in neurones A + B + I | ✗ |

*(Total 8 marks)*

**10** Classify the following chemicals as agonistic or antagonistic by placing a tick in the appropriate column.

| Effect | Agonistic | Antagonistic |
| --- | --- | --- |
| neostigmine is a drug that prevents the action of cholinesterase enzyme | | |
| botulinum toxin prevents release of acetylcholine from vesicles | | |
| curare blocks the action of acetylcholine | | |
| nicotine mimics the action of acetylcholine | | |
| amphetamines stimulate the release of neurotransmitters from vesicles | | |
| chlorpromazine blocks receptors | | |
| cocaine causes neurotransmitters to linger in the receptors | | |
| sarin nerve gas prevents the breakdown of acetycholine after transmission | | |

*(Total 8 marks)*

# Receptors and the nervous system

The nervous system is organised into a number of parts. This chapter will consider the function of **receptors** (particularly the **rods** and **cones** in the eye) in collecting information and the functions of the **spinal cord** and **brain** (the central nervous system or **CNS**) in processing this information and coordinating responses to it. The structure and function of nerve cells (neurones) and the nature of nerve impulses are considered in Chapter 6.

**This chapter includes:**
- the organisation of the nervous system
- the main features of receptors
- the structure and functions of the parts of the eye
- detection of light by rods and cones
- the structure and function of the spinal cord
- nature of reflexes and the reflex arc
- the functions of the main parts of the brain
- the functions of the autonomic nervous system.

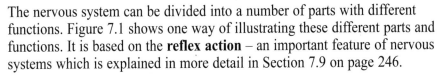

## 7.1 The organisation of the nervous system

The nervous system can be divided into a number of parts with different functions. Figure 7.1 shows one way of illustrating these different parts and functions. It is based on the **reflex action** – an important feature of nervous systems which is explained in more detail in Section 7.9 on page 246.

The collecting of information is carried out by **receptors**, e.g. temperature receptors in the skin or light-sensitive cells in the eye. These are able to detect **stimuli** such as light. Receptors respond to stimuli by producing nerve

**Figure 7.1** *The organisation of the nervous system*

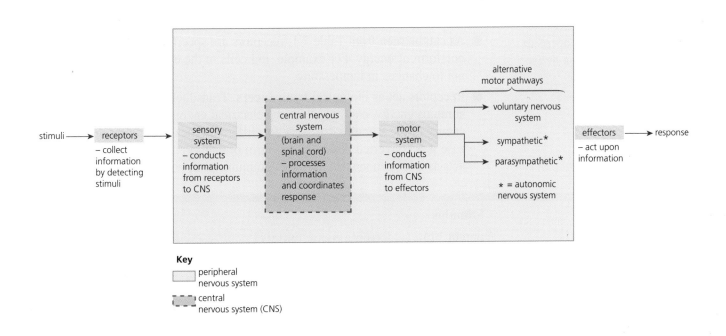

stimuli → receptors
– collect information by detecting stimuli

sensory system
– conducts information from receptors to CNS

central nervous system
(brain and spinal cord)
– processes information and coordinates response

motor system
– conducts information from CNS to effectors

alternative motor pathways

voluntary nervous system

sympathetic*

parasympathetic*

* = autonomic nervous system

effectors
– act upon information

→ response

**Key**

peripheral nervous system

central nervous system (CNS)

*Figure 7.2 Prey and predator*

*Table 7.1 Types of receptor*

impulses; these electrical signals are the universal form of information within the neurones of the nervous system. The information is now carried as nerve impulses within the nervous system by the pathways shown in Figure 7.1. Note that the motor system is sub-divided with three different possible routes to the **effectors**. This is explained on page 259. Eventually the effector carries out a response appropriate to the original information received (the stimulus). The main effectors, muscles, are considered in Chapter 8.

## 7.2 Receptors

Mammals need constant information about changes in their internal and external environments to enable them to respond in an appropriate way for survival. Detecting and responding to internal changes (within the body) is mainly dealt with in Chapter 9, Section 9.1 on **homeostasis**, e.g. the regulation of blood glucose levels. An example of an important change in the external environment is a prey animal detecting the appearance of a predator by sight – by the use of its eyes (Figure 7.2).

Changes in the environment act as stimuli and they are detected by specialised cells called **receptors** (or sensory receptors). The ability to detect these stimuli is a property possessed by living things called **sensitivity**.

There are many different types of receptor and they can be classified in various ways. One classification based on the type of stimulus detected is given in Table 7.1.

Receptors can exist as single cells, e.g. the **Pacinian corpuscle** – a pressure receptor (see Box 7.1) or grouped together in large numbers with accessory structures to form a **sense organ**, e.g. the eye.

### Features of receptors

- As can be seen from Table 7.1, receptors are specific – they only respond to one form of energy. For example, rod cells in the eye respond to light and not to changes in temperature.
- Receptors act as biological **transducers**. Transduction is the conversion of stimulus energy, e.g. light, into the electrical energy of a nerve impulse. The energy of the stimulus will vary for different types of receptor, but in each case it is converted into only one form of energy – the nerve impulse. This acts as a 'common currency' for communication within the body.

| Type | Stimulus |
|------|----------|
| photoreceptors | detect light (electromagnetic energy) |
| mechanoreceptors | detect touch, pressure or sound (mechanical energy) |
| thermoreceptors | detect temperature changes (thermal energy) |
| chemoreceptors | detect chemicals, e.g. taste, smell (chemical energy) |
| electroreceptors | detect electrical fields (electromagnetic energy) |

- The stimulus causes a change in the permeability of the receptor membrane to sodium ions. This in turn creates a **generator potential**.
- The frequency of the nerve impulses produced by the receptor varies with the intensity of the stimulus.

The last two features will be dealt with in relation to the function of **rod cells** – see Section 7.5, page 237. The function of receptors will be illustrated by reference mainly to the mammalian eye in Section 7.3, page 230. The Pacinian corpuscle is described in Box 7.1 and another receptor, the **muscle spindle** is covered in Section 8.8, page 281.

**Extension**

### Box 7.1  The Pacinian corpuscle

The Pacinian corpuscle is an example of a mechanoreceptor – it is sensitive to touch and pressure. Pacinian corpuscles are found in the skin and in joints. They are one of the simplest types of receptor consisting of the ending of a sensory neurone which is enclosed in a capsule made up of many layers of connective tissue (Figure 7.3).

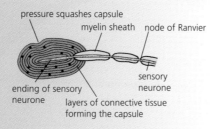

*Figure 7.3  The Pacinian corpuscle – a mechanoreceptor*

The capsule acts as a pressure-sensitive pad. When pressure is applied to the skin the capsule becomes squashed. The deformation of the capsule presses on the sensitive nerve ending at the centre and stretches the nerve membrane. This stretching acts as a stimulus to the membrane and opens the gates of sodium channel proteins. These channels are known as stretch-mediated channels. This sudden increase in permeability of the membrane to sodium ions ($Na^+$) allows $Na^+$ to diffuse rapidly into the nerve ending, depolarising the membrane and producing the generator potential. If the generator potential reaches a threshold value an action potential is generated. The size of the generator potential varies with the size of the stimulus (see Box 7.4, page 239).

The Pacinian corpuscle illustrates one of the features of all receptors – they only respond to specific stimuli. Pacinian corpuscles are sensitive to pressure – they do not respond to other stimuli, e.g. temperature. The skin has separate receptors to detect temperature changes.

**Did You Know**

Transduction can be very efficient. Rods in the eye can detect a very weak stimulus – just 2–3 **photons** (units of light energy) of light, but this is converted into a nervous impulse travelling to the brain which has about 100 000 times as much energy. The stimulus is amplified.

**7.3** **The mammalian eye – an example of a sense organ**

The mammalian eye is complex, consisting of very large numbers of receptor cells together with accessory structures such as the lens which help the receptor cells work. The structure of the eye and brief details on the functions of the main parts are given in Figure 7.4 and Table 7.2.

1 Distinguish between the following:
  (a) pupil and iris
  (b) retina and choroid.
                                     *(2 marks)*

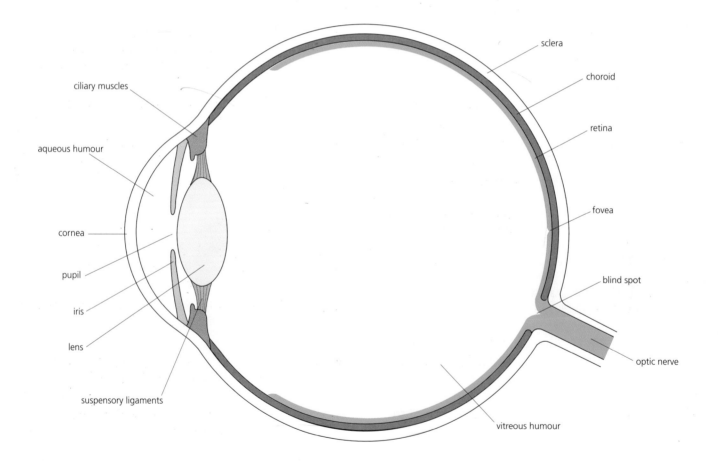

**Figure 7.4** *The structure of the human eye*

| Part of eye | Location and composition | Function |
|---|---|---|
| sclera | tough connective tissue which encloses the eye | protects the eye and maintains its shape |
| cornea | the front part of the sclera which is transparent to allow light to enter the eye | bends (refracts) the light to help focus it onto the retina |
| aqueous humour | a transparent watery solution; fills the front part of the eye | maintains the shape of the front chamber of the eye |
| iris | found in front of the lens, it contains circular and radial muscle and is pigmented | controls the size of the pupil to adjust the amount of light entering the eye |
| pupil | a hole in the centre of the iris | allows light to enter the eye; its size can be varied |
| ciliary muscles | circular and radial muscles which form a ring around the lens towards the front of the eye | adjust the focusing of light by changing the shape of the lens |
| suspensory ligaments | strong ligaments which connect the ciliary muscles to the lens | contraction and relaxation of the ciliary muscles tightens or loosens the ligaments; this changes the shape of the lens |
| lens | flexible, transparent disc | its shape can be changed to adjust focusing |
| vitreous humour | transparent, jelly-like material; fills the rear part of the eye | maintains the shape of the rear chamber and supports the lens |
| retina | found at the back of the eye, it contains light sensitive cells – the rods and cones; these are the photoreceptors | the detection of light |
| fovea | a small area at the centre of the retina containing only cones | most light is focused onto here; it gives the most detailed vision (greatest acuity) |
| blind spot | a small part of the retina where the optic nerve leaves the eye; contains no rods or cones | not sensitive to light because no rods or cones are present |
| choroid | a layer of pigmented cells at the back of the eye behind the retina | prevents reflection of light; contains blood vessels which supply the retina |
| optic nerve | a bundle of sensory nerve fibres which leave from the back of the eye | carries nerve impulses from the retina to the brain |

*Table 7.2  The main parts of the eye*

**Definition**

Refraction is the bending of light rays which occurs when the rays pass from one medium to another of a different density.

As a sense organ, the eye is sensitive to both the intensity and the wavelength of light. It works on the same fundamental principles as a camera:

● the amount of light entering is controlled
● the light is focused by a lens to give a clear image
● the image is detected by a sensitive surface (Figure 7.5).

We are able to see by detecting the light rays reflected from an object. From each point of the object light rays are reflected in all directions. Only some of these rays will enter the eye and the rays from each point of the object need to be focused (brought together) to produce a single point of light on the retina. The retina contains the receptors – the rods and cones. In this way each point on the object is reproduced as a point of light on the retina, forming an image. This image is reduced in size and it is inverted (upside down) but the brain learns to adjust to this and we 'perceive' the image as being the correct size and the correct way up.

### Accommodation

The focusing of the light rays onto the retina from objects at different distances is known as **accommodation**. Focusing involves the bending or **refraction** of the light rays. Refraction occurs when light rays pass from one medium to another which has a different density, e.g. from the air into the cornea.

Most refraction occurs at the cornea but the degree of refraction here cannot be adjusted. Refraction also occurs at the lens and this can be adjusted. The shape of the lens can be changed, altering the amount of refraction to bring the light rays to a sharp focus on the retina. Changing the shape of the lens is regularly required as we alter between looking at near or distant objects. These changes occur automatically as a reflex action called accommodation (Figure 7.6).

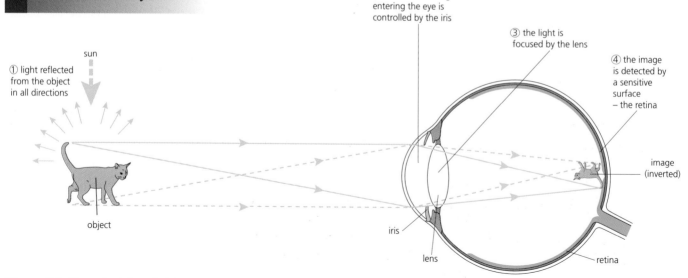

*Figure 7.5  The principles of vision*

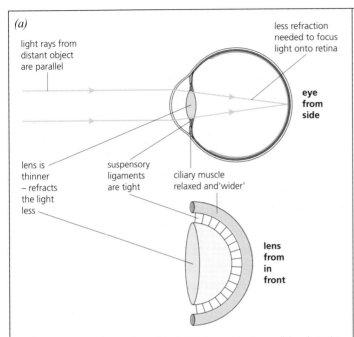 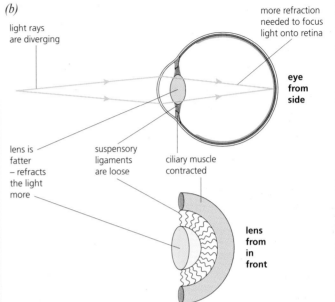

*(a)*

Light rays entering the eye from distant objects are virtually parallel and require little refraction (bending) to bring them to a focus. The circular ciliary muscles relax, forming a wider 'ring' and this causes them to pull on the suspensory ligaments, making them tight. The ligaments in turn pull on the lens, stretching it and making it flatter. This less rounded (less convex) lens causes less refraction. Accommodation to long distance is aided by the internal pressure of the eye caused by the vitreous humour. This helps to widen the ciliary muscles and tighten the ligaments.

*(b)*

If we now change to looking at a near object, the light rays entering the eye from the object are spreading out (diverging) and require more refraction to bring them to a focus. The circular ciliary muscles contract, forming a narrower 'ring'. The suspensory ligaments become loose as they are not being pulled by the muscles. The lens is not stretched and because it is elastic it reverts to its natural rounded shape. This fatter (more convex) lens causes more refraction, allowing a sharp focus to be achieved.

**Figure 7.6** *Accommodation (a) viewing a distant object, (b) viewing a near object*

2 (a) People who are long sighted cannot focus on objects which are close to them. Explain how the lens in their glasses helps them to focus.
*(1 mark)*

(b) Some people need thicker lenses in their glasses than others – suggest why this might be the case.
*(1 mark)*

### Remember

**Exam hint**

The changes that occur in focusing can be difficult to remember. Try the following:

Our eyes get tired when we look at near objects, e.g. when revising by reading our notes. They don't get tired when we look into the distance as when on a beach holiday and we look at the sea. Tiredness is caused by the contraction of a muscle, *not* by it being relaxed. We contract the circular ciliary muscles for near vision and relax them for far vision.

## The functions of the iris

The iris adjusts the amount of light which enters the eye by means of a **reflex mechanism** – the **iris reflex** (see Section 7.9, page 246). This is controlled by the **autonomic nervous system** (see Section 7.11, page 259) and the brain, and is an example of a **cranial reflex**. As with many other reflexes this one is protective. Very bright light can damage the sensitive rods and cones by over-stimulating them. The amount of light entering the eye needs to be reduced quickly. In contrast, in dim light it is important to allow as much of the available light as possible into the eye to stimulate the rods and cones.

The iris consists of a diaphragm containing circular and radial muscles surrounding a small hole – the pupil. The pupil is the only route for light to enter the eye (see Figure 7.4 on page 230). In bright light, nervous impulses are sent from the brain along **parasympathetic** nerves to the iris. These impulses stimulate the circular muscles to contract and the radial muscles to relax. This makes the pupil smaller so that less light enters the eye. In dim light, nervous impulses are sent from the brain along **sympathetic** nerves to the iris. These impulses stimulate the circular muscles to relax and the radial muscles to contract. This **dilates** the pupil, making it wider so that more light is able to enter the eye. These changes are shown in Figure 7.7. (Further details on sympathetic and parasympathetic nerves are given in Section 7.11 on page 259.)

*Figure 7.7  The iris reflex – controlling the size of the pupil*

*Figure 7.8  Human eyes in close up to show variation*

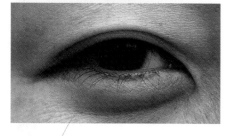

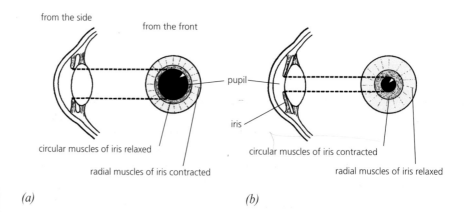

Dim light – wider pupil

Bright light – narrower pupil

from the side

from the front

pupil

iris

circular muscles of iris relaxed

radial muscles of iris contracted

circular muscles of iris contracted

radial muscles of iris relaxed

*(a)*

*(b)*

The circular and radial muscles work **antagonistically** – they are an example of an **antagonistic** pair of muscles, having the opposite effects to each other. (see Section 8.3, page 270). A second function of the iris is to prevent any light entering other than through the pupil. The iris is pigmented, giving the eye its particular colour (brown, blue, etc.) and blocking out light (Figure 7.8).

## The transmissive properties of the eye

It is important to note that the function of the eye depends upon many of its parts being **transmissive** (allowing light to pass through). The structures involved in the light path – the cornea, the lens and the aqueous and vitreous humours – are all transparent. They allow light to pass through them so it can reach the retina.

## 7.4 The structure of the retina

*Figure 7.9* *The structure of the retina*

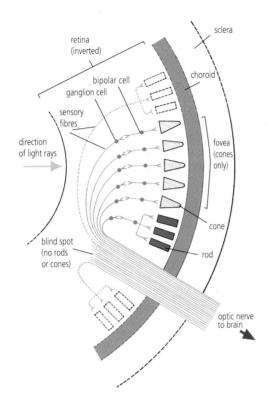

The retina can be considered as consisting of two layers (Figure 7.9). Firstly there is a layer of photoreceptors – the rod and cone cells. This is the crucial layer where the light is actually detected. The rods and cones are not evenly distributed in the retina (see Section 7.6, page 242). The second layer is complex, containing different types of nerve cells and large numbers of fibres from sensory neurones which collect to form the optic nerve. the rods and cones form synapses with cells in this layer. An outline of the pathway is as follows: The rods and cones synapse with **bipolar cells** (bipolar neurones) and the bipolar cells connect by synapses to **ganglion cells**. It is the axons leading from these ganglion cells which collect to form the optic nerve, taking information to the brain. This pathway can be shown as a flowchart:

rods and cones → (synapse) → bipolar cells → (synapse) → ganglion cells
→ axons from ganglion collect together → optic nerve → brain

Behind the retina, at the back of the eye and furthest away from where light enters, is the **choroid**. This is a layer of pigmented cells containing the black pigment **melanin**. The melanin absorbs light rays, preventing them from being reflected back through the rods and cones and causing 'fuzzy' distorted vision.

### ! Did You Know

Cats' eyes shine in the dark! Unlike humans where the choroid prevents reflection, cats and some other nocturnal animals have a reflective layer in the choroid called the tapetum. In members of the cat family this is made of shiny crystals of guanine. In dim light conditions the light is reflected back through the rods so that they receive maximum stimulation, helping the cat to see at night.

*Figure 7.10* *Eyes reflecting light*

It should be noted that the retina is **inverted** – that is the photoreceptor cells are in the outer layer of the retina, at the back of the eye, rather than nearer the source of the light rays on the inside of the retina. Light has to pass through the inner layer of bipolar, ganglion and other cells before it reaches the photoreceptors. This would appear to be an inefficient arrangement but it occurs in all vertebrates, including birds such as hawks and eagles which have outstandingly good eyesight – about 10 times better than humans in terms of the detail seen. For example, when hunting, a Peregrine falcon can see a pigeon from a distance of 1066 m.

This inversion of the retina also explains the occurrence of the **blind spot**. The fibres forming the optic nerve start off on the inside of the retina and then cross the surface of the retina to form the optic nerve, which then goes to the brain. At the point where the optic nerve passes through the retina there is no room for any rods or cones. This small region is called the blind spot because as there are no photoreceptors here it is not sensitive to light. Box 7.2 shows how to find the blind spot (see also Figure 7.16 on page 242).

**Box 7.2  Finding your blind spot**

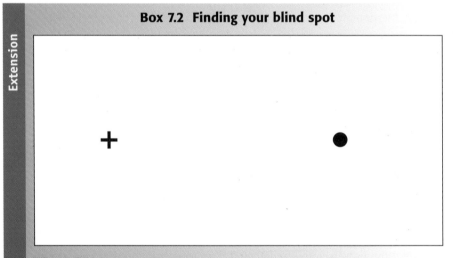

*Figure 7.11*

You may find it useful to copy Figure 7.11 to use for this demonstration.

1  Cover your left eye with your hand.

2  Holding Figure 7.11 at arms length, stare at the cross with your right eye.

3  Keeping the right eye focused on the cross, now slowly move the book towards you. You should notice that at some point the circle disappears. At this point the image of the circle is falling directly on the blind spot of the right eye – the eye cannot see it. Normally we use two eyes so we do not notice the blind spot because it is made up for by vision from the other eye. Repeat the test with both eyes and check for yourself.

**Extension**

## 7.5 Photoreception – the detection of light by rods and cones

The mammalian eye is an example of a sense organ. The receptor cells are found in the retina at the back of the eye and are of two types, rods and cones. These are present in very large numbers, with approximately 120 million rods and 6 million cones per eye. The structures of rods and cones are shown in Figure 7.12 and the differences between them are summarised in Table 7.3 on page 240.

Rods and cones have three main regions. The outer segment is the light-sensitive region containing photosensitive pigments. Here light energy is converted into a **generator potential**. In rods this region contains up to 1000 membrane-lined vesicles. The photosensitive pigment **rhodopsin** is embedded in the membranes of the vesicles. Vesicles are also present in cones but they consist of infoldings of the outer membrane. The cones have a different photosensitive pigment, **iodopsin**, embedded in the membranes. This structural arrangement of the membranes gives a very large surface area of pigment for the absorption of light.

The inner segment of both rods and cones contains many mitochondria to provide the energy to resynthesise the visual pigments after they have been broken down by light. It also contains the nucleus and many ribosomes to synthesise proteins such as those needed to make the vesicles.

The synaptic region is the equivalent of the synaptic bulb in neurones and allows the formation of synapses with bipolar cells which in turn link to the optic nerve and then to the brain.

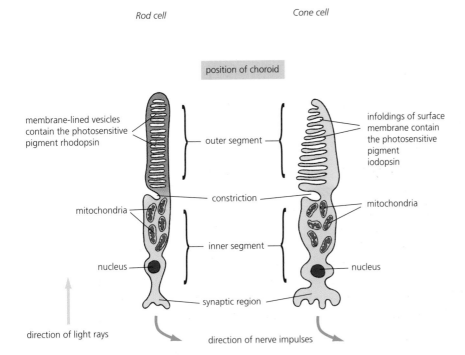

*Figure 7.12 The structure of rods and cones*

## Photoreception in a rod cell

Light detection depends on the light-sensitive pigment rhodopsin. This pigment is a pink/purple colour and it is therefore sometimes known as visual purple. Rhodopsin consists of a protein, **opsin**, joined with **retinal** (retinene) which is derived from vitamin A. Retinal can exist in two forms. When rhodopsin absorbs light, the energy from the light causes the retinal to change to a different shape which can no longer attach tightly to the opsin, resulting in the splitting of the rhodopsin molecule into retinal and opsin. (This breakdown is often referred to as **bleaching**.)

> ### Did You Know
>
> Carrots do help you see in the dark! Carrots are a good source of vitamin A which is needed to make rhodopsin. Too little vitamin A causes 'night blindness' where a lack of rhodopsin means that the rods cannot function. We use rods to see in dim light conditions. However, vitamin A is also found in other vegetables and particularly in liver which has much higher quantities than carrots. In excessive quantities vitamin A can be poisonous and some arctic explorers have died from eating too many polar bear livers!

> ### Definition
>
> The generator potential is the change in the potential difference of a receptor membrane caused by a stimulus. If the generator potential reaches threshold it produces an action potential.

The chemical breakdown of the rhodopsin changes the permeability of the plasma membrane of the rod to sodium ($Na^+$) ions (Figure 7.13). This causes a change in the distribution of $Na^+$ ions and this in turn changes the potential difference across the rod membrane. The alteration in the potential difference is called the generator potential. If the generator potential is large enough to reach a threshold size it causes an action potential and therefore a nerve impulse in the sensory neurone leading away from the receptor. The size of the generator potential is proportional to the size of the stimulus and this influences the frequency of the impulses. See Box 7.4 on page 239 for more detail.

It should be noted that the production of the generator potential in rods in fact involves a **hyperpolarisation** (inside of cell becoming more negative) unlike in most other receptors (e.g. the Pacinian corpuscle, Box 7.1, page 229) where it involves a depolarisation. For further details on rod hyperpolarisation see Box 7.3.

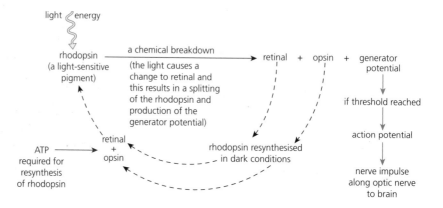

***Figure 7.13*** *Rhodopsin and light*

## Box 7.3  Rod cells and hyperpolarisation

In the dark, when the rod is not being stimulated, sodium pumps continuously pump Na$^+$ ions out of the rod. Unlike a neurone membrane the rod membrane is very permeable to Na$^+$ so that it easily diffuses back into the rod. When stimulated by light the chemical breakdown of rhodopsin causes the rod membrane to become less permeable to Na$^+$ so that less can diffuse in. As the pumps continue to pump Na$^+$ out, the inside of the rod becomes more negative. This hyperpolarisation is the generator potential. If there is a large enough change to reach threshold, it fires off action potentials in the sensory neurone that leads via the optic nerve to the brain.

## Resynthesis of rhodopsin

Before it can be used again the rhodopsin must be resynthesised from the opsin and retinal. This requires energy from the hydrolysis of ATP provided by the large numbers of mitochondria in the inner segment of the rod cells. In normal daylight the rhodopsin is almost entirely broken down (the cones are used for

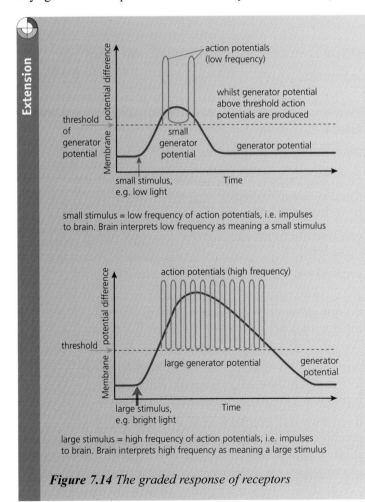

*Figure 7.14 The graded response of receptors*

## Box 7.4  The graded response of receptors

The size of the generator potential produced depends on the size of the stimulus (Figure 7.14). Generator potentials *do not* obey the 'all or nothing' law (see Section 6.9, page 203). Very low light intensity causes a small generator potential in rod cells whereas brighter light causes a larger generator potential. If the generator potential is larger than the threshold then an action potential is produced. Action potentials will continue to be produced for as long as the generator potential remains above threshold. A large generator potential will stay above threshold for longer, producing more action potentials than a small generator potential. More importantly the larger generator potential produces a higher frequency of action potentials – they are fired off more rapidly. Different sizes of stimuli (varying brightness of light) therefore cause a different frequency of nerve impulses travelling to the brain. This is crucial to enable the brain to differentiate between varying light levels. Action potentials *do* obey the 'all or nothing' law – they are always the same size, and therefore relative size of nerve impulse cannot be used to distinguish between a smaller and larger stimulus. Frequency of impulse is used instead.

vision) and the eye is said to be **light adapted**. After about 30 minutes in complete darkness nearly all the rhodopsin has reformed and the rods are sensitive – they can be used again. The eye is said to be **dark adapted**. This explains why we have very poor vision when we first go into a dark area from somewhere brightly lit. It takes quite a few minutes before we can see more clearly because it takes time to resynthesise enough rhodopsin to use our rods effectively – for our eyes to become dark adapted. When we return to the brightly lit area our rhodopsin is very quickly broken down again.

### Photoreception in cones

Light reception is very similar in cones to that in rods. The pigment involved is iodopsin. However, iodopsin exists in three different forms and there are three different types of cone, each with a different type of iodopsin (see Section 7.7, page 243). This forms the basis of colour vision. Iodopsin is less easily broken down by light and cones only produce a generator potential above a certain light intensity (i.e. in relatively bright light). Unlike rhodopsin, the iodopsin is not immediately all broken down and it can be resynthesised in the light, enabling us to continue to see in good light conditions. The differences between rods and cones are summarised in Table 7.3.

3  Give one similarity and two differences between rods and cones.

*(3 marks)*

*Table 7.3 The differences between rods and cones*

| Rods | Cones |
|---|---|
| outer segment shaped like a rod | outer segment shaped like a cone |
| contain the photosensitive pigment rhodopsin | contain the photosensitive pigment iodopsin |
| only one type of rod because only one type of rhodopsin | three types of cone because three types of iodopsin |
| do not detect colour | can detect colour (because of the three types of iodopsin) |
| distributed fairly evenly over the retina but none at the fovea | located mainly at the fovea |
| show convergence, i.e. many rods share a single neurone connection to the brain (see Figure 7.15) | do not show convergence, i.e. each cone has its own neurone connection to the brain |
| due to convergence they are sensitive to dim light (high sensitivity); used for night vision | due to the 1:1 connection with the neurones to the brain they are not sensitive to dim light (low sensitivity); used for day vision |
| due to convergence, they produce an image which lacks detail (low visual acuity) | due to the 1:1 connection with the brain they produce images with a high level of detail (high visual acuity) |

## 7.6  Sensitivity and acuity: the importance of convergence

- **Sensitivity** is the ability to detect low light levels – to be able to see in conditions of low light intensity. Rods are more sensitive than cones.
- **Visual acuity** is the degree of detail which can be seen – the ability to distinguish between two points which are close together. Cones have better visual acuity than rods.

This difference in sensitivity and acuity between rods and cones is largely explained by the connections they make with the neurones of the optic nerve via the bipolar cells (Figure 7.15).

*Sensitivity* (ability to detect low light levels)

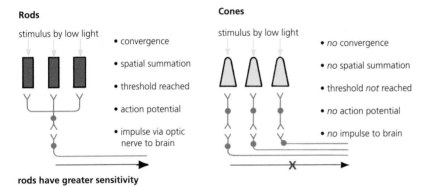

(a)

*Figure 7.15  Sensitivity and acuity – the importance of convergence*

*Acuity of vision* (in good light) – degree of detail which can be seen; ability to distinguish two points close together

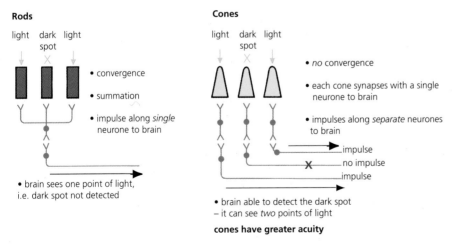

(b)

## Sensitivity

Several rods synapse with a single bipolar cell (see Figure 7.15a). This is called synaptic or retinal **convergence** (see also Section 6.11, page 211). The number varies from two or three rods to several hundred rods per bipolar cell. This arrangement allows **spatial summation** (see Section 6.11, page 212). Stimulations from a number of different rods at the same time are added together (summated) to reach threshold and produce an impulse in a neurone of the optic nerve. In effect one rod on its own cannot produce sufficient generator potential to reach its threshold but when the generator potentials from a number of rods are added together threshold is reached. This arrangement gives rods increased sensitivity because the effects of very low light levels can be added together so the light can be detected.

Impulses from cones do not undergo convergence. One cone synapses with one bipolar cell so there is no summation and cones have low sensitivity.

## Acuity

As described above, cones do not show convergence and one cone synapses with one bipolar cell – there is a 1:1 relationship (see Figure 7.15b). Each single point of bright light on the retina produces an action potential and an impulse in one neurone which goes to the brain via the optic nerve. This gives high acuity – the brain is able to distinguish between two points which are close together.

Impulses from rods do undergo convergence and a number of separate points of light are summated to produce an impulse in one neurone to the brain. The brain is not able to distinguish this impulse as separate points of light. Information has been combined to give a 'blurring' of the image. Therefore rods have low visual acuity.

## The distribution of rods and cones in the retina

Not all mammals have cones and those which don't, e.g. cattle, can only see in black and white as cones are required for colour vision. In mammals with colour vision, e.g. humans, the cones are mainly located at the fovea with only 10% found elsewhere. As the fovea is small (0.5 mm in diameter) the concentration of cones is high, with approximately 50 000 per square mm. Rods are more numerous than cones but they are distributed fairly evenly throughout the retina except at the fovea where there are none (see Figure 7.16 and Figure 7.9 on page 235).

### Did You Know

Waving a red rag at a bull is probably not a good idea. However, waving a rag of any other colour is equally undesirable. Bulls (and cows) are colour blind. The bull is only likely to be attracted – and perhaps made angry – by the movement of the rag. Its colour is irrelevant.

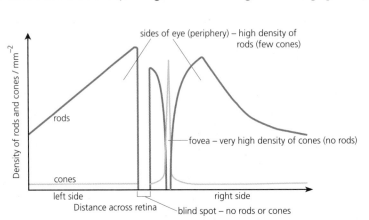

*Figure 7.16* *The distribution of rods and cones*

In bright light visual acuity is therefore greatest at the fovea due to the cones, and we move our eyes so as to keep the image we are interested in focused on the fovea. However, in dim light acuity is poorest at the fovea as the cones have low sensitivity. We see best if we look slightly to one side – out of the 'corners' of our eyes. Light from the object we are interested in then falls mainly on the rods which are away from the fovea. The rods can be stimulated by the dimmer light. Rods cannot detect colour so we lose colour vision in dim light.

To summarise, in bright light the cones at the fovea are stimulated, giving colour vision with high acuity, whereas in dim light the rods away from the fovea are sensitive enough to be stimulated, giving black and white vision with low acuity. At the blind spot there are no rods or cones (see page 236) and therefore light cannot be detected at all.

**4** Give one difference and one similarity between the fovea and the blind spot.
*(2 marks)*

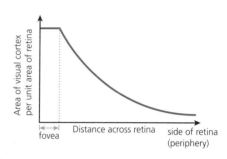

**Note:** a much bigger area of the visual cortex is allocated to the impulses arriving from the fovea. Each cone supplies a separate neurone to the visual cortex. Rods have convergence so fewer neurones supplied to visual cortex.

*Figure 7.17 Allocation of area of visual cortex per unit area of retina (only half of the retina is shown)*

### Interpretation by the brain

Information from the eyes is carried as nerve impulses by the thousands of neurones of the optic nerve to a part of the brain called the **visual cortex**, each part of which represents a part of the retina. Impulses from the different parts of the retina stimulate the corresponding part of the visual cortex. Because of convergence, rods have fewer neurones going to the brain and therefore a smaller area of cortex is devoted to impulses from rods (Figure 7.17). The 'pattern' of stimulation at the visual cortex is interpreted and we can 'see'. The significance and meaning of what we see depends upon reference to other parts of the brain where previous visual information is stored (see page 251 for further information on brain function and Box 7.9 on page 255 for more detail on the brain and vision).

### 7.7 Colour vision – the trichromatic theory

There are three types of cone cell, each containing a different form of iodopsin. Each is sensitive to different wavelengths of light corresponding to the colours blue, green and red (Figure 7.18). In effect the three types of cone are blue cones, green cones and red cones. The names refer to the colour of light absorbed and *not* to the colour of their appearance.

Pure red light will only break down the red iodopsin and only the red cones will fire impulses to the brain. This is interpreted by the brain as red. However, yellow light will break down some of the red iodopsin and some of the green iodopsin and so both red and green cones will fire impulses to the brain. This is interpreted by the brain as yellow. In this way the full range of colours can be seen depending on the relative proportions of the different cones which are stimulated. White light stimulates all three types of cone equally, all three types of cone fire and the brain interprets this as the colour white. Some of the colour combinations are given in Table 7.4 on page 244 and Figure 7.18.

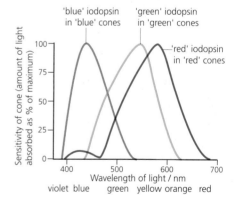

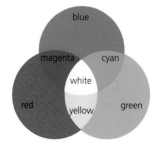

*Figure 7.18 Cones and the wavelength of light*

*Table 7.4* Cones and the perception of colour

| Cones stimulated by the light | | | The colour perceived by the brain |
|---|---|---|---|
| Green | Red | Blue | |
| yes | no | no | green |
| no | yes | no | red |
| no | no | yes | blue |
| yes | yes | yes | white |
| yes | no | yes | cyan (light blue) |
| no | yes | yes | magenta (purple/mauve) |
| yes | yes | no | yellow |

Colour as such is an invention of the brain. Cones enable us to be sensitive to and to distinguish between different *wavelengths* of light. This information is interpreted by the brain as different colours. Rods have only one type of rhodopsin so they cannot distinguish between different wavelengths of light and cannot detect colour.

**Extension**

### Box 7.5  Colour blindness

There are various types of defective colour vision. For example, people with red-green colour blindness are lacking either green cones or red cones. They cannot distinguish between red and green. Colour blindness is inherited – it is a sex-linked, recessive characteristic (see Section 3.7, page 82 for further details). This means that it is more common in men, with a frequency of about 8% for red-green colour blindness for men compared to a frequency of only 0.5% for all types of colour blindness in women.

### 7.8  The spinal cord

In mammals, the **central nervous system (CNS)** consists of the brain and the spinal cord. It is responsible for the control and coordination of the activity of the nervous system. During development in the embryo the brain and spinal cord arise from the same structure. A hollow neural (nerve) tube develops, running the length of the mammal. The front end of this tube enlarges to form the brain and the rest of the tube becomes the spinal cord (see Figure 7.27 on page 252).

The spinal cord is a tube of nervous tissue which extends from the base of the brain down the body. It is enclosed within the vertebral column which protects it from damage (Figure 7.19).

The spinal cord has a small **central canal** containing **cerebrospinal fluid (CSF)**. CSF also circulates within spaces in the brain (see page 252) and its functions include protective cushioning for the delicate brain and spinal cord.

### Remember

**Exam hint**

Note that the vertebral column is also known as the backbone or spinal column and it is important not to confuse the terms spinal *cord* and spinal *column*. At A level the terms vertebral column and spinal cord should be used.

Surrounding the central canal is an area of **grey matter** forming a distinctive 'H' or 'butterfly' shape, and around the grey matter is an outer layer of **white matter**. The white matter consists mainly of myelinated nerve fibres with the white colour being due to the myelin sheaths. The grey matter consists mainly of the cell bodies of nerve cells and unmyelinated, **relay neurones**. The absence of myelin produces the grey colour.

At intervals down the body spinal nerves connect with the spinal cord. As with other nerves these contain many separate nerve fibres (Figure 7.20). Spinal nerves contain both **sensory and motor neurones**.

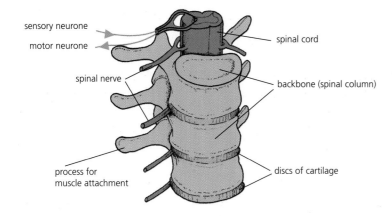

(a)

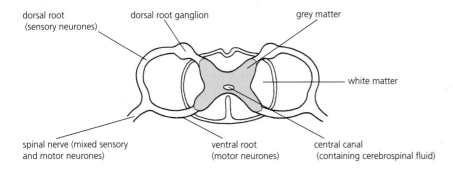

(b)

*Figure 7.19  The spinal cord (a) position of spinal cord enclosed by the backbone, (b) TS spinal cord*

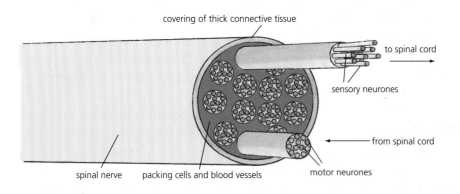

*Figure 7.20  Section through a spinal nerve*

Near to the spinal cord the spinal nerve divides to form two branches. Sensory neurones enter the spinal cord via the **dorsal (upper) root** and motor neurones leave the spinal cord to join the spinal nerve via the **ventral (lower) root**. A swelling on the dorsal root – the **dorsal root ganglion**, encloses the cell bodies of the sensory neurones. Also found within the spinal cord are nerve fibres which conduct information to and from the brain. These are known as ascending and descending fibres or tracts.

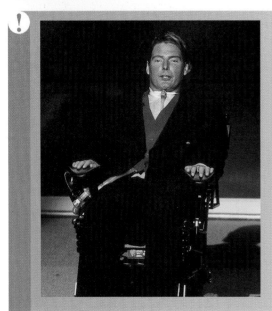

Figure 7.21 Christopher Reeve

**Did You Know**

Damage to the spinal cord or brain is particularly serious because nerve cells cannot replace themselves. If we cut our skin the cells can be replaced by new skin cells. If we damage our spinal cord the body cannot usually repair itself by producing new nerve cells. A possible solution may be found in research involving embryonic stem cells. These are 'young' unspecialised cells which have the potential to develop into any type of cell, including nerve cells. If implanted into a damaged area of the spinal cord or brain they may be able to repair the injury. This research holds out hope for those with Huntington's disease where brain cells die, and people like the 'superman' actor Christopher Reeve (Figure 7.21) who is paralysed because of spinal damage caused by a riding accident.

## 7.9 Reflexes and the reflex arc

As seen earlier it is essential for survival that mammals respond to changes in their internal and external environments. In general terms these responses can be described as behaviours, and the simplest form of behavioural response is the **reflex** or **reflex action**.

In an active mammal a very large number of reflexes are occurring almost continually, helping the body adjust to change and maintaining the conditions of the body in a constant state. These include reflexes which are specifically protective, such as the iris reflex protecting against bright light, and others such as those involved in ongoing adjustments of balance and posture. The advantages for survival (**adaptive value**) of reflexes include the following:

● Reflex actions produce a fast response which is important as many reflexes prevent damage to the body. Reflexes are rapid because only two or three neurones are involved in the pathway, reducing the number of synapses which slow down nervous conduction.

● Reflex actions allow a protective response to occur the *first* time a dangerous stimulus occurs. There is no learning needed as there is no time – the first time might be lethal.

● Reflex actions are automatic and involuntary – the same stimulus produces the same response every time without the need for a conscious decision by the brain. This reduces the demands on the voluntary parts of the brain and helps prevent overloading. It is automatic because the same restricted pathway (reflex arc) of neurones is used each time. Although the voluntary parts of the brain are not involved in making the actual response, they are often 'informed' of the stimulus (e.g. the awareness of pain from a burnt finger) by a separate nerve pathway.

5 For each of the following, state whether it involves a reflex action or a voluntary action:

(a) blinking in response to dust in the eye

(b) catching a ball

(c) writing

(d) shivering when cold. *(4 marks)*

**Definitions**

- A **reflex action** can be defined as a rapid, automatic response to a stimulus which is **involuntary**, i.e. not under conscious control.

- A **reflex arc** is the specific pathway taken by the nerve impulses in a reflex action. This involves a series of structures including the receptor, the neurones and the effector.

## Spinal and cranial reflexes

Spinal reflexes are coordinated by the spinal cord and include the knee jerk reflex and the **withdrawal reflex** (withdrawing the hand from danger). Cranial reflexes are coordinated by the involuntary parts of the brain. Examples are the iris reflex, blinking and coughing.

6 Suggest why the coughing reflex and the iris reflex are coordinated by the brain and not by the spinal cord. *(1 mark)*

Conditioned reflexes are a different type of reflex. These are simple forms of learning where the reflex is modified by past experience. The response of an

**Did You Know**

Some of the voluntary movements that we learn become so automatic and so fast that they appear to be reflex actions. For example, when a goalkeeper saves a penalty in football (usually against England), we may talk about a reflex save but it is a voluntary response.

Similarly, as part of driving lessons we become so quick at the 'emergency stop' that this voluntary response seems like a reflex.

*Figure 7.22 Saving a penalty is not a reflex*

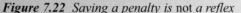

existing reflex becomes associated with a new and different stimulus. A well-known example is when we learn to respond to the sight or smell of food by producing saliva (see Section 9.7, page 310).

### The reflex arc

There are two main types of reflex arc. Some only involve two neurones, for example the knee jerk reflex described in Box 7.6, and some involve three neurones, e.g. the withdrawal reflex described below. The terms monosynaptic reflex (one synapse) and polysynaptic reflex (more than one synapse) are sometimes used to describe the two and three neurone reflex arcs respectively.

### The withdrawal reflex – an example of a simple spinal reflex

If we accidentally touch something hot with our hand we automatically move our hand away quickly due to a spinal reflex called the withdrawal reflex. The response protects us from damaging our hand by burning (Figure 7.23). This example illustrates most of the features of reflexes. The reflex action involves the following pathway:

stimulus → receptor → sensory neurone → CNS (relay neurone)
→ motor (effector) neurone → effector (muscle) → response

*Figure 7.23  The withdrawal reflex*

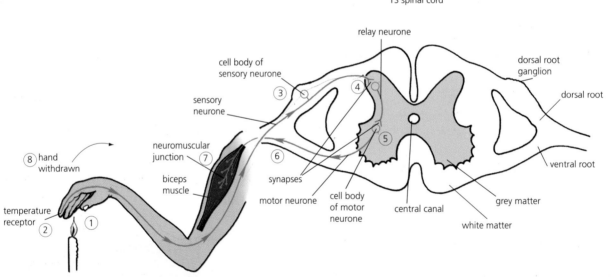

1. The stimulus is the heat of the object.
2. This is detected by temperature receptors in the skin of the fingers.
3. Nerve impulses are sent from the receptor along a sensory neurone to the spinal cord where it enters by the dorsal root.
4. Within the grey matter of the spinal cord the sensory neurone forms a synapse with a relay (connector) neurone. The nerve impulses are transmitted across the synapse.
5. From the relay neurone the impulses are transmitted across another synapse to a motor neurone.
6. The axon of the motor neurone leaves the spinal cord by the ventral root and conducts the impulses to muscles in the arm, e.g. the biceps.
7. The arrival of the impulses at the muscle causes it to contract, moving the hand away from the hot object.
8. The muscle is the effector and the withdrawal of the hand is the response.

The role of the spinal cord in coordinating this response is to ensure that the incoming information from the sensory neurones is switched via the relay neurone to the correct outgoing motor neurone, which goes to the appropriate effector.

The structures of sensory and motor neurones are described in Section 6.1, page 189. Their functions can be summarised as follows:

- Sensory neurones – conduct nerve impulses from receptors to the central nervous system (CNS).
- Motor neurones – conduct nerve impulses from the CNS to effectors.
- Relay neurones – conduct nerve impulses from sensory neurones to motor neurones.

**7** Suggest an advantage of the monosynaptic reflex (involving only two neurones – see Box 7.6) compared to a polysynaptic reflex (involving three neurones) such as the withdrawal reflex described on page 248.

*(1 mark)*

### Box 7.6  The human knee jerk reflex

This is an example of the simplest type of reflex – a monosynaptic reflex. It only involves two neurones and one synapse. It also provides an example of the role of **muscle spindles** – see Section 8.8, page 281. This reflex is commonly used by doctors to test that the nervous system is working properly.

Whilst the patient is sitting, the knee tendon is given a light tap with a rubber hammer. This stretches the muscle attached to the tendon above the patella. Muscle spindles are small stretch receptors found in muscles. The spindle detects the stretching and sends impulses along a sensory neurone to the spinal cord. This synapses directly with a motor neurone (not via a relay neurone) and impulses pass along the motor neurone to the muscle that extends the leg. The muscle is stimulated to contract, straightening the leg – the knee jerk (Figure 7.24).

It may seem strange that nature has equipped us to respond to a blow on the knee by a hammer, but the hammer just provides an easy way of stretching the muscle. This type of reflex action occurs naturally all the time. For example, when a person stumbles this reflex straightens the leg so as to regain our balance.

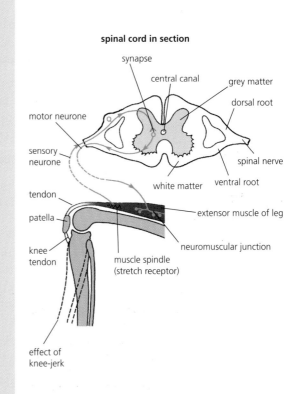

*Figure 7.24  The human knee jerk reflex*

## Box 7.7  Taxes and kineses

A reflex action is one example of a simple behavioural response. **Taxes** and **kineses** are two other types of simple behaviour patterns found in animals (singular: taxis and kinesis). For both taxes and kineses the stimuli that trigger the responses can be varied, e.g. temperature, light intensity, humidity, chemicals and gravity. The nature of the stimuli can be used as an addition to the name, e.g. phototaxis (light) and chemotaxis (chemicals).

The function of both types of response is to maintain the animal in a 'favourable environment'. This needs to be defined quite widely, referring to almost anything that helps the animal to survive and reproduce. In exam questions you may be provided with information or you may have to give your own examples. The commonest themes involve keeping the animal in the right place: for the finding of food; for the avoiding of predators and for finding a mate. All of these can be seen as 'maintaining the animal in a favourable environment'.

Examples of taxes:

- Female moths secrete chemicals (pheromones) which attract male moths to move towards them for mating (see Did You Know, page 229).

- Earthworms move away from the light and towards the dark. This helps to keep them underground where they are safer from predators and where they can get food.

- Mosquitoes move towards higher temperatures which make them more likely to find a human so that they can obtain a blood meal! (The human body tends to be warmer than the surrounding air.)

Examples of kineses:

- Flour beetles are pests of grain stores. In low humidity (dry air) they move less and change direction less often. As the humidity increases they move at a faster rate and change direction more often. This behaviour means that they tend to spend more time where conditions are very dry, such as the conditions found in grain stores. This keeps them in the store where food is available. In an experiment, flour beetles were observed in Petri dishes kept at different conditions of relative humidity. The distance moved and the number of turns were recorded (Figure 7.25 and Table 7.5).

- In contrast to flour beetles, woodlice display a kinesis which tends to keep them in moist conditions. This helps them survive because they are very vulnerable to dehydration and cannot afford to lose too much water by evaporation. As the humidity increases their rate of movement decreases.

**Definition**

A taxis is a type of behaviour involving the movement of an animal towards or away from a stimulus. It is a *directional* response – the *direction* of the stimulus rather than the *intensity* determines the response.

**Definition**

A kinesis is a type of behaviour involving the animal changing its speed of movement and/or its rate of turning (changing direction) as a result of the intensity of a stimulus. It is a *non-directional* response – the response depends upon the *intensity* of the stimulus rather than the *direction*.

- *Entobdella* is a parasitic worm which lives attached to the outside of fish. It obtains oxygen from the water through its body surface. As the oxygen concentration in the water decreases the worm increases its wriggling movements. This increases the circulation of water over its body, bringing fresh water with more oxygen to the body surface. Note that in this example the movements of the worm do not cause it to actually travel anywhere, it remains attached to the fish!

The changes in movement produced by kineses seem random and aimless compared to the directional nature of taxes. In fact they are very effective. In general terms, kineses cause the animal to move less in a favourable environment so that it tends to remain there. If it happens accidentally to move to an unfavourable environment its rate of movement increases and this means that it is more likely to 'escape' back to the favourable conditions where it will slow down again (see Figure 7.25).

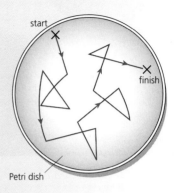

*Figure 7.25  Kinesis in the flour beetle*

| Relative humidity/ percentage | Distance moved/ cm per minute | Turns/ number per hour |
|---|---|---|
| 20 | 11.0 | 24 |
| 40 | 16.0 | 31 |
| 60 | 20.5 | 41 |
| 80 | 21.5 | 53 |
| 100 | 22.0 | 61 |

*Table 7.5  The results of experiments with flour beetles*

## 7.10  The brain

The brain is the most complex structure in a mammal's body and we are still some way from fully understanding how it works. Together with the spinal cord it controls and coordinates the activity of the nervous system but it has many more specific functions than the spinal cord and has a more dominant role. Briefly, the functions of the brain include:

- receiving sensory information from both inside and outside the body
- processing and coordinating the response to this information
- maintaining involuntary activities such as heart rate
- initiating voluntary activities such as locomotion
- reasoning, learning and memory (the 'higher' mental functions).

Many of these functions are carried out by specific parts of the brain but others, particularly some of the higher functions, involve complex relationships between different areas of the brain and are also influenced by our environment. For example, there is no agreement yet as to the exact location or even the exact nature of 'consciousness' – our sense of self.

### Did You Know

We may have parts of the brain that 'coordinate' being in love! Researchers have found that when people were shown photographs of their 'true loves' there was an increased flow of blood to four specific parts of the brain. The significance of these results is not yet known.

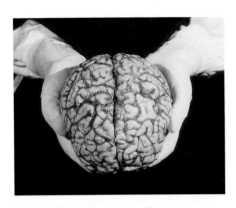

*Figure 7.26  The human brain – external view*

The brain is not the largest organ in the body (it is only approximately 2% of total body weight) and it is remarkable what is achieved within a relatively small space (Figure 7.26). A key feature is the complexity of the interconnections between nerve cells (particularly in the **cerebral cortex** – the outer layer of the **cerebral hemispheres**) allowing the formation of huge numbers of **neural networks**. Overall the human brain contains an estimated 100 thousand million nerve cells and each nerve cell in the cerebral cortex makes connections (synapses) with up to 10 000 other nerve cells. Changes to these synapses and networks may form the basis of memory.

**Structure and function of the human brain**

Although the parts vary in their proportions, all mammals have three main parts to their brain – **forebrain**, **midbrain** and **hindbrain**. In humans these three parts are seen most clearly during the early stages of embryonic development. In adults they are not so obvious as separate divisions (Figure 7.27). In particular the forebrain becomes much larger and covers the midbrain so that it is not visible from above. In the adult the hindbrain includes the **cerebellum** and the **medulla oblongata**. The forebrain includes the **hypothalamus** and the **cerebral hemispheres**. The midbrain changes least during development. External and sectional views of the adult human brain are given in Figure 7.28.

Within the brain are fluid-filled spaces called **ventricles** which contain cerebrospinal fluid (CSF). CSF is secreted by the ventricles and circulates through them and along the central canal of the spinal cord before being reabsorbed back into the blood. Amongst other functions the CSF provides protective cushioning for the delicate brain tissue. Some protection is also provided by three layers of tough membranes called **meninges** which surround the brain and the spinal cord. CSF also circulates in a space between the meninges. Meningitis is an infection of the meninges.

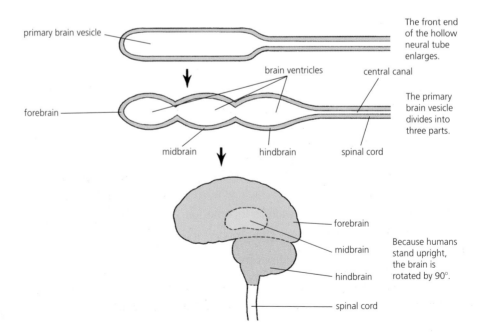

*Figure 7.27  The development of the human brain (not to scale)*

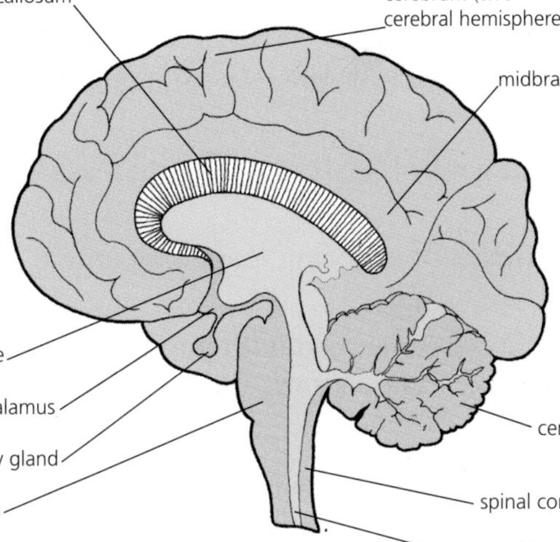

*(a)*

vertical section

corpus callosum

cerebrum (two
cerebral hemispheres)

midbrain

**FRONT**

**REAR**

ventricle

hypothalamus

pituitary gland

medulla

cerebellum

spinal cord

central canal

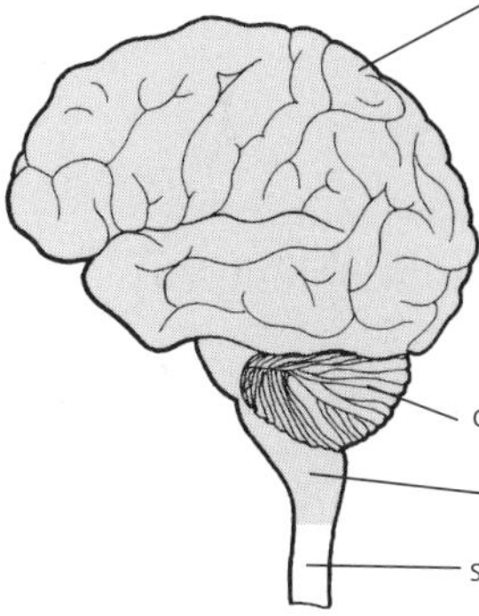

*(b)*

external view from side

left cerebral hemisphere

cerebellum

medulla

spinal cord

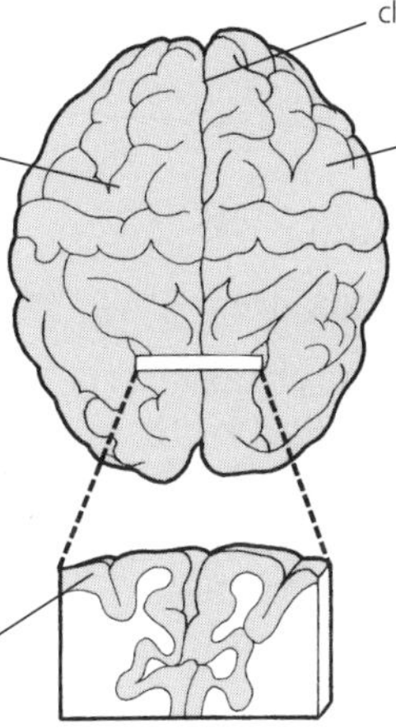

*(c)*

external view of cerebral hemispheres from above

cleft

left cerebral
hemisphere

right cerebral
hemisphere

section through
part of cerebrum
(note: highly folded)

cerebral cortex
(the surface layer)

***Figure 7.28*** *The structure of the
human brain*

## Box 7.8  Phineas Gage

**Extension**

Phineas P. Gage worked in a quarry in Vermont in the USA. In 1848 there was an accidental explosion and a large iron bar was blasted through his skull and brain. Amazingly he was not killed and he was even able to walk to seek help. He recovered physically and most of his specific brain functions, e.g. speech, hearing, sight, were unaffected. What did seem to be affected was his personality. Before the accident he was a calm, quiet man but afterwards he was prone to violent mood swings. However, he was still alert enough to sell his skull, cash in advance, before he died. The skull is kept in the museum of the Harvard Medical School (Figure 7.29).

Whilst probably an extreme example, what this case may demonstrate is that the brain has some capacity to compensate, with undamaged parts 'taking over' the functions of damaged parts.

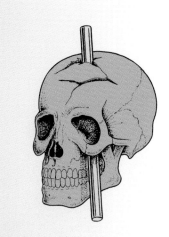

***Figure 7.29*** *The skull of Phineas Gage*

The functions of the main parts of the brain are summarised in Table 7.6.

| Part | Function |
| --- | --- |
| **hindbrain**<br>cerebellum | • controls balance and posture<br>• together with the cerebrum controls precise voluntary movements |
| medulla (medulla oblongata) | • contains the centres that control heart rate, blood pressure and rate of breathing<br>• controls the cranial reflexes of sneezing, coughing and salivation |
| **midbrain** | • controls the cranial reflexes concerned with pupil size (iris reflex) and lens shape (accommodation) |
| **forebrain**<br>hypothalamus | • contains centres which control many aspects of homeostasis, e.g. body temperature and osmoregulation<br>• production of the hormone ADH which is then stored in the posterior pituitary gland |
| corpus callosum | • connects left and right cerebral hemispheres |
| cerebral hemispheres<br>(There are two cerebral hemispheres, left and right. Together they make up the cerebrum. Further details on the functions of the cerebral hemispheres are given in Box 7.9.) | • receive sensory information from receptors<br>• interpret and analyse this information and are the site of 'higher' mental activities, e.g. memory and learning<br>• send out motor impulses to control the action of voluntary muscles |

*Table 7.6  The functions of the main parts of the brain*

**Remember**

**Exam hint**

Be very careful not to confuse the following terms as they look very similar:

• Cerebral hemispheres – largest parts of the brain; two present – left and right.

• Cerebrum – overall name for the two cerebral hemispheres taken together; main structure in the forebrain.

• Cerebral cortex – the outer 3 mm of the cerebral hemispheres; highly folded; the site of many functions, e.g. memory.

• Cerebellum – a separate structure (not part of the cerebrum); found in the hindbrain; has its own functions, e.g. balance.

**8** When we have a fever the body temperature is much hotter than usual as if the body's own 'thermostat' had been reset. Using Table 7.6, suggest in which part of the brain the thermostat might be located. *(1 mark)*

*Figure 7.30* *London taxi*

All that studying makes your brain larger! London 'black cab' taxi drivers go through rigorous studying called 'the knowledge' where they learn the detailed geography of London so that they know how to drive almost anywhere within London without using a map. They develop a detailed mental map of the streets in their head. Researchers have found that part of the brain of a 'cabby' is significantly increased in size. The area enlarged is a part of the brain which is involved in spatial memory.

Extension

### Box 7.9 The cerebral hemispheres

An outline of the functions of the cerebral hemispheres is given in Table 7.6. These are the largest part of the human brain. They are composed of white matter and grey matter as in the spinal cord, but with the white matter on the inside, not the outside. A thin (approx. 3 mm) outer layer is called the **cerebral cortex** and this consists of grey matter – millions of unmyelinated nerve cells (see Figure 7.28d). Most of the functions of the cerebral hemispheres are located in this thin surface layer, which is highly folded, giving a greater surface area for the huge cell numbers.

Bemeath the cortex the bulk of the cerebral hemispheres consists of white matter – myelinated neurones which connect the cortex with other parts of the brain and with the spinal cord. The myelin sheaths produce the white colour. Some of the main functions of the cerebral hemispheres are listed below. For the sake of brevity the term **cortex** will be used (rather than cerebral cortex or cerebral hemispheres). The location of some of these functions is shown in Figure 7.31.

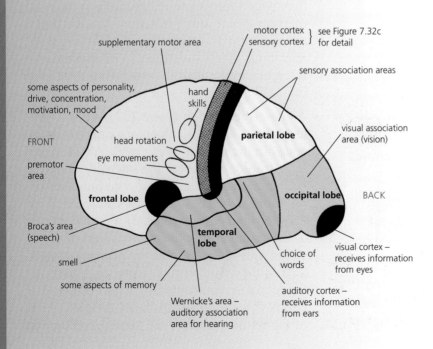

*Figure 7.31* *Location and function in the cerebral hemispheres*

**Box 7.9  The cerebral hemispheres continued**

### Sensory areas

These receive nerve impulses from receptors all over the body. Different parts of the cortex receive impulses from specific receptors. For example the **primary visual area** at the back of the brain receives sensory information from the eyes. Impulses from touch receptors in the skin and from pressure receptors in the muscles and joints are received by a narrow 'strip' of cortex at the top of the brain called the **sensory cortex**. This is like a map of the body as perceived by the brain. Different parts of the strip receive impulses from receptors in different parts of the body, e.g. impulses from the toes at one end of the strip and impulses from the tongue at the other end. The size of the area of the cortex devoted to a particular part of the body is proportional to the number of the receptors found there (how sensitive that part is). For example, within the strip, the largest areas are allocated to impulses from the fingers and the mouth as these have the greatest density of sensory receptors. Elsewhere, the primary visual area is also large, reflecting the huge numbers of rods and cones which send it information.

These arrangements are shown well in 'homunculus' diagrams and models where the parts of the human body are shown in proportion to the area of the cortex devoted to them (Figure 7.32a and c).

Damage to a specific part of the sensory cortex, for example by a stroke, causes a loss of 'feeling' from the matching part of the body and can also lead to the person being unaware even of the existence of that part. This is because although nervous information is sent from the receptors in that part of the body, the damaged brain cannot interpret this information when it arrives.

### Motor areas

A similar arrangement exists for motor areas which control the movement of voluntary muscles by sending out nerve impulses along motor nerves. There is another 'strip' – the **motor cortex** – alongside the sensory cortex. This strip is divided up according to the part of the body that nerve impulses are sent to.

The proportions are rather different to the sensory cortex although the areas for the fingers and mouth are also large (Figure 7.32b and c). There are also motor areas outside of the motor cortex, in particular the **supplementary motor area** which controls a variety of motor activities including voluntary eye movements.

One unusual aspect of motor control is that motor areas in the right cerebral hemisphere control muscles on the left-hand side of the body and vice versa. Damage to a motor area in the right hemisphere will therefore cause paralysis of the corresponding muscles on the left-hand side of the body.

### Association areas

These interpret and analyse incoming impulses from the sensory areas in the light of previously received information (memory) and then pass appropriate instructions to the relevant motor area for action. **Association areas** are involved in memory, learning and reasoning. Their function can be illustrated by reference to the **visual association area** (see Figure 7.31).

As described earlier (page 237), we 'see' by using light-sensitive receptors called rods and cones in the retinas of our eyes. All of these light stimuli, whatever the intensity, shape or colour, are converted into nerve impulses of the same size. These impulses are carried by the thousands of neurones of the optic nerve to the visual cortex (a sensory area). It is sometimes referred to as the primary visual area of the visual cortex. Here, each part of the retina is represented by a part of the visual cortex. The impulses from the different parts of the retina stimulate the corresponding part

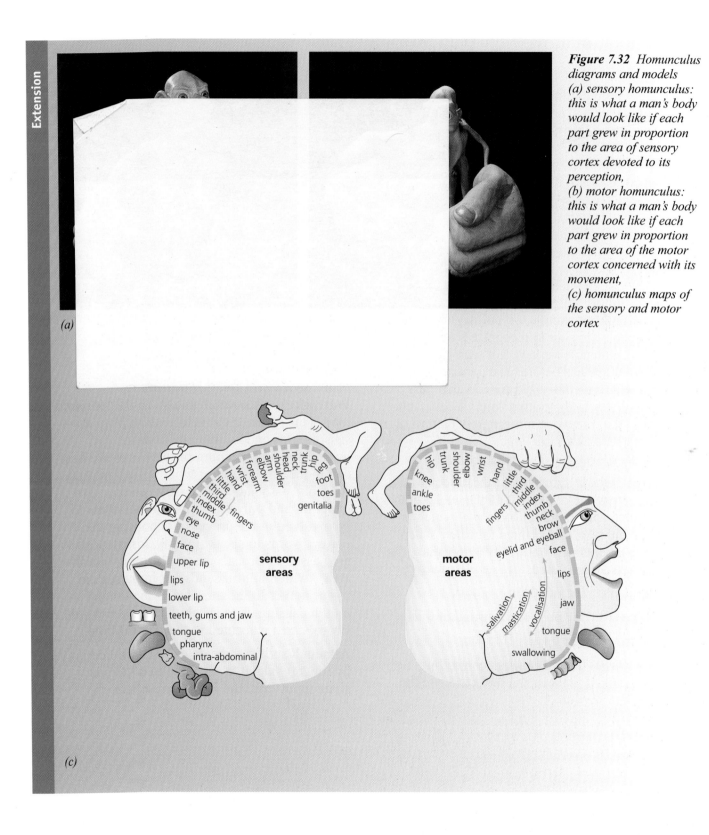

**Figure 7.32** Homunculus diagrams and models
(a) sensory homunculus: this is what a man's body would look like if each part grew in proportion to the area of sensory cortex devoted to its perception,
(b) motor homunculus: this is what a man's body would look like if each part grew in proportion to the area of the motor cortex concerned with its movement,
(c) homunculus maps of the sensory and motor cortex

(a)

(c)

sensory areas

motor areas

Extension

### Box 7.9 The cerebral hemispheres continued

of the visual cortex. This produces a 'pattern' of stimulation at the cortex. This pattern must now be analysed and interpreted to give it meaning and this is the function of the visual association area. Although the incoming sensory impulses are all the same size, their frequency varies, as does the location and type of receptor where the impulse originated, for example whether a green cone or a blue cone. All of this information is compared with visual information (images) previously received and stored in the association area as memory. It may also be compared with the sensory information arriving from other receptors. For example, on holiday in another country we are able to recognise a bird as a bird even though we have not seen that particular species before. The general shape matches our memory and we can also hear noises that sound like bird song. We are able to 'see' the otherwise meaningless pattern of sensory data as a bird.

As well as memory, learning and problem solving are also functions of the association areas. When we are taught maths we learn to recognise what numbers look like, which is which and how to add them together. When we see the maths problem of 9 + 6 we use the visual association area:

• to identify the numbers

• to refer to our previous learning of maths

• to solve the problem

• to instruct the motor areas controlling the muscles of the arm and hand to write the correct answer.

We then are aware of whether the answer looks correct or not!!

Association areas do not work in isolation, and analysis and interpretation of sensory data will often involve co-operation between different areas of the cortex and other parts of the brain.

### Speech

In most people the coordination of speech and language is located in the left cerebral hemisphere, e.g. **Broca's area** – a motor area for speech – is found on the left side. When we listen to speech, information arrives from the ears at the **auditory cortex** (the sensory area for sound). This is located in the **temporal lobe** (lobe at the base) of the left cerebral hemisphere (see Figure 7.31). The main association area for hearing is **Wernicke's area** which is also in the temporal lobe. This receives the information from the sensory area and is involved both in interpreting the meaning of the words we hear and in the production of what we say in reply. A second association area between the temporal and **parietal** (upper) **lobe** seems to be involved in the process of choosing the correct words to speak. Finally, instructions are passed to Broca's area – the main motor area – which in turn sends out impulses via motor nerves to the muscles of the tongue and mouth causing us to speak the selected words. Broca's area is located in the **frontal lobe**.

Although speech appears to be controlled only by the left side, it is probable that there is some input from the right hemisphere.

**Did You Know**

For most people, the left side of the brain seems to be more important and is referred to as dominant. Louis Pasteur, the famous French scientist, suffered a stroke in the right side of his brain. He had some paralysis on the left side of the body but was able to continue to work successfully. When he died years later it was found that there had in fact been extensive brain damage. If a stroke of the same severity had damaged the left side of his brain, he would have been completely incapacitated or might even have died as a result.

*Figure 7.33  Louis Pasteur*

## 7.11  The autonomic nervous system

As seen earlier the parts of the nervous system other than the CNS make up the **peripheral nervous system** (see Figure 7.1 on page 227). The peripheral nervous system can in turn be divided into the **voluntary (somatic) nervous system** which controls activities that are usually voluntary, e.g. contraction of leg muscles for walking, and the **autonomic nervous system** (**ANS**). The ANS controls the activities which are normally involuntary, e.g. heart rate, breathing rate and sweating. These are the crucial 'basic maintenance' functions that continue and are adjusted without us having to think about them.

The voluntary nervous system usually supplies skeletal (striated) muscles while the ANS usually supplies involuntary (smooth) muscles and glands. The ANS also differs from the voluntary nervous system in that it consists only of motor neurones. It does not have its own system of sensory neurones. The motor neurones of the ANS originate from the brain and spinal cord. Some carry impulses via a separate chain of ganglia (swellings) lying close to the spinal cord.

9  Give one structural difference between the ANS and the voluntary (somatic) nervous system. *(1 mark)*

The ANS has two divisions – the **sympathetic** and the **parasympathetic**. Organs receive nerve impulses from both divisions – there is **dual control** over the activity of the organ (Figure 7.34).

Impulses from sympathetic nerves tend to have an excitatory effect and sympathetic responses are more important when the body is active or stressed. The parasympathetic tends to inhibit or decrease activity and its responses are more important when the body is at rest. The two divisions work antagonistically rather like an accelerator and a brake – one speeds up the organ and the other slows it down. The reason for their different effects is that the two divisions have different neurotransmitters at the synapse (neuromuscular junction) made with the

**Remember**

**Exam hint**

Think *auto*nomic = *auto*matic /involuntary.

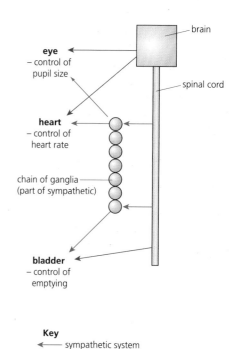

*Figure 7.34  The autonomic nervous system – a simplified outline with only some of the structures shown*

muscle of the supplied organ. The parasympathetic uses acetylcholine whereas the sympathetic uses noradrenaline. Noradrenaline is similar chemically to the hormone adrenaline so that they can both fit into the same receptor sites. Therefore adrenaline tends to have the same effect on an organ as the sympathetic nervous system, e.g. they both speed up heart rate in times of stress.

Some of the effects of the sympathetic and the parasympathetic systems are summarised in Table 7.7. This also illustrates the antagonistic dual control that occurs in most cases.

The responses controlled by the ANS are types of reflex action. As with the reflexes discussed earlier (e.g. withdrawal reflex, page 248) most of these are automatic and are not under conscious control. However, a few, such as the emptying of the bladder (see below) are under voluntary control which has to be learnt. Some specific examples of activities controlled by the ANS will now be considered.

### Control of heart rate

As can be seen from Table 7.7 the sympathetic speeds up heart (rate of beating of the heart) and the parasympathetic slows down heart rate. Full details of this can be found in *AS Biology*, Section 10.6, page 284.

### Control of pupil diameter – the iris reflex

This has been covered earlier on page 234. It is another typical ANS-controlled response with antagonistic dual control. The sympathetic causes dilation of the pupil and the parasympathetic causes constriction.

### Control of tear production

This is unusual in that it does not involve dual control – only the parasympathetic has an effect. Tears consist of a watery solution of salts, mucus and the enzyme **lysozyme** which kills bacteria. They are produced from a **tear (lachrymal) gland** above each eye. Tears are necessary to keep the front of the eye moist, clean and free from bacteria. They are drained away into the nasal passage by ducts in the corner of the eye.

| Sympathetic nervous system | Parasympathetic nervous system |
|---|---|
| increases heart rate | decreases heart rate |
| increases blood pressure | decreases blood pressure |
| increases ventilation (breathing) rate | decreases ventilation (breathing) rate |
| dilates pupils | constricts pupils |
| no effect on secretion of tears | increases secretion of tears |
| contracts bladder sphincter | relaxes bladder sphincter |
| relaxes bladder wall | contracts bladder wall |
| increases sweat production | no effect on sweat production |

**Table 7.7** *The effects of the sympathetic and parasympathetic systems*

**10** Name two protective mechanisms for the eye and state whether each is under voluntary or involuntary control. *(2 marks)*

Tears are produced constantly, but certain stimuli including dirt particles, cold air and infections can trigger increased secretion, as can the 'crying' caused by emotional stress. Stimulation of the tear glands by nerve impulses from parasympathetic nerves causes an increase in tear secretion. However, stimulation by sympathetic nerves has no effect. This is one of the few ANS responses which does not have dual control. In the absence of stimulation by the parasympathetic, tear production soon returns to normal levels without requiring antagonistic stimulation by the sympathetic.

### Control of emptying of the bladder (urination or micturation)

This involves antagonistic dual control but is unusual in that it is one of the few ANS responses that is under voluntary control in adult humans and has to be learned. This response involves different parts of the bladder. When the bladder is full the stretch receptors trigger nervous impulses along parasympathetic nerves which cause the ring of muscle (**sphincter**) at the exit of the bladder to relax. This opens the sphincter allowing urine to leave. Impulses from the parasympathetic also cause the muscles in the wall of the bladder to contract, reducing the volume of the bladder and pushing the urine out through the open sphincter. This is the automatic reflex found in all mammals and in young humans.

In childhood, humans learn to 'override' this involuntary response by a process of conditioning. We impose conscious voluntary control so that urination is not automatic but occurs when we decide it should. This is achieved mainly by voluntary control of a second sphincter muscle in the tube (**urethra**) which leads from the bladder. This sphincter is opened and closed voluntarily. There are limits of course. We can only delay the emptying of a very full bladder for a certain length of time!

The sympathetic nervous system has the antagonistic effect, providing dual control. As the bladder starts to fill, stimulation from sympathetic nerves causes the sphincter to contract, shutting the exit, and causes the wall muscles to relax, allowing the bladder to stretch and increase in volume as it fills with urine.

### Box 7.10  Imposing voluntary control

*Extension*

Humans routinely impose voluntary control on the involuntary autonomic reflexes involved in urination and defecation. It is now thought that other autonomic reflexes can also be brought under conscious control by a process of learning. Some forms of meditation involve such processes. Progress has already been made in people learning to regulate their own heart rate and blood pressure by conscious control. Apart from stress reduction and general well-being such techniques could prove very helpful for people at risk due to high blood pressure or overactive hearts.

### Did You Know

Before his discovery of penicillin, Alexander Fleming investigated the ability of the lysozyme enzyme in tears to kill bacteria. To obtain enough tears for his research he squeezed lemon juice into the eyes of student volunteers! The acidity of the lemon juice stimulated the production of the tears.

*Figure 7.35 Alexander Fleming*

**Summary – ⑦ Receptors and the nervous system**

- The nervous system is organised into a central nervous system, consisting of the brain and spinal cord, and the peripheral nervous system. The peripheral system in turn is divided into the autonomic system and the voluntary system.

- Receptors collect information, converting stimuli into nerve impulses.

- The eye is a sense organ which detects light. It is able to focus light rays onto the retina which contains photoreceptor cells called rods and cones.

- Rods and cones contain a photosensitive pigment which is broken down by light to produce a change in potential difference. If this is large enough it generates an action potential and a nerve impulse passes to the brain.

- Rods and cones have differences in the connections that they make with the neurones of the optic nerve. Rods show convergence whereas cones do not. Because of this, rods are more sensitive to dim light. Cones are only sensitive to higher light levels but produce an image with greater detail.

- There are three types of cone: red, green and blue. These are sensitive to different wavelengths of light and this forms the basis of the trichromatic theory of colour vision. Rods are equally sensitive to all wavelengths of light.

- The spinal cord coordinates spinal reflex actions such as the withdrawal of the hand when it touches something hot. A pathway of three neurones (the reflex arc) is involved.

- Reflexes help survival because they are fast and automatic – not involving conscious decisions.

- The brain coordinates the whole nervous system and has a wide range of functions, including controlling voluntary actions and the higher mental abilities of memory and learning. It is a very complex organ with many different parts responsible for specific functions.

- The autonomic nervous system controls essential involuntary activities such as heart rate and breathing rate. These also are reflexes.

# ? Answers

**1 (a)** The iris is the circular and radial muscles surrounding the pupil. The pupil is a hole in the centre of the iris *(1)*.

**(b)** The retina consists of the rod and cone cells – the photoreceptors. The choroid is a layer of pigmented cells which prevent reflection *(1)*.

**2 (a)** It helps by bending (refracting) the light rays more so that they can be focused onto the retina *(1)*.

**(b)** Those needing thicker lenses cannot refract the light as much as other people. They need thicker lenses to provide greater refraction *(1)*.

**3** Similarity: both contain light sensitive pigment/both are receptors which detect light *(1)*. Differences: any two from Table 7.3, page 240 *(2)*.

**4** Similarity: neither have rods *(1)*. Difference: cones present at fovea and not at blind spot *(1)*.

**5 (a)** Reflex, (b) voluntary, (c) voluntary, (d) reflex *(4)*.

**6** The structures involved are nearer to the brain therefore response is quicker *(1)*.

**7** Synapses slow down the response – the monosynaptic has fewer synapses and therefore is quicker *(1)*.

**8** The hypothalamus *(1)*.

**9** ANS has no sensory neurones and somatic has /ANS supplies involuntary muscles and somatic supplies voluntary muscles *(1)*.

**10** Iris reflex – involuntary; production of tears (crying) – mainly involuntary; blinking – can be either voluntary or involuntary; closing the eyes – voluntary (any 2) *(2)*.

## End of Chapter Questions

**1 (a)** State which structure or substance within the mammalian eye is referred to in each of the following statements.

(i) There are no light sensitive cells in this area as all the nerve fibres converge here to form the optic nerve. *(1 mark)*

(ii) This layer is highly vascularised and its chief function is nutritive. *(1 mark)*

(iii) When light strikes a rod cell this reddish pigment is broken down. *(1 mark)*

**(b)** The sketches A and B below represent the iris of a mammalian eye in two different situations.

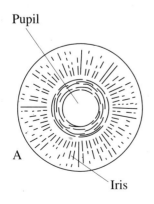

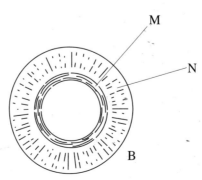

(i) Name M and N. *(1 mark)*

(ii) Briefly explain why and how the form of the iris might change from A to B during normal vision. *(2 marks)*

(c) Complete and annotate the diagram below to explain what is meant by *inversion of the image*. *(2 marks)*

*Oxford 1998*

*(Total 8 marks)*

2 The diagram below shows a rod cell from the retina of a mammal.

(a) Name the regions labelled A and B. *(2 marks)*

(b) State the location of most of the rod cells in the human retina. *(1 mark)*

(c) (i) Give the name of the light sensitive pigment contained in the rod cells. *(1 mark)*

(ii) Use the letter P to label on the diagram the region of the rod cell in which this pigment is located. *(1 mark)*

*Edexcel 1996*

*(Total 5 marks)*

**3** Read through the following passage, which refers to the detection of light in the mammalian eye, and then write on the dotted lines the most appropriate word or words to complete the passage.

Light is detected by specialised cells in the retina of the eye. The rod cells

detect light at ........................ intensity. They contain a photosensitive

pigment called ........................, which consists of a protein molecule,

..............................., joined to a molecule of retinal. Light causes the pigment

to separate into its two component molecules. When we move from the light

to the dark the pigment is re-synthesised. This process is known as

............................ adaptation.

*Edexcel 1998*                                              *(Total 4 marks)*

**4** The diagram below represents an enlarged section of part of the retina of a human eye.

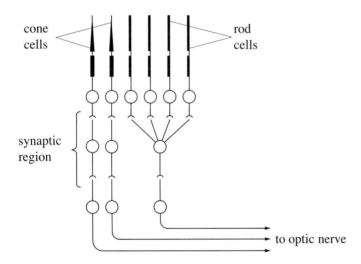

**(a)** Draw an arrow on the diagram to represent the direction of incoming light.                                                                       *(1 mark)*

**(b)** Explain how, in the synaptic region:

   (i) the connections of the rod cells enable us to see in conditions of low light intensity.                                                        *(2 marks)*

   (ii) the connections of the cone cells enable us to distinguish between objects close together.                                                   *(2 marks)*

*NEAB (AQA B) 1995*                                        *(Total 5 marks)*

**5 (a)** Explain how light energy falling on a rod cell in the retina of the eye is converted to electrical energy. *(2 marks)*

**(b)** Suggest why rod cells contain large numbers of mitochondria. *(2 marks)*

**(c)** Explain how the possession of different types of cone cell helps us to see:

(i) blue light; *(1 mark)*

(ii) white light. *(1 mark)*

*AQA A 1996*                                                      *(Total 6 marks)*

**6** The diagram below shows a transverse section of the spinal cord of a small mammal.

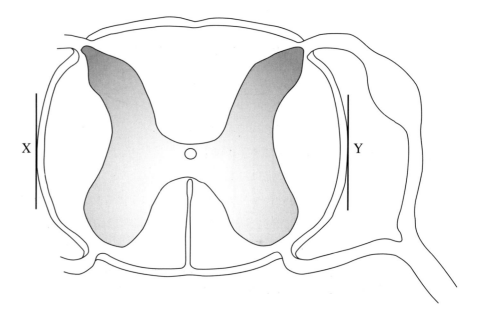

**(a)** Name the substance present in the central canal. *(1 mark)*

N3.2 **(b)** The magnification of the diagram is ×25. Calculate the actual diameter of the spinal cord at the position between X and Y. *(2 marks)*

**(c)** On the diagram, draw and label the neurones involved in a simple spinal reflex. *(2 marks)*

*Edexcel 1999*                                                   *(Total 5 marks)*

7  The graph shows the effect of light on the locomotion of a free-living flatworm.

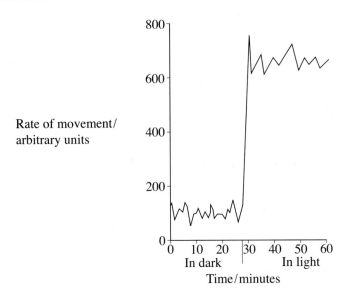

(a)  Describe the effect of light on the locomotion of the flatworm.  *(1 mark)*

(b)  (i)  Name this type of behavioural response.  *(1 mark)*

(ii) Give a reason for your answer.  *(1 mark)*

(c)  Suggest how the pattern of behaviour shown in the graph helps to keep the flatworm in favourable conditions.  *(2 marks)*

*AQA A 1998*  *(Total 5 marks)*

8  The table below refers to the functions of some regions of the brain. Complete the table by inserting the correct word or words.

| Region of the brain | *One* function |
| --- | --- |
|  | Control of voluntary movement |
| Medulla |  |
|  | Control of balance and fine movement |
| Hypothalamus |  |

*Edexcel 1998*  *(Total 4 marks)*

9 The table compares some effects of the sympathetic and parasympathetic systems.

| Feature | Sympathetic | Parasympathetic |
|---|---|---|
| Pupil of eye | Dilates | Constricts |
| Salivary gland | Inhibits secretion of saliva | Stimulates secretion of saliva |
| Lungs | Dilates bronchi and bronchioles | Constricts bronchi and bronchioles |
| Arterioles to gut and smooth muscle | Constricts | No effect |
| Arterioles to brain | Dilates | No effect |
| Heart rate | | |
| Stroke volume | | |

(a) Complete the table by filling in the spaces to suggest the effects of these two systems on heart rate and stroke volume. *(2 marks)*

Use information in the table and your own knowledge to answer the following questions.

(b) In giving dental treatment, it is important that any local anaesthetic stays close to the site of its injection. Explain why dental anaesthetics usually contain adrenaline. *(2 marks)*

(c) Many people suffer from motion sickness when travelling in cars. A number of drugs are used to control this and some work by inhibiting the parasympathetic system. Explain why side-effects of such drugs may include:

(i) dryness of the mouth;

(ii) blurred vision. *(2 marks)*

*AQA A 1992* *(Total 6 marks)*

# 8 ) Muscle: structure and function

Muscle tissue consists of specialised cells which can contract when stimulated. This contraction causes movement in animals, for example the movement of bones as in walking, or the movement of materials within the body such as moving food through the gut. There are three different types of muscle: skeletal, smooth and cardiac. Skeletal muscles are attached to bones and are involved in the movement of limbs and in locomotion. Skeletal muscle consists of two types of protein filaments which slide over each other when the muscle contracts. This causes the muscle to get shorter and pull on a bone causing it to move.

**This chapter includes:**
- the three types of muscle tissue
- antagonistic muscle action
- the detailed structure of skeletal (striated) muscle
- the sliding filament hypothesis of muscle contraction
- motor units and muscle spindles
- the effects of exercise on muscles.

## 8.1 ) Types of muscle

In mammals, there are three types of muscle tissue: **skeletal** (or striated), **smooth** and **cardiac**. The characteristics of these three types are summarised in Table 8.1. Further details of cardiac muscle are given in *AS Biology*, Section 10.5, page 283.

The main feature of all types of muscle cells is the ability to **contract** by up to half their resting length and then to **relax** ready for the next contraction. Muscles are made largely of protein and in mammals muscles can make up as much as 40% of body weight. Muscles are one of the main examples of **effectors**.

*Table 8.1 The characteristics of muscle types in mammals*

| Muscle type | Skeletal (striated, voluntary) | Smooth (unstriated, involuntary) | Cardiac |
|---|---|---|---|
| **Distribution** | attached to bones | present in walls of tubular organs, e.g. gut, uterus, bladder, blood vessels, and diaphragm | only found in the walls of the heart |
| **Function** | movement of part of body and locomotion | movement of materials within the body | pumping of the heart to maintain blood circulation |
| **Stucture of cells** | long fibres; distinctive cross-striations (banding) | shorter, spindle-shaped cells; no cross-striations | cells are branched forming a linked network; toughened cell membranes between cells |
| **Control of contraction** | by voluntary nervous system | by involuntary (autonomic) nervous system | contraction is myogenic (self-initiated) but rate can be changed by autonomic nervous system |
| **Muscle stamina** | contracts rapidly but fatigues easily | contracts slowly but can be maintained for long periods | contraction maintained without fatigue |

## 8.2 Effectors

Effectors are structures which, as a result of a nervous or hormonal stimulus, produce a response. Effectors include muscles and secretory glands. Skeletal muscles are attached to the bones of the skeleton. When they contract they move the bones, for example as in locomotion. Muscles therefore are effectors which facilitate movement.

## 8.3 Antagonistic muscle action

In mammals, movements of the body are caused by the contraction of skeletal muscles working across a joint. The ends of the muscle are attached by **tendons** to different bones. Tendons do not stretch, so contraction of the muscle will pull on the bone. When the muscle contracts, one of the bones stays still so that the shortening of the muscle causes the other bone to move.

Muscles can only exert force when they contract – they can only pull on a bone. When they relax they cannot push the bone back again. Because of this skeletal muscles are usually found in pairs where one pulls the bone in one direction and the other pulls the bone back to its original position. While one of the pair is contracting the other is relaxed. The movement caused by the contracting muscle extends the relaxed muscle so that it is returned to its resting length ready for its next contraction.

Because they have opposite effects to each other these muscles are said to be **antagonistic**. The biceps and triceps found in the arm form an **antagonistic pair**. Some of the antagonistic pairs found in the leg are shown in Figure 8.1. Some smooth muscles also form antagonistic pairs, e.g. the circular and longitudinal muscles of the human gut and in the iris of the eye.

*Figure 8.1* *Movement in the human leg*

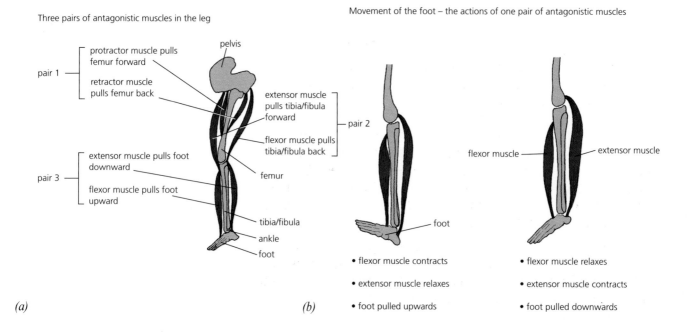

Three pairs of antagonistic muscles in the leg

pair 1 — protractor muscle pulls femur forward / retractor muscle pulls femur back

pelvis

extensor muscle pulls tibia/fibula forward — pair 2

flexor muscle pulls tibia/fibula back

pair 3 — extensor muscle pulls foot downward / flexor muscle pulls foot upward

femur

tibia/fibula

ankle

foot

*(a)*

Movement of the foot – the actions of one pair of antagonistic muscles

foot

- flexor muscle contracts
- extensor muscle relaxes
- foot pulled upwards

flexor muscle

extensor muscle

- flexor muscle relaxes
- extensor muscle contracts
- foot pulled downwards

*(b)*

**1** Many muscles form antagonistic pairs, both moving the same bone. How does this help explain the fact that the human body has more than twice as many muscles as bones? *(2 marks)*

**Extension**

### Box 8.1  The synovial joint

This type of joint is found in the skeletons of mammals, for example the knee, hip and elbow. These joints are found where one bone moves relative to another (Figure 8.2). The joint is enclosed in a **capsule** formed by **ligaments**. Ligaments also hold the bones together at the joint. The ends of the bones are covered with smooth **cartilage** (articular cartilage), which reduces friction between the bones. Lining the joint is a **synovial membrane** which secretes **synovial fluid**. The fluid fills the joint capsule and acts as a lubricant, further reducing friction.

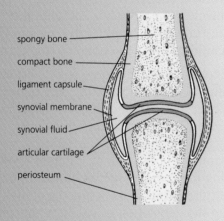

spongy bone
compact bone
ligament capsule
synovial membrane
synovial fluid
articular cartilage
periosteum

***Figure 8.2*** *The structure of a synovial joint*

**Did You Know**

Researchers are investigating the use of coral as a substitute for bone tissue in the repair of damaged bones. Some species of coral have a structure similar to bone, and coral tends not to be rejected by the immune system. New human bone tissue grows around and within the coral.

**Arthritis** (or osteoarthritis) is a common condition in which joints, including synovial joints, become damaged. The protective layers of cartilage on the ends of the bones are worn away by wear and tear and the joint becomes swollen and painful. Movement of the joint becomes difficult. Treatment to reduce the pain and swelling may help but sometimes the joint becomes so damaged that a replacement joint is needed. Replacement of worn out hip joints with artificial ones is now common practice. The new joint is made of metal and plastic and totally replaces the damaged parts.

**Rheumatoid arthritis** produces similar symptoms but has a different cause. It is an auto-immune disease meaning that the person's own tissues are attacked by their immune system. In this case the cartilage of the joint is damaged.

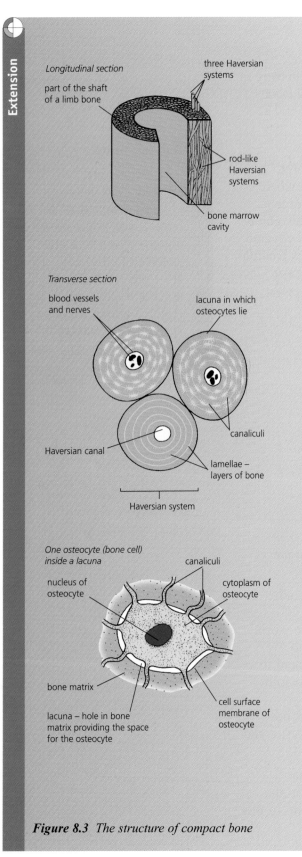

*Longitudinal section*

part of the shaft
of a limb bone

three Haversian
systems

rod-like
Haversian
systems

bone marrow
cavity

*Transverse section*

blood vessels
and nerves

lacuna in which
osteocytes lie

Haversian canal

canaliculi

lamellae –
layers of bone

Haversian system

*One osteocyte (bone cell)
inside a lacuna*

canaliculi

nucleus of
osteocyte

cytoplasm of
osteocyte

bone matrix

cell surface
membrane of
osteocyte

lacuna – hole in bone
matrix providing the space
for the osteocyte

*Figure 8.3  The structure of compact bone*

## Box 8.2  The structure of compact bone

To allow movement of the body, skeletal muscles need a firm structure to pull against when they contract. This strength and resistance to compression is provided by bones. A typical long bone such as the femur in the leg consists of a number of types of tissue including **compact bone**.

Compact bone is a calcified connective tissue composed of bone cells called **osteocytes** embedded in a solid matrix of **collagen fibres** (30%) and **inorganic salts** (70%) (Figure 8.3). The inorganic salts are a complex mixture of calcium, phosphate and hydroxyl ions called **hydroxyapatite**. The osteocytes lie in cavities in the matrix called **lacunae**, and cause the matrix to be deposited in concentric layers called **lamellae**. These build up to form rod-like **Haversian systems**. In the centre of each Haversian system is a **Haversian canal** containing blood vessels and nerves. The living osteocytes are connected to each other and with the blood vessels in the Haversian canal by thin channels in the matrix called **canaliculi** which contain cytoplasm.

**Osteoporosis** is a condition where there is a deterioration in bone structure due to a loss of calcium. The bones can get thinner and weaker and are more prone to fracture. The level of calcium in the blood is kept in balance homeostatically by the action of two hormones. **Parathormone** (parathrin) raises blood calcium levels partly by stimulating the removal of calcium from bones. **Calcitonin**, on the other hand, lowers blood calcium levels by increasing the deposition of calcium phosphate in bone tissue. These two hormones have antagonistic effects and work in a similar way to insulin and glucagon in the control of blood glucose levels. (see Section 9.3, page 295). **Oestrogens** (hormones) inhibit parathormone. In females oestrogen secretion declines after the menopause and therefore inhibition of parathormone also declines. The effects of parathormone are subsequently increased, with increased removal of calcium from bone causing osteoporosis. This explains why osteoporosis is more common in females after the menopause. One treatment is **hormone replacement therapy** (**HRT**) where females are given oestrogens during and after the menopause. Parathormone is inhibited and calcium removal from bone is minimised. Other less harmful side-effects of the menopause are also reduced by this treatment.

## 8.4 The structure of skeletal muscle

Skeletal muscle is also known as striated or striped muscle because of its banded or striped appearance. It can also be referred to as voluntary muscle because its contraction is under voluntary control. Each complete skeletal muscle consists of large numbers of specialised cells called **muscle fibres**, together with connective tissue and many blood vessels (Figure 8.4). The muscle fibres are grouped into bundles. They are very long cells, many centimetres in length, with each cell having large numbers of nuclei which lie close to the cell surface.

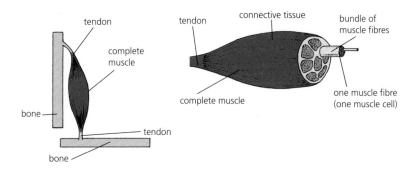

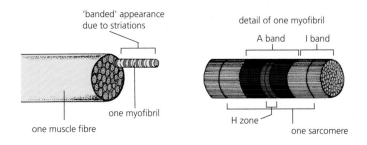

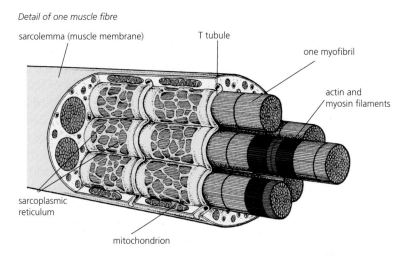

*Figure 8.4* *The structure of skeletal muscle*

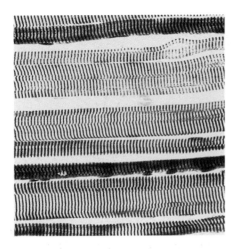

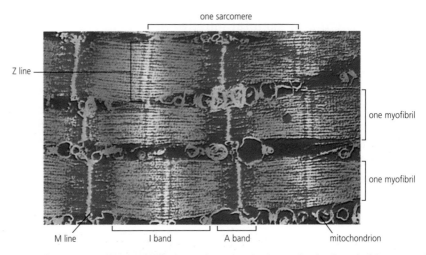

*Figure 8.5* *Photomicrograph of skeletal muscle (× 500)*
*Figure 8.6* *Electronmicrograph of part of a skeletal muscle fibre*

When viewed with a light microscope the muscle tissue looks banded but no detail can be seen (Figure 8.5). With the electron microscope the detailed structure of the muscle fibre is visible (Figure 8.6).

The muscle fibre (the muscle cell) is covered by a cell surface membrane – the **sarcolemma** (see Figure 8.4). At intervals infoldings of the sarcolemma form transverse or **T tubules** which penetrate into the cell. The contents of the cell are known as the **sarcoplasm** and there are various component structures. Just under the surface of the sarcolemma are large numbers of **mitochondria** to provide the high levels of energy needed for muscle contraction. There is also an extensive network of endoplasmic reticulum called the **sarcoplasmic reticulum**. This is a system of flattened membrane-lined tubes or cisternae which store **calcium ions**. The release of these ions triggers contraction (see page 277).

### 8.5 The structure of myofibrils

The main component of each muscle fibre is large numbers of **myofibrils**. It is the myofibrils within a muscle fibre which make it look banded under the microscope. The dark and light bands on all the myofibrils within one cell are lined up with each other.

A myofibril contains two types of filament. The thicker filaments are made of a protein called **myosin** whereas the thinner filaments are made of a protein called **actin**. The myofibrils run the whole length of the long muscle cells and are divided into many repeating units called **sarcomeres** (Figure 8.7). At intervals the actin filaments are attached to cross bands of connective material called **Z lines** (or Z discs). One sarcomere is the part of a myofibril between two Z lines. The myosin filaments are found in the central part of each sarcomere. In the middle of their length, the myosin filaments are attached to different cross bands of connective material called **M lines** (or M discs). The Z and M lines help to space out the actin and myosin filaments so that they can slide easily between each other (Figure 8.7b).

For part of their lengths the actin and myosin filaments lie between each other (they overlap) and with both filaments present, this part of the sarcomere looks the darkest. Nearest to the Z lines the sarcomere is made up of the thin actin filaments

> **! Did You Know**
>
> A baby at 6 months has a much smaller volume of muscle and a much larger volume of bone compared to the proportion in an adult. For the first 3 years of life muscle grows twice as fast as bone until the adult proportions are reached.

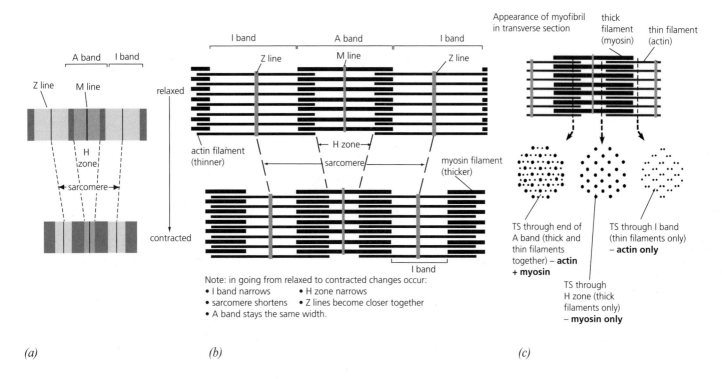

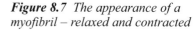

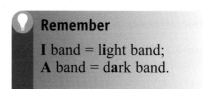

(a)    (b)    (c)

only and this part looks lightest in colour. In the very centre of the sarcomere only myosin filaments are present and this part is intermediate in colour. Each of these sections of the sarcomere has its own name (see Figure 8.7b). The light part with only actin is called the **I band**; the darker part with both actin and myosin is the **A band**; the **H zone**, with myosin only, lies in the centre of the A band.

In two-dimensional diagrams it appears that each myosin is associated with two actin filaments. In fact in transverse section it can be seen that there is a hexagonal lattice arrangement (see Figure 8.7c and Figure 8.8d). Each myosin filament is associated with six actin filaments which surround it, and each actin filament is linked with more than one myosin filament. Each myosin forms cross-bridges with the six actins, producing an overall contraction which is more coordinated, smoother and stronger.

*Figure 8.7* *The appearance of a myofibril – relaxed and contracted*

**Remember**

**I** band = **l**ight band;
**A** band = **d**ark band.

2   Some skeletal muscle tissue was treated with a solvent which dissolved the myosin filaments but left the actin filaments. As seen with an electron microscope, which of the following would still be visible: (a) H zone, (b) Z line, (c) I band, (d) A band?    *(1 mark)*

### Changes in appearance during contraction

When the muscle contracts the actin filaments are pulled between the myosin filaments towards the centre of the sarcomere. There is now a greater area of overlap between the actin and myosin, producing characteristic changes of appearance compared to the relaxed sarcomere (see Figure 8.7). In the contracted sarcomere the I bands and H zones become narrower, the overall sarcomere becomes shorter, the Z lines become closer together but the A bands stay the same length. It should be noted that although contraction involves a shortening of the sarcomere, this is achieved by the filaments sliding between each other rather than individual filaments changing in length. The individual filaments stay the same length whether contracted or relaxed.

3 Skeletal muscle has a characteristic appearance of I bands, H zones, A bands and sarcomeres. How does the appearance of each of these alter when a muscle changes from being contracted to being relaxed? Does it become longer, shorter or stay the same? *(4 marks)*

## The structure of myosin and actin

Myosin filaments are made up of several hundred molecules of the protein myosin. It is an unusual molecule with a long rod shaped 'tail' made of fibrous protein and two roundish 'heads' made of globular protein. The heads face outwards and link up with the actin during contraction. Actin filaments are made up of two long chains of small molecules of the globular protein actin. The chains are arranged as two helical strands twisted around each other (Figure 8.8).

Each actin filament is associated with two other proteins (see Figure 8.10). **Tropomyosin** forms long, thin strands wound around the actin so as to cover or block the sites where the myosin heads attach. Tropomyosin is attached to a different protein, **troponin**. **Calcium ions** bind to the troponin during the initiation of contraction. Tropomyosin and troponin are sometimes known as the blocking or switching proteins due to their role in contraction (see page 277).

*Figure 8.8 The structure of actin and myosin filaments*

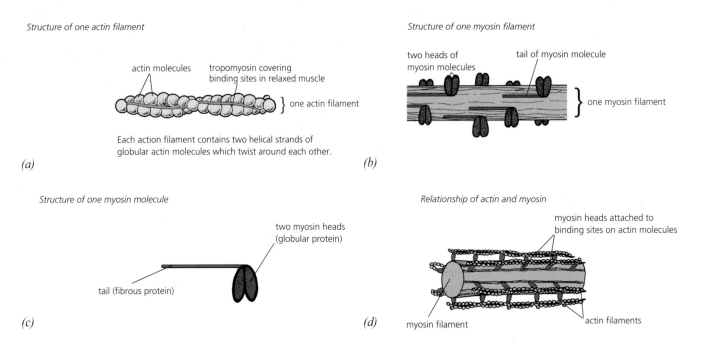

Structure of one actin filament

actin molecules    tropomyosin covering binding sites in relaxed muscle

} one actin filament

Each action filament contains two helical strands of globular actin molecules which twist around each other.

*(a)*

Structure of one myosin filament

two heads of myosin molecules          tail of myosin molecule

} one myosin filament

*(b)*

Structure of one myosin molecule

two myosin heads (globular protein)

tail (fibrous protein)

*(c)*

Relationship of actin and myosin

myosin heads attached to binding sites on actin molecules

myosin filament          actin filaments

*(d)*

## 8.6  Muscle contraction

Muscle contraction can be explained by the **sliding filament hypothesis**. Muscles contract as the actin and myosin filaments slide between each other. This shortens each sarcomere and therefore shortens the muscle fibre. This complex process can be summarised in three stages as follows:

### Initiation of contraction:

● The muscle is *stimulated* to contract by the arrival of a nerve impulse (Figure 8.9). Note: for details of the junction between nerve and muscle (neuromuscular junction) see Box 6.7, page 210.

● An action potential spreads along the sarcolemma and down the T tubules.

● This causes the membranes of the sarcoplasmic reticulum to have an increased permeability to *calcium ions* which *diffuse out* rapidly into the myofibrils.

● The *calcium ions* cause a change in position of the tropomyosin, *unblocking the binding sites* on the actin filaments (Figure 8.10).

*Figure 8.10 Muscle contraction – the roles of tropomyosin, troponin and calcium ions*

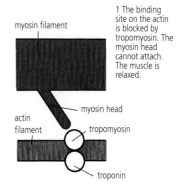

1 The binding site on the actin is blocked by tropomyosin. The myosin head cannot attach. The muscle is relaxed.

myosin filament

myosin head

actin filament

tropomyosin

troponin

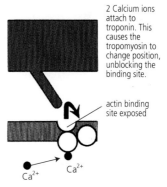

2 Calcium ions attach to troponin. This causes the tropomyosin to change position, unblocking the binding site.

actin binding site exposed

$Ca^{2+}$    $Ca^{2+}$

3 The myosin head can now attach to the actin binding site. This forms an actomyosin cross bridge and allows contraction to occur.

actomyosin cross-bridge

$Ca^{2+}$

*Figure 8.9  Initiation of muscle contraction*

1 A nerve impulse arrives in a motor neurone.

2 The nerve impulse is transmitted to the muscle at a neuromuscular junction.

3 This sets up an action potential in the sarcolemma (muscle membrane).

4 The action potential spreads down the membranes of the T tubules.

5 The action potential causes the membranes of the nearby sarcoplasmic reticulum to become much more permeable and calcium ions diffuse rapidly out.

6 Calcium ions quickly reach the actin filaments in the myofibrils, unblocking the binding sites and causing contraction.

7 When nervous stimulation stops, the calcium ions are pumped back into the reticulum and contraction ceases.

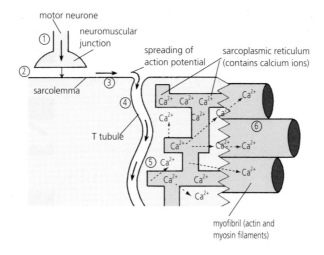

motor neurone

neuromuscular junction

spreading of action potential

sarcoplasmic reticulum (contains calcium ions)

sarcolemma

T tubule

myofibril (actin and myosin filaments)

*Figure 8.11 The sliding filament hypothesis*

**Contraction:**

- The myosin heads are now able to attach to the binding sites forming **actomyosin cross-bridges** between the actin and myosin filaments (Figure 8.11).
- The myosin heads now *change in angle, pulling the actin* over the myosin towards the centre of the sarcomere.
- *ATP is needed to break the links between actin and myosin* and allow the cycle to be repeated.

1 Muscle – binding sites on actin blocked by tropomyosin.

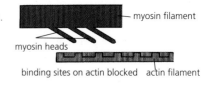

2 Nervous stimulation releases Ca$^{2+}$ ions which unblock the binding sites. Myosin heads can now attach to actin binding sites to form cross-bridges.

3 Myosin heads change angle, sliding the actin filament past the myosin towards the centre of the sarcomere.

4 ATP is needed to detach the myosin heads from the actin.

5 Energy from ATP is used to reposition the heads so that they attach further along the actin.

6 Myosin heads change angle, sliding the actin again.

- A molecule of ATP attaches to each myosin head.
- Hydrolysis of the ATP *releases energy* which is used to *detach* the myosin head from the actin *and to reposition* the head further along the actin.
- The myosin heads change in angle pulling the actin again and the cycle is repeated.
- To fully shorten (contract) the sarcomere, this cycle must be *repeated* rapidly *many times* in a **'ratchet'** mechanism. In effect the myosin heads 'walk' along the actin filaments until they reach the end.

**Relaxation:**

- When nervous stimulation of the muscle stops, the cycle of contraction also *stops*.
- The calcium ions are *pumped* by active transport out of the myofibrils and *back into* the sarcoplasmic reticulum where they are stored. This also requires energy from the hydrolysis of ATP.
- With no calcium ions present, the tropomyosin changes position and *blocks* the binding sites on the actin.
- The myosin heads are *unable* to bind with the actin.
- The muscle is now *relaxed* and can be extended by the contraction of the *antagonistic muscle.*

## Remember

The functions of key components in the sliding filament hypothesis:

- **Tropomyosin** is a blocking protein. When the muscle is relaxed it blocks the binding sites on the actin filaments. For the muscle to contract the tropomyosin must be moved, unblocking the binding sites and allowing the attachment of the myosin heads to form actomyosin cross-bridges.

- **Calcium ions:** unblock the binding sites on the actin filaments by moving the tropomyosin. This initiates muscle contraction by allowing actomyosin cross-bridges to form.

- **ATP** when hydrolysed provides the energy to detach the myosin heads from the actin filaments and to reposition the head further along the actin. ATP also provides the energy to pump the calcium ions back into the sarcoplasmic reticulum to allow relaxation of the muscle at the end of contraction.

4 After contraction, why is it necessary to break the cross-links between the actin and myosin filaments? *(2 marks)*

## The strength of contraction

The cycle of the myosin heads binding, tilting, detaching and repositioning is repeated between 50 and 100 times per second. Each time it happens the actin filaments are pulled about 10 nm towards the centre of the sarcomere. This is what generates the force that the whole muscle exerts.

Each myofibril contains thousands of myosin heads forming thousands of cross-bridges. Although the tilting of just one myosin head can only produce a tiny force, the action of all the myosin heads in a complete muscle can add up to a very large force. The greater the number of actomyosin cross-bridges that can be formed, the greater the force produced by each sarcomere and by the whole muscle (Figure 8.12). As the sarcomere contracts there is more overlap between the actin and myosin, more cross-bridges can be formed and more force can be generated. However, when the sarcomere is fully contracted the actin filaments meet and they cannot slide any further. At this point the force generated drops to zero.

*Figure 8.12* *The relationship between the force generated by the muscle and the number of actomyosin cross-bridges*

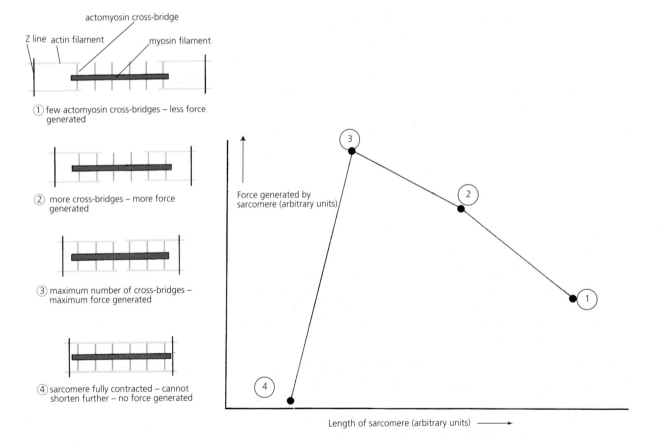

**5** Suggest whether each of the following situations would cause stronger or weaker muscle contraction than usual.

(a) a shortage of calcium ions in the body

(b) increased permeability of the sarcoplasmic reticulum to calcium ions.

*(2 marks)*

## 8.7 Motor units

Muscle fibres are divided up functionally into groups which work together called **motor units** (Figure 8.13). One motor unit consists of one motor neurone and all the muscle fibres it supplies. Note how the axon of the motor neurone divides up to supply a number of muscle fibres. The number of muscle fibres in a single motor unit varies from 100–200 in a large muscle to as few as 1 or 2 in the human eye muscles. Large muscles such as the biceps are supplied by many motor neurones and therefore consist of many motor units.

The tension in a particular muscle depends upon how many motor units are contracted at one time. In muscles involved in maintaining posture, only a few motor units contract at any one time, producing a continuous low tension, e.g. in the leg muscles when standing still. When greater force is required, e.g. for running, most of the motor units in the appropriate leg muscles will contract at the same time.

From this it can be seen that *whole* muscles contract in a 'graded' manner, producing a range of tensions. In contrast, both individual muscle fibres and single motor units contract in an '**all or nothing**' manner – they either contract fully or not at all. This is similar to the response of nerve cells (see Section 6.9, page 203).

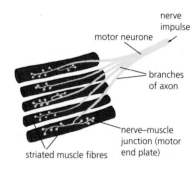

*Figure 8.13 A single motor unit – motor nerve plus the muscle fibres it innervates (stimulates)*

## 8.8 Muscle spindles

When a skeletal muscle is stretched it responds by contracting. This is controlled by many small sense organs found in skeletal muscle called **muscle spindles**. These are stretch receptors (sometimes called proprioceptors). Muscle spindles continually monitor the degree of tension in muscles. The structure of the spindle is shown in Figure 8.14.

When the muscle spindle is stretched nerve impulses are sent via sensory neurones to the central nervous system. This triggers impulses to return via motor neurones to the muscle, causing it to contract. This is an example of a **reflex action** – the stretch reflex. One well-known example of a stretch reflex is the knee-jerk reflex used as a test by doctors (see Box 7.6, page 249).

The more the spindle is stretched, the greater the muscle contraction caused by the reflex as a response. This is important as the skeletal muscles automatically adjust to the load placed on them. Adjusting our balance on a slightly tilting train requires a small degree of contraction, whereas lifting a heavy bag requires much greater contraction by the muscles.

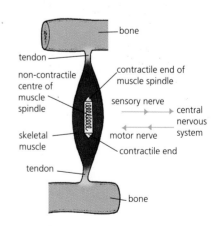

*Figure 8.14 A muscle spindle*

### Box 8.3  Muscles and exercise

**Fast twitch and slow twitch muscle fibres**

Striated muscle fibres are of two main types. Most whole muscles have a mixture of both types of fibre but the proportion varies. The characteristics of these two types are given in Table 8.2.

| Fast twitch | Slow twitch |
|---|---|
| also known as 'white' muscle because little or no myoglobin present (myoglobin acts as an oxygen store – see page 284 for details) | also known as 'red' muscle because large quantities of myoglobin are present, giving a reddish brown colour |
| rapid, short-term contraction | slower, sustained contraction |
| energy provided mainly by anaerobic respiration | energy provided mainly by aerobic respiration |
| few mitochondria present | many mitochondria present |
| large amounts of creatine phosphate present (acts as a store of phosphate and energy – see page 284 for details) | little creatine phosphate present |
| less extensive blood supply | more extensive blood supply |
| fatigues quickly | fatigues more slowly |

*Table 8.2  Features of fast and slow twitch muscle*

**Athletes and muscle type**

In humans the relative proportions of fast and slow twitch fibres present is largely inherited but can also be influenced by training. Top-class sprinters are likely to have been born with a naturally high proportion of fast twitch muscle fibres, whereas marathon runners will have inherited a higher proportion of slow twitch fibres. For short periods sprinters can produce much greater power and speed but they fatigue quickly, lacking the endurance of the marathon runners to keep running at slower speeds but for far longer.

In one study sprinters were found to have 62% fast twitch and 38% slow twitch fibres compared to marathon runners with 18% fast twitch and 82% slow twitch fibres (Figure 8.15).

**Muscles and training**

Although the relative proportions of fast and slow twitch fibres can probably not be changed much by training, other aspects of physiology can be. Regular training programs where the muscle is made to do work against a load, e.g. weight training, stimulates muscle growth producing a larger cross-sectional area of muscle. Individual muscle cells (muscle fibres) increase in diameter, partly due to an increase in the number of myofibrils in each cell. The amounts of mitochondria, creatine phosphate and ATP present in each cell increases and the cells also become more tolerant to the build up of **lactate (lactic acid)**.

(a)

This type of training is useful to those involved in power events such as throwers, e.g. javelin or shotput, but also to sprinters who need fast, powerful contraction for a short time. Training for endurance events, e.g. marathon running, involves steadier work over longer periods of time. Muscle bulk does not tend to increase but other changes take place. More glycogen can be stored in the muscles, the heart increases in size allowing blood to be delivered to the muscles more quickly and efficiently, and the overall potential for aerobic respiration increases.

### Supplying the energy for muscle contraction (Figure 8.16)

* **ATP:** the immediate source of energy for muscle contraction is ATP, but a muscle usually contains only enough ATP for around 3 seconds of activity. More ATP must be made while the muscle is active. With gentle activity this ATP can be provided by the aerobic respiration of glucose.

(b)

*Figure 8.15* (a) Sprinters, (b) marathon runner

*Figure 8.16* Some of the pathways involved in supplying the energy for muscle contraction

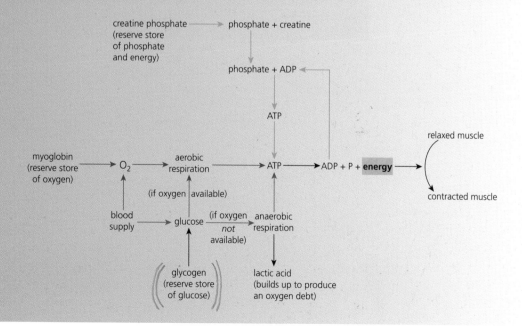

## Box 8.3  Muscles and exercise continued

- **Creatine phosphate:** this substance, found in muscle, acts as a reserve store of phosphate and energy. The ADP produced when ATP is hydrolysed to release energy is reconverted to ATP. This uses phosphate from the creatine phosphate (CP). CP also releases energy as it is broken down which is used in the regeneration of the ATP. When the muscle is relaxed, phosphate from ATP is used to regenerate the CP to be stored again.

- **Glycogen:** the muscle contains a store of glycogen which is broken down to glucose. The glucose is respired aerobically. With more strenuous and long-lasting exercise the supplies of glycogen may soon get used up.

- **Myoglobin:** is a protein found in muscles. It has a similar structure to haemoglobin. It can attach to oxygen and so store it in the muscles as a 'reserve supply'. Oxygen can be released for aerobic respiration when oxygen supplies from the blood get very low (see also *AS Biology*, Box 11.3, page 320).

- **Anaerobic respiration:** muscle cells change to anaerobic respiration if their activity continues at a level where oxygen cannot be supplied at a sufficient rate to maintain aerobic respiration. Anaerobic respiration is covered in detail in Section 1.11, page 19. Anaerobic respiration causes a build up of lactate. This build up causes muscle fatigue and cramp, and means that anaerobic respiration cannot continue for long. To remove the lactate and restore normal conditions requires large amounts of oxygen – **the oxygen debt**. This can only be repaid when muscle activity reduces and aerobic respiration is restored.

## Summary – ⑧ Muscle: structure and function

- Mammals have three types of muscle: smooth, cardiac and skeletal (striated).

- The main feature of muscle cells is the ability to contract and shorten.

- Skeletal muscles are attached to bones. When they contract they cause a bone to move, e.g. as in walking.

- Most skeletal muscles are found in antagonistic pairs. One of the pair pulls a bone one way and the other muscle pulls the bone back to its original position.

- Skeletal muscle consists of cells called muscle fibres with characteristic features including large numbers of mitochondria, extensive sarcoplasmic reticulum for storage of calcium ions and infoldings of the cell surface membrane called T tubules.

- The main component of muscle fibres is large numbers of protein filaments called myofibrils. These are of two types: thinner actin and thicker myosin.

- The myofibrils run the whole length of the long muscle cells and are divided into repeating units called sarcomeres.

- Each sarcomere has distinctive regions which give it a characteristic striped (banded) appearance. The regions of the sarcomere include A bands, I bands and H zones.

- When the muscle fibre contracts the I bands and H zones become shorter but the A band stays the same length.

- Stimulation of the muscle causes the myosin filaments to form actomyosin cross-bridges with the actin filaments and to pull the actin filaments towards the centre of the sarcomere.

- The actin and myosin filaments slide between each other, shortening the muscle fibre. This is known as the sliding filament hypothesis.

- Both ATP and calcium ions are essential for this process. ATP provides the energy to break the cross-bridges. Calcium ions unblock the binding sites on the actin, initiating contraction.

- Each whole muscle, e.g. the biceps, is divided functionally into many motor units consisting of a number of muscle fibres (muscle cells) and the motor nerve that supplies them.

- Whole muscles can contract in a graded manner. If only a few motor units are stimulated to contract a small force is produced but contraction of most of the motor units at the same time will produce a much stronger force.

- Muscle spindles are small sense organs found within muscles. They are stretch receptors and they constantly monitor the degree of tension in the muscle.

- Supplying the energy for muscle contraction involves a variety of pathways. ATP can be produced by both aerobic and anaerobic respiration. Creatine phosphate acts as a reserve store of phosphate and energy. Myoglobin is able to store oxygen to help maintain aerobic respiration.

## ? Answers

1 With antagonistic muscle action each bone is usually associated with at least two muscles; one to move it one way and another to return it to its original position. Therefore there will be more muscles than bones *(2)*.

2 Visible – I band and Z line. Not visible – A band and H zone *(2)*.

3 The muscle is relaxing, i.e. getting longer. H zones, I bands and the sarcomeres get longer; A bands stay the same length. None of them get shorter *(4)*.

4 Either: to allow repositioning of the myosin heads further along the actin filaments; so that further sliding of the actin can occur. Or: to allow the muscle to relax and to be extended; so that it is ready for contraction the next time the muscle is stimulated *(2)*.

5 (a) A weaker contraction because fewer calcium ions therefore fewer actin binding sites unblocked and fewer cross-bridges formed *(1)*.

   (b) A stronger contraction because calcium ions released more easily, more actin binding sites unblocked and more cross-bridges formed; the greater the number of actomyosin cross bridges that can be formed the greater the force that can be exerted by the muscle *(2)*.

## End of Chapter Questions

1 (a) The drawing shows some of the muscles used to open and close the mouth on a human skull.

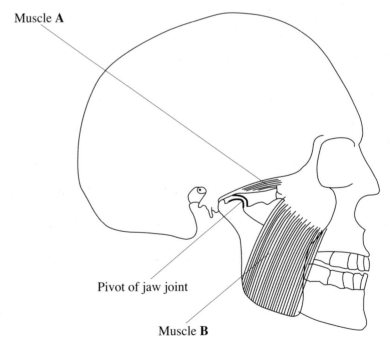

Muscle **A**

Pivot of jaw joint

Muscle **B**

   (i) Use your knowledge of antagonistic muscle action to explain how muscles **A** and **B** may be used to open and close the mouth. *(3 marks)*

   (ii) Give an advantage of the difference in length of the two muscles, **A** and **B**. *(1 mark)*

**(b)** The diagram shows the appearance of a sarcomere from a relaxed muscle fibril, as seen with a light microscope.

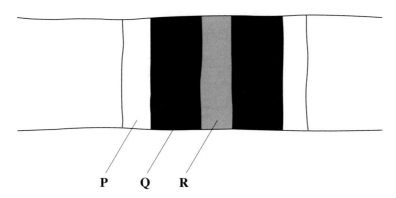

P      Q      R

(i) Use your knowledge of the sliding filament hypothesis to explain the appearance of each of the bands **P**, **Q** and **R**. *(3 marks)*

(ii) Draw a similar diagram to show the appearance of the sarcomere when the fibril is contracted. *(1 mark)*

**(c)** Muscles use energy from respiration for contraction. Describe how the energy released in mitochondria during respiration produces contraction of a muscle fibril. *(4 marks)*

*AQA B 1999*                                             *(Total 12 marks)*

**2** Read through the following passage, which refers to the structure and function of striated muscle, then write on the dotted lines the most appropriate word or words to complete the account.

A striated muscle fibre contains many strands, called ........................., which run the length of each fibre. Within these strands there are thick filaments made of the protein ......................, and thin filaments which contain the protein ............................ . When a muscle contracts, the two kinds of filament slide past each other. Energy to enable the sliding filaments to move is provided by molecules of ................................... attached to the head of each thick filament.

*Edexcel 1999*                                                  *(Total 4 marks)*

**3** **(a)** Describe the role played by calcium ions in muscle contraction. *(2 marks)*

**(b)** The graph shows the force generated by a muscle plotted against the mean sarcomere length. The diagrams show the appearance of the sarcomere at points **A** and **B**.

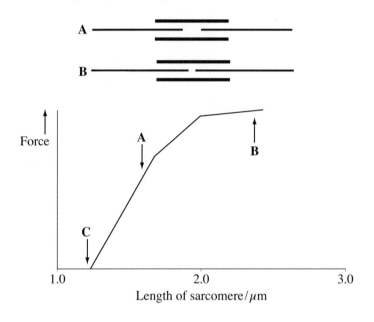

(i) What is the relationship between the force generated by the muscle and the number of actomyosin bridges? Explain your answer.
*(2 marks)*

N2.1

(ii) Draw a simple diagram to show the appearance of the sarcomere at point **C** on the graph. *(1 mark)*

*AQA B 2000*

*(Total 5 marks)*

**4** The diagram below shows part of a myofibril of a striated muscle fibre.

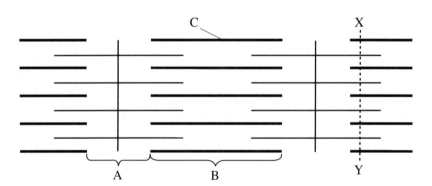

(a) Name the regions labelled A and B. *(2 marks)*

(b) Name the protein which makes up part C. *(1 mark)*

(c) Make a drawing to show the appearance of a transverse section of a myofibril as it would be seen in a section across XY. *(2 marks)*

(d) Describe the effect of training on muscle size. *(3 marks)*

*Edexcel 2000* **(Total 8 marks)**

5   The diagram represents actin and myosin in a muscle cell.

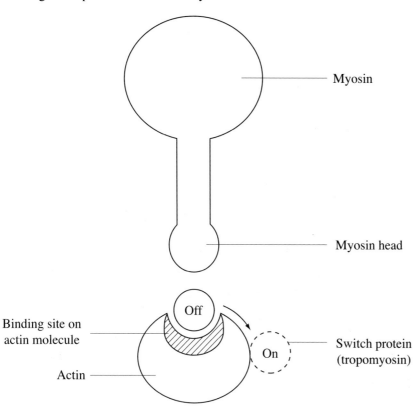

(a) With reference to the diagram:

  (i)  describe the part played by calcium ions in muscle contraction; *(2 marks)*

  (ii) explain how the muscle cell contracts. *(2 marks)*

(b) Describe how ATP is used in muscle contraction. *(1 mark)*

*AEB 1999* **(Total 5 marks)**

6  The drawing has been made from a slide of skeletal muscle tissue seen with a light microscope at a magnification of 800 times. It shows parts of two motor units.

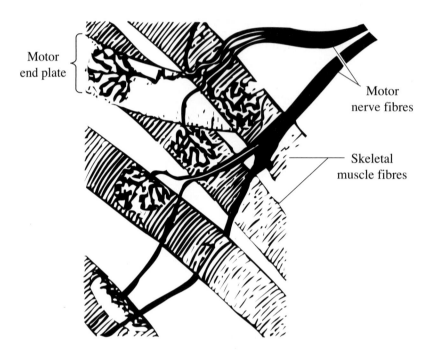

Motor end plate

Motor nerve fibres

Skeletal muscle fibres

(a) Use evidence from the drawing to suggest:

   (i)  a meaning for the term *motor unit*;                                  *(1 mark)*

   (ii) why all the muscle fibres shown will not necessarily contract at the same time.                                                        *(1 mark)*

(b) Briefly describe the sequence of events at the motor end plate which leads to an action potential passing along the muscle fibre.      *(3 marks)*

The diagram shows the pathways by which energy is produced for muscle contraction. Numbers **1** to **3** indicate the order in which the various pathways are called on to supply ATP as muscular effort increases.

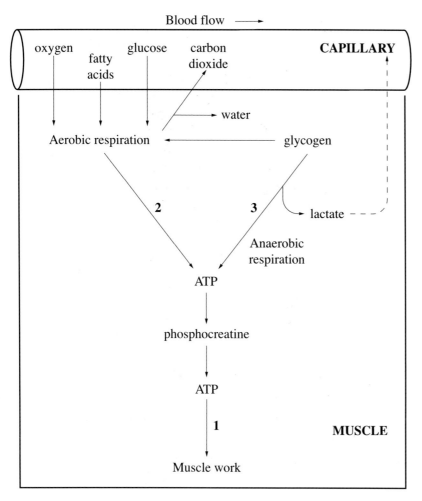

(c) (i) What happens to the lactate produced in pathway **3**? *(3 marks)*

(ii) Explain the part played by phosphocreatine in supplying energy to the muscle. *(2 marks)*

Most muscles contain both slow twitch and fast twitch fibres but the proportion of each depends on the function of the muscle as a whole. The table lists some differences between slow twitch and fast twitch muscle fibres.

| Characteristic | Slow twitch fibre | Fast twitch fibre |
|---|---|---|
| Contraction time/milliseconds | 110 | 50 |
| Mitochondria | Many present | Relatively few |
| Glycogen store | Low | High |
| Myosin ATPase activity | Low | High |
| Capillaries | Many present | Fewer present |
| Sarcoplasmic reticulum | Poorly developed | Well developed |
| Rate of fatigue | Slow | Fast |

**(d)** Suggest why muscles concerned with maintaining the posture of the body might be expected to have a large proportion of slow twitch fibres.
*(2 marks)*

**(e)** Explaining your answer in each case, give **two** pieces of evidence from the table which:

(i) suggest that fast twitch fibres may easily build up an oxygen debt;
*(4 marks)*

(ii) might account for the difference in speed of contraction of the two types of fibre.
*(4 marks)*

*AEB 1996*

*(Total 20 marks)*

# 9 Homeostasis

Environmental conditions in which organisms live are very variable. Temperatures can fluctuate enormously, especially in terrestrial environments. The chemical composition and water content of the surroundings and food supply are always changing. Living organisms usually maintain constant internal conditions, to a certain extent, in spite of these changes. This process is called **homeostasis**.

Constant internal conditions are essential for enzymes to work efficiently and for maintaining the correct water balance between cells and surrounding fluids. Mammals have many very efficient homeostatic mechanisms.

**This chapter includes:**

- the concept and importance of homeostasis
- the principles and role of negative feedback mechanisms
- control of blood glucose concentration: insulin and glucagon and their role in the regulation of blood glucose concentration
- diabetes mellitus and its control
- control of body temperature (thermoregulation): ectothermic and endothermic animals; heat gain and heat loss; the mechanisms of temperature regulation in endothermic animals
- control of digestive secretions in mammals
- a comparison between nervous and hormonal communication.

## 9.1 Homeostasis and its importance

The chemical reactions which make up the metabolism of all living organisms are controlled by enzymes. Enzymes operate most efficiently at a particular optimum temperature and at an optimum pH (see *AS Biology*, Section 4.4, page 104). A steady supply of substrate and removal of waste products is also necessary in order that metabolic reactions can continue to take place efficiently. Homeostatic mechanisms provide the constant optimum conditions essential for the chemical reactions of life. In animals, the cells are bathed in tissue fluid, the contents and temperature of which are regulated homeostatically so that the cells are always provided with their optimum conditions.

It is also necessary to maintain the correct water potential of body fluids in animals at all times (see *AS Biology*, Section 2.7, page 49). If the water potential of tissue fluid is higher (less negative) than that of the cytoplasm of cells, water enters the cells and eventually they may burst. Alternatively, if the water potential of the fluid is lower (more negative) than that of the cytoplasm, water leaves the cells, they shrink and may die. Homeostatic mechanisms maintain the correct water balance within animals. Plant cells have cell walls which resist the excessive intake of water that might cause a cell to burst. So although plants have mechanisms which prevent the excessive loss of water, they do not have the efficient homeostatic control of water balance which is found in animals.

Some of the factors in mammals which are regulated at a constant level homeostatically are:

- concentration of glucose in the blood
- concentration of individual inorganic ions in the blood (e.g. calcium ions)
- water potential of blood and body fluids
- pH of the blood and body fluids
- core body temperature
- hydrostatic pressure of the blood (blood pressure).

The ability to maintain a favourable constant internal environment when external conditions are unfavourable or fluctuating gives organisms

### Definition

**Homeostasis** is the maintenance of a constant internal environment. This includes maintaining the chemical composition and temperature of body fluids at constant levels (within narrow limits).

*(a)*

*(b)*

**Figure 9.1** *Living in hostile environments (a) Male emperor penguins (*Aptenodytes forsteri*) maintain a body temperature of 38 °C in the Antarctic winter by a mixture of huddling behaviour, physiological mechanisms and structural adaptations, (b) The kangaroo rat (*Dipodomys sp.*) maintains a water balance in its body although there is no available liquid water in the desert. It obtains water from its food and, internally, from the process of respiration. Water loss is kept to a minimum by behavioural and physiological mechanisms together with structural adaptations*

**Figure 9.2** *A generalised negative feedback system*

independence from the external conditions. This allows them to lead active lives at different times of day and in different seasons. It allows them to inhabit parts of the Earth where the climate provides conditions which are very different from those which are essential internally. For example, penguins maintain an internal body core temperature of 38 °C in Antarctic conditions where temperatures may fall as low as –60 °C. Kangaroo rats maintain the essential water content and water potential of body tissues in desert conditions where no drinking water is available (Figure 9.1).

### 9.2 The principles and role of negative feedback

Homeostasis is brought about by **negative feedback** systems. As seen in Figure 9.2, a negative feedback system usually involves the following stages:

- Conditions cause the level of a particular factor (e.g. temperature) to change from the **norm**.
- This change is detected by **receptors**.
- The receptors communicate information by means of **hormones** or the **nervous system** to **effectors**. These bring about a **corrective mechanism** which restores the factor towards its normal level. Effectors are often referred to as **target organs** when communication is hormonal.
- The return to the norm is detected by the receptors and the corrective mechanism is switched off.

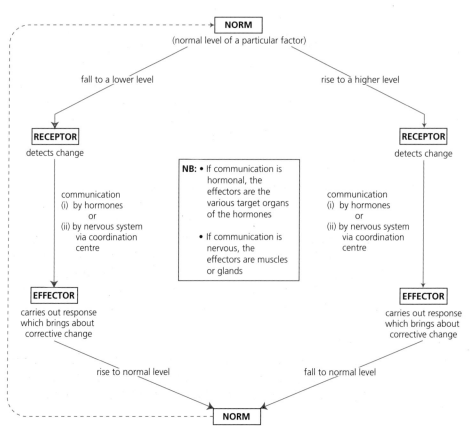

- The corrective mechanism, however, may cause the level of the factor to overshoot the norm, triggering a mechanism to restore the norm in the opposite direction. For this reason the level of the factor is not absolutely constant but oscillates about the norm. This is called **dynamic equilibrium**.

Two further important points about negative feedback systems in animals should be noted.

- When nervous communication is involved, it is usual for the information to be transmitted from the receptors to a **coordination centre** in the brain and then relayed to the effectors.

- It is usual to have **dual control** of one factor. Different mechanisms restore the norm from deviations in different directions from the normal level. For example, if the temperature rises above normal in a mammal, **sweating** is triggered which helps to cool the body. However, if the temperature falls below normal, **shivering** is triggered which helps to warm the body – there is not *just* reliance on ceasing to sweat. This operation of different mechanisms gives more precise control and restores the norm quickly, keeping the fluctuation from the normal conditions to a minimum.

### 9.3  The control of glucose concentration in the blood

It is essential that the blood provides the cells with a constant supply of glucose for respiration. Brain cells are especially sensitive to lowered glucose levels as glucose is the only respiratory substrate which they can use.

The normal blood glucose concentration in humans is about 90 mg of glucose in $100 \text{ cm}^3$ of blood. If the level falls much below this concentration, then the cells are not supplied with sufficient glucose for respiration. However, if the level rises above this level then the high concentration of glucose, as it is a solute, lowers the water potential of the blood and affects the water balance between cells and tissue fluid. The level of glucose is regulated around this optimum homeostatically by negative feedback. Communication is hormonal. The main hormones involved are **insulin** and **glucagon**.

#### The production of insulin and glucagon

Insulin and glucagon are produced in a gland called the pancreas. Most of the cells of the pancreas secrete the enzymes and other components of pancreatic juice and the secretions are poured directly down the pancreatic duct into the duodenum. In this respect, the pancreas acts as an **exocrine** gland. However, small clusters of cells, called **Islets of Langerhans**, secrete hormones which pass directly into the blood. In this way the pancreas acts as an **endocrine** gland. There are two types of cell within the Islets of Langerhans, alpha cells (α-cells) which secrete glucagon and beta cells (β-cells) which secrete insulin (Figure 9.3).

#### The function of insulin

Insulin has several effects, all of which help to *reduce* the level of glucose in the blood. When insulin binds to a receptor molecule in the surface membrane of a cell, it triggers responses in the cell. Examples of such responses are given below.

---

**Definition**

A **negative feedback system** is one in which a change from the normal level or composition (**norm**) is detected and triggers a response which opposes the change and restores the norm. The return to the norm is then detected and this corrective response is switched off.

**Definitions**

- A **hormone** is a chemical substance which is secreted into the blood by special cells or glands (endocrine glands). It is transported to other parts of the body where it has an effect on particular target cells or organs.

- A **gland** is an organ which synthesises and secretes a chemical substance.

- An **exocrine gland** is a gland whose secretions are transported from the organ along a duct. Most of the digestive glands, such as the salivary glands and gastric glands, are exocrine.

- An **endocrine (ductless) gland** produces hormones which are released directly into the blood.

**Figure 9.3** *(a) Photomicrograph of pancreatic tissue showing cells which secrete pancreatic juice surrounding an Islet of Langerhans, (b) An Islet of Langerhans and surrounding cells*

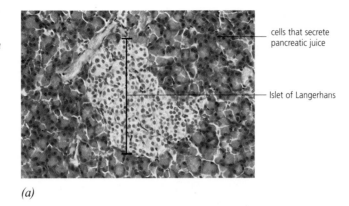

cells that secrete pancreatic juice

Islet of Langerhans

*(a)*

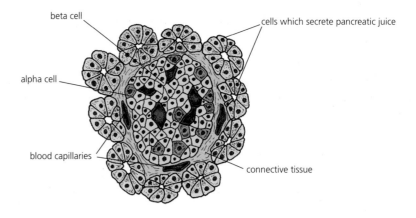

beta cell

alpha cell

blood capillaries

cells which secrete pancreatic juice

connective tissue

*(b)*

- There are protein molecules in cell membranes which transport glucose from the blood into the cells by facilitated diffusion (see *AS Biology*, Section 2.5, page 47). Insulin causes more of these protein molecules to be sent into the membrane from vesicles in the cytoplasm of the cells (Figure 9.4.) This results in cells taking in glucose more rapidly from the blood.

- In some cells, especially liver and muscle cells, there are enzymes which speed up the conversion of glucose into glycogen. This process is known as **glycogenesis** (= *glycogen forming*). The insoluble glycogen is then stored as granules in the cytoplasm – a human liver can store up to 75 g of glycogen. Insulin activates these enzymes. This also results in the cells taking up more glucose from the blood, as the conversion to glycogen lowers the concentration of glucose within the cell which increases the diffusion gradient of glucose.

- Insulin also causes other intracellular enzymes to become active. These speed up the synthesis of proteins and lipids. Although glucose is the main substrate for respiration, some cells also respire the components of proteins and lipids. If fewer of these are available, then cells use more glucose and take up more from the blood.

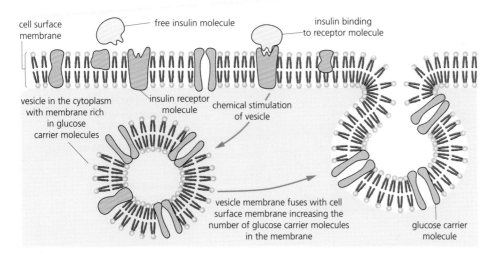

**Figure 9.4** *Insulin stimulates cells to take in more glucose from the blood by increasing the number of glucose carrier molecules in the cell surface membrane*

## The function of glucagon

The effects of glucagon *raise* of the level glucose in the blood. The main effects are as follows:

● Glucagon activates enzymes in the liver cells which speed up the conversion of the insoluble stored glycogen molecules into glucose. This process is known as **glycogenolysis** (= *glycogen splitting*). The glucose molecules then diffuse out of the cells into the blood because of the high concentration now present in the liver cells.

● Glucagon increases the production of glucose from other nutrients such as amino acids or fatty acids in liver cells. This process is known as **gluconeogenesis** (= *new glucose formation*). Glucose produced in this way also enters the blood.

Figure 9.5 summarises the ways in which glucose enters and leaves the blood.

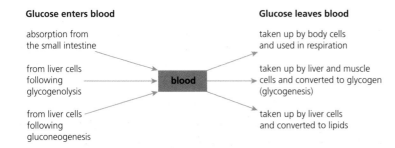

**Figure 9.5** *Entry and exit of glucose into and out of the blood*

## The mechanism of control of blood glucose concentration

Figure 9.6 illustrates the negative feedback system which is involved in the control of blood glucose concentration. Following a meal, glucose produced by the digestion of carbohydrates is absorbed into the blood capillaries in the intestine wall, raising the level of glucose in the blood. The high concentration of glucose stimulates the β-cells of the Islets of Langerhans in the pancreas to secrete insulin. Insulin circulates in the blood and binds to the receptor sites on cell surface membranes, causing the effects listed above. As a result the blood glucose level falls back towards the norm and so the secretion of insulin also falls, in turn stopping these corrective processes.

*Figure 9.6 The negative feedback systems involved in the regulation of the glucose concentration in the blood (compare with Figure 9.2)*

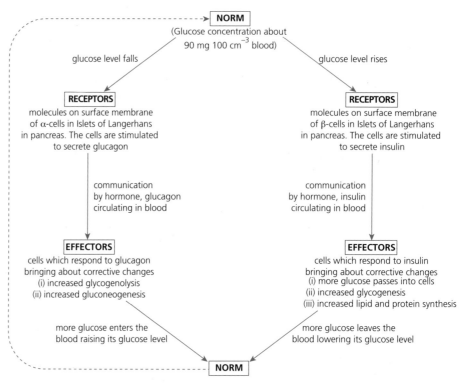

**NORM**
(Glucose concentration about
90 mg 100 cm$^{-3}$ blood)

glucose level falls                                    glucose level rises

**RECEPTORS**
molecules on surface membrane
of α-cells in Islets of Langerhans
in pancreas. The cells are stimulated
to secrete glucagon

**RECEPTORS**
molecules on surface membrane
of β-cells in Islets of Langerhans
in pancreas. The cells are stimulated
to secrete insulin

communication
by hormone, glucagon
circulating in blood

communication
by hormone, insulin
circulating in blood

**EFFECTORS**
cells which respond to glucagon
bringing about corrective changes
(i) increased glycogenolysis
(ii) increased gluconeogenesis

**EFFECTORS**
cells which respond to insulin
bringing about corrective changes
(i) more glucose passes into cells
(ii) increased glycogenesis
(iii) increased lipid and protein synthesis

more glucose enters the
blood raising its glucose level

more glucose leaves the
blood lowering its glucose level

**NORM**

**NB:** There are two separate feedback mechanisms operating in glucose regulation.
For example, if glucose level falls, glucagon stimulates responses which raise the
glucose level. There is not just reliance upon the fact that less insulin is secreted.

The body cells are constantly taking up glucose from the blood and respiring it. If a long period has elapsed since the last meal and no more glucose is absorbed from the intestines, then the blood glucose level begins to fall below the norm. The lowering of the level of glucose stimulates the α-cells of the Islets of Langerhans to secrete glucagon. This passes into the blood and binds to the receptor sites on the surface membranes of liver cells, causing the cells to carry out the functions listed above. This raises the glucose level back towards the norm. So control is by negative feedback systems with separate mechanisms for controlling the return to the norm from different directions.

1 (a) Explain why it is true to say that blood glucose is controlled by two separate negative feedback systems (dual control). *(2 marks)*

(b) Why is this important? *(1 mark)*

2 Blood flowing from the pancreas passes into the hepatic portal vein which carries blood directly to the liver. Why is this important in the control of the level of blood glucose? *(2 marks)*

### The role of other hormones in glucose metabolism

The level of glucose in the blood is largely maintained by the homeostatic mechanisms described above. However, other hormones can sometimes be involved in glucose metabolism and can override these homeostatic mechanisms at times of stress. For example, a group of hormones including **cortisol** can be released from the **adrenal cortex** glands. Their function is to ensure that enough energy is available to carry out normal metabolism in times of physiological stress. For example, if the body's reserves of glycogen become very low (normally the liver has enough glycogen to supply the body with glucose for about 12 hours), these hormones increase the formation of glucose from amino acids in the liver (gluconeogenesis).

The **adrenal medulla** glands also release hormones at times of emergency. These hormones include **adrenaline**. They are not as essential as cortisol but prepare the body for rapid physical action such as running away from danger. They have many effects including promoting the breakdown of glycogen to glucose in liver cells (glycogenolysis) which results in more glucose being released into the blood. This extra glucose is respired rapidly by cells such as muscle and brain cells and releases the extra energy needed for emergency activity.

## 9.4   Diabetes mellitus and its control

Over one million people in the UK are unable to control their blood glucose concentration effectively. They are in danger of suffering the effects of a high blood glucose level (**hyperglycaemia**) or a low blood glucose level (**hypoglycaemia**). This condition is known as **diabetes mellitus**.

There are two types of diabetes mellitus, Type I and Type II.

### Type I diabetes mellitus (juvenile-onset diabetes or insulin-dependent diabetes)

About 10% of all cases of diabetes mellitus are of this type. It usually arises rapidly and before the age of 20. The β-cells of the Islets of Langerhans are destroyed, possibly by an auto-immune reaction. (This occurs when the body's immune response is activated against its own tissues, see *AS Biology*, Section 11.9, page 324). They are no longer able to secrete insulin, therefore the blood glucose rises in an uncontrolled way after a meal until the levels become so high that total reabsorption of glucose cannot take place in the nephrons of the kidney (see Section 10.4, page 330) and the excess glucose is excreted in the urine (**glycosuria**). A test for the presence of glucose in the urine can be used in the diagnosis of diabetes mellitus of either type (see *AS Biology*, Section 4.5, page 110). The lack of insulin means that no glucose will have been stored as glycogen in the liver.

Later, if no further meal is taken, the glucose levels fall as glucose is used up in respiration. The release of glucagon may be stimulated from α-cells but there are no glycogen reserves in the liver from which glucose can be released. Glucose can be produced by gluconeogenesis, and fats and proteins can be respired. A marked loss of weight can occur within a few weeks. The presence of glucose in the urine prevents the kidney from reabsorbing as much water as usual (see Section 10.4, question 6, page 334) and the sufferer usually passes excessive amounts of urine, leading to dehydration and thirst. There are many other symptoms. For example, the excessive breakdown of fats can lead to a condition known as **ketosis** which lowers the pH of the blood and can result in death.

## Remember

**Diabetes mellitus**, the inability to control blood glucose levels, must not be confused with the condition **diabetes insipidus**. In this condition the body cannot produce the hormone ADH (see Box 10.2, page 336). The similarity in the name arises because diabetes (= *overflow*) refers to the flow of urine. In diabetes mellitus (= *honey*) there is glucose in the urine. In diabetes insipidus (= *tasteless*) there is a copious flow of very watery urine.

Since 1921 this condition has been treated by injections of insulin. Originally, the insulin was extracted from the pancreases of pigs. Insulin is a protein molecule and human insulin differs slightly in its amino acid sequence from pig insulin. Pig insulin was, nevertheless, quite effective. However, the development of genetic engineering techniques has led to the gene for human insulin being introduced into microorganisms (see *AS Biology*, Figure 16.1, page 482) and human insulin (humulin) is now produced on a large scale by genetically modified bacteria or yeast. There are several advantages to producing insulin this way:

● It is easier to produce insulin on a large scale.
● Human insulin does not produce an immune reaction, which was sometimes experienced by diabetics after being injected with pigs' insulin.
● It avoids the risk of transferring microorganisms from the pig which could be dangerous to humans.

A person with Type I diabetes must also carefully manage his intake of carbohydrates in the diet. The overall aim should be to absorb glucose steadily and slowly to balance the use by respiration. It is therefore preferable to take carbohydrates in the form of starch as in potatoes, pasta or bread. The time taken to digest the starch into glucose means that the glucose is absorbed gradually. Although the intake of sugars can be useful in emergencies, in general, this should be avoided as their rapid absorption could lead to a rapid rise in blood glucose level. This may be difficult to balance with the correct timing of insulin injections.

Most people with Type I diabetes lead a normal and very active life and control their condition by a combination of insulin injections and taking care with their diet (Figure 9.7).

(a)

(b)

*Figure 9.7 (a) Young people can be taught to inject themselves with insulin, (b) It is possible for people with diabetes to lead a very active life – Sir Steve Redgrave, who has won five Olympic gold medals for rowing, has diabetes mellitus*

**?**

**3** (a) Why is it necessary to inject insulin (a protein) rather than give it by mouth?                                                            *(1 mark)*

(b) Why should insulin be injected into muscle or fat rather than directly into a vein?                                                                *(2 marks)*

(c) Suggest what might happen if far too much insulin were injected into the body?                                                                    *(3 marks)*

### Type II diabetes mellitus (mature-onset diabetes or non insulin-dependent diabetes)

Ninety per cent of cases of diabetes mellitus are of this type. It mostly arises in people who are over 40 and who are overweight. The symptoms are the same as those of type I diabetes but they are much milder. The sufferer may secrete normal amounts of insulin but the insulin receptors in the membranes of body cells are either fewer in number or less sensitive than normal. Therefore the insulin is less effective in reducing the blood glucose level. This condition can usually be controlled by carefully regulating carbohydrate intake in the ways suggested above, and by exercise. Usually, insulin injections are not required, although some elderly people with cells which have a low sensitivity to insulin do boost their insulin levels with injections. Sometimes drugs which stimulate insulin secretion are used.

## 9.5  The control of body temperature (thermoregulation)

The rate of all metabolic reactions is controlled by enzymes, and enzymes only operate efficiently at temperatures close to their optimum. Therefore, it follows that temperatures within the bodies of organisms should be kept as close to that optimum as possible. The optimum temperature for the human body core (all areas more than 2.5 cm below the body surface) is about 36.9 °C.

Many animals cannot regulate their body temperatures at all, but those that do use one of the following methods of temperature control: **ectothermy** (= *outside heat*) or **endothermy** (= *inside heat*).

### Ectothermic animals

Ectothermic animals control their body temperatures by their behaviour. As the name suggests, most of their body heat is derived from their immediate environment and their body temperature follows that of their surroundings. Air temperatures can show marked variations daily and seasonally. The behaviour of ectotherms which live on land is therefore restricted by the need to maintain a suitable body temperature. Examples of such behaviour patterns can be seen in desert lizards which are active by day (diurnal). Figure 9.8 shows how the behaviour of a North American horned lizard is governed by the temperature of its surroundings on summer days.

### Endothermic animals

All mammals and birds are endotherms. As their name suggests, most of their body heat is derived internally from respiration. They maintain a constant core body temperature by controlling the amount of heat produced and lost by physiological and behavioural means. They maintain a body temperature independent of their immediate environment and do not have to remain in areas with suitable temperatures in the way that ectotherms do (Figure 9.9).

### Definitions

- An **ectothermic** animal derives most of its body heat from its environment and regulates its body temperature by its behaviour.

- An **endothermic** animal generates most of its body heat internally and controls its body temperature by regulating the production and loss of this heat energy. It does this by physiological and behavioural means. Only birds and mammals are endothermic.

**Figure 9.8** *(a) A North American horned lizard* (Phrynosoma *sp.), (b) The regular pattern of life during the summer*

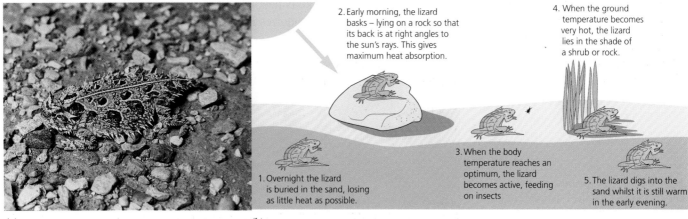

2. Early morning, the lizard basks – lying on a rock so that its back is at right angles to the sun's rays. This gives maximum heat absorption.

4. When the ground temperature becomes very hot, the lizard lies in the shade of a shrub or rock.

1. Overnight the lizard is buried in the sand, losing as little heat as possible.

3. When the body temperature reaches an optimum, the lizard becomes active, feeding on insects

5. The lizard digs into the sand whilst it is still warm in the early evening.

*(a)*                              *(b)*

> **Did You Know**
>
> Temperature intolerance is remarkably similar throughout the living world. Almost all eukaryotic organisms (animals, plants, fungi and simple organisms like algae) are not able to tolerate prolonged body temperatures above 40 °C. It has been suggested that enzymes of some vital metabolic pathway common to all of these organisms are denatured at this temperature.
>
> Only a few prokaryotic organisms can tolerate higher temperatures. Some of these inhabit hot springs and deep vents in the ocean floor where temperatures can reach 350 °C. It is possible that these organisms may have arisen before the evolution of the heat-sensitive metabolic pathways which are found in other organisms.

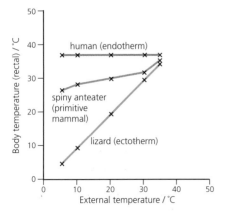

*Figure 9.9 The relationship between body temperature (as measured in the rectum) and external temperature in three different animals (from CJ Martin, 1930)*

## Advantages and disadvantages of endothermy

Endotherms can maintain a body temperature which is markedly different from that of their environment. This allows them to remain active at all times and in all seasons and to live and exploit areas of the Earth in which ectotherms could not survive. However, as most of their body heat is derived from respiration, they have a higher metabolic rate than ectotherms and use much of their food intake in respiration. They usually live in areas with temperatures lower than their body temperature and are constantly losing heat to the environment. Producing enough heat energy to balance this can use up to 80% of their food intake. This means that an endotherm must eat much more then an ectotherm of similar size and so spend much more time searching for food.

The fact that ectotherms require less food than endotherms makes them more successful at surviving in desert habitats, where food is scarce.

**4** Figure 9.10 shows an ectothermic desert iguana and an endothermic brown rat. Both animals are about 45 cm long (including tail). If 80% of the endotherm's food intake is needed just to supply body heat, how does the amount of food the rat requires in a day compare with that needed by an iguana?
*(1 mark)*

*Figure 9.10 (a) The desert iguana,* Dispsosaurus dorsalis *(ectothermic), (b) The brown rat,* Rattus norvegicus *(endothermic)*

(a)                    (b)

## Box 9.1  Size and shape in endotherms

For a given shape, a smaller object has a larger surface area to volume ratio than a bigger object (see *AS Biology*, Section 9.2, page 232). As heat is lost from the body surface and is produced by the volume of respiring tissues, it follows that a smaller endotherm has a larger heat loss to heat production ratio than a larger one of the same shape. In other words, smaller animals must respire more rapidly to release enough heat energy to compensate for the rapid loss of heat. The small animals need to take in enough food to maintain this high metabolic rate. Even in warm climates, there is a minimum limit to the size of endotherms below which it is impossible to eat enough food to maintain body temperature. The smallest bird is the bee hummingbird which weighs about 2 g. The smallest living mammal is usually thought to be Savi's pygmy shrew which weighs between 1.2 g and 2.7 g (Figure 9.11).

For endotherms of a certain volume it is possible to have different surface areas determined by body shape. In general, the rounder the shape, the smaller the surface area to volume ratio and the easier it is to conserve heat. Elongated extremities have large surface areas and lose heat rapidly. The shapes of animals are adapted to help them survive in extreme conditions. The polar bear is large and is as round as it is possible for a four-footed creature to be! Animals which live in warmer regions tend to have long noses, big ears or flaps of skin which allow them to lose heat (Figure 9.12).

*Figure 9.11* (a) The smallest bird – the bee humming bird (Mellisuga helenae) is found in Cuba, (b) The smallest mammal – Savi's pigmy shrew (Suncus etruscus) is found around the Mediterranean and across Asia

(a)                    (b)

*Figure 9.12* The zebu cow (a) which lives in tropical regions, has a larger surface area to volume ratio than that of its European relatives (b) as a result of its longer ears, longer nose and dewlap

## 9.6  Thermoregulation in mammals

### Heat gain and heat loss in mammals

The methods by which the body of a mammal can gain or lose heat are summarised below. It is by regulating heat gain and heat loss that mammals maintain a constant body temperature.

**Heat gain:**

- Respiration by all tissues. When at rest, most heat is produced by respiration in liver cells. During exercise, most heat is produced by respiration in muscle cells. Respiration in **brown adipose tissue** cells is important at certain times (see Box 9.2).

- Intake of hot food and drink. The heat energy in the food and drink molecules is transferred to the molecules of the cells of the body.

- Absorption from surroundings. On the rare occasions when environmental temperatures are higher than body temperature there is a net gain of radiant heat, and some heat is gained by conduction. This explains why you get so hot when relaxing in a hot bath or sauna.

### Box 9.2  Brown adipose tissue (brown fat)

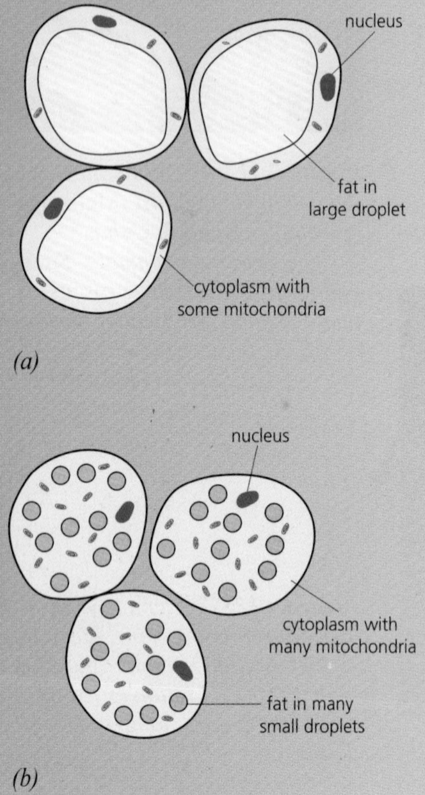

**Adipose tissue** consists of cells which store lipid in the form of fat. Most adipose tissue, which is found under the skin (**subcutaneous fat**) or around organs such as the kidneys, is white. It is therefore known as **white** adipose tissue. However, in new-born and hibernating mammals it is common to find adipose tissue which is a pink-brown colour. This is called **brown** adipose tissue.

Brown adipose tissue occurs in the neck region and around the aorta near the heart. When stimulated by nerve impulses, it respires its fat stores rapidly. The type of respiration carried out releases a higher proportion of the energy in the form of heat than normal respiration. The heat energy warms the blood as it emerges from the heart and is thus distributed rapidly around the body – a method of producing heat known as **non-shivering thermogenesis**.

Much brown adipose tissue is found in mammals which hibernate and in newborn mammals. Non-shivering thermogenesis warms up the bodies of hibernating mammals as they come out of hibernation. This form of rapid heat production can also warm up the bodies of newly born mammals very quickly in emergencies. This may be necessary because young mammals often have inefficient thermoregulatory systems at first and can also lose heat quickly owing to their small size.

**Figure 9.13** Cells of (a) white and (b) brown adipose tissue

labels:
(a) nucleus; fat in large droplet; cytoplasm with some mitochondria
(b) nucleus; cytoplasm with many mitochondria; fat in many small droplets

**?**

5  Describe two features of brown adipose tissue cells which indicate that they could metabolise fats more rapidly than white adipose tissue cells.

*(2 marks)*

## Heat loss:

● From the body surface when, as is usual, the temperature of the surroundings is cooler than body temperature, by conduction of heat energy to surrounding molecules and by radiation of heat energy. The rate of heat loss by these methods is faster if there is a steeper temperature gradient between the body surface and the surroundings. Convection currents in the air can increase the rate of heat loss by conduction and radiation by replacing warmed air with cooler air, so increasing this temperature gradient.

● By the evaporation of water from the surface of the animal, as this absorbs latent heat of vaporisation. Such water can be produced by sweating. Water also evaporates from the lungs and this can be increased by panting.

● By passing out hot faeces and urine.

### The regulation of heat gain and heat loss

Heat energy can be produced in the body by respiration, when conversion from chemical energy in a respiratory substrate, such as glucose, takes place. Heat energy is lost mainly through the body surface. The skin therefore plays an important part in thermoregulation. Figures 9.14 and 9.15 explain the role of some of the structures in the skin of a mammal in thermoregulation. The main methods of regulating heat production and heat loss are summarised in Table 9.1.

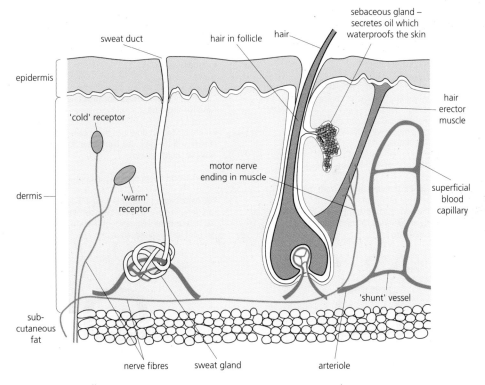

Mammals with thick fur do not usually have sweat glands. Sweat only cools the body as it evaporates and this would not be effective below a layer of fur. Furry mammals lose heat more effectively by the evaporation of water from the lungs and tongue during panting.

(a)

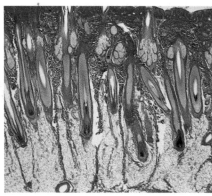

(b)

**Figure 9.14** *The structure of mammal skin and its role in thermoregulation. (a) A simplified vertical section through the skin of a mammal showing the structures involved in thermoregulation, (b) Photomicrograph of a vertical section through human skin*

*(a)*

**In a cold environment**

- hair erector muscles are stimulated to contract
- hairs are pulled erect

**In a warm environment**

- hair erector muscles relax
- hairs lie flat

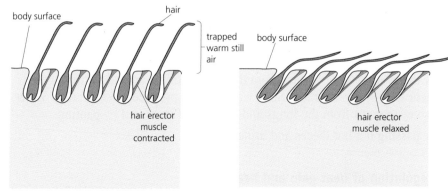

body surface          hair

trapped warm still air

body surface

hair erector muscle contracted

hair erector muscle relaxed

- air is trapped next to the body surface. Convection currents cannot occur. Warm air is not replaced with cooler air. As the trapped air, which is a poor conductor of heat, warms up, the temperature gradient between the body and environment becomes less steep and heat loss is reduced.

- convection currents can replace warm air close to the body surface with cooler air. This maintains a steep temperature gradient between the body and the environment. Heat loss by conduction and radiation is increased.

**NB:** In humans, owing to lack of hair, this mechanism is of little use. The action of the hair erector muscles can be seen by the goose pimples produced when we are cold.

*(b)*

**In a cold environment**

- vasoconstriction occurs

- sphincter muscles around arterioles leading to superficial capillaries contract

- this constricts the passage into these capillaries and more blood flows through deeper shunt vessels

- less blood flows close to the body surface

- as most blood is diverted further from the body surface, the temperature gradient between the body surface and the environment is less steep, so heat loss by conduction and radiation is reduced.

**In a warm environment**

- vasodilation occurs

- sphincter muscles around arterioles leading to superficial capillaries are not stimulated to contract and therefore relax

- more blood can flow into these capillaries, dilating them with the pressure; less blood flows through deeper shunt vessels

- more blood flows close to the body surface

- as more blood flows close to the body surface, the temperature gradient between the body surface and the environment becomes steeper, so heat loss by conduction and radiation is increased.

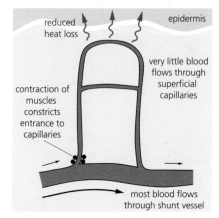

reduced heat loss          epidermis

very little blood flows through superficial capillaries

contraction of muscles constricts entrance to capillaries

most blood flows through shunt vessel

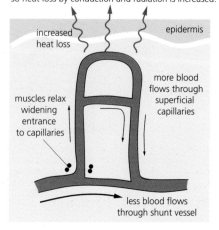

increased heat loss          epidermis

more blood flows through superficial capillaries

muscles relax widening entrance to capillaries

less blood flows through shunt vessel

**Figure 9.15** *(a) The role of hairs in thermoregulation,*
*(b) The role of superficial blood capillaries in thermoregulation*

| In a cold environment | In a warm environment |
|---|---|
| **Increase in heat production** | **No increase in heat production** |
| • exercising increases respiration rate of muscle cells | • remaining as still as possible |
| • shivering increases respiration rate in muscle cells | • no shivering |
| • raising metabolic rate increases respiration rate in all cells | • no raising of metabolic rate |
| • increasing respiration rate in brown fat under special circumstances | • no activation of brown fat |
| **Decrease in heat loss (heat conservation)** | **Increase in heat loss** |
| • vasoconstriction of arterioles in skin | • vasodilation of arterioles in skin |
| • reducing sweating and panting | • increasing sweating and panting |
| • erecting hairs | • not erecting hairs (they remain flat) |
| • curling up, reducing surface area | • spreading out, increasing surface area |

*Table 9.1 Methods of regulating heat production and heat loss*

## The mechanism of thermoregulation

The homeostatic mechanisms involved operate by means of negative feedback systems where, unlike the mechanisms controlling blood glucose levels, communication is by the nervous system and involves receptors, effectors and a coordination centre. However, once again there are separate mechanisms for restoring the norm, allowing heat to be lost or gained as appropriate.

The receptors are **thermoreceptors** which detect changes in temperature. Receptors in the skin are important for detecting changes in environmental temperature and can act as an 'advance warning' system. Some are stimulated by a decrease in temperature (**cold receptors**) and others are stimulated by an increase in temperature (**warm receptors**). There are also similar cold and warm thermoreceptors in the **hypothalamus** region of the brain which detect changes in the temperature of the blood flowing near this region. These are important for detecting changes in the body **core** temperature. The central part of the body, the body core, surrounds the vital organs and it is the temperature of this region which must be kept as constant as possible (close to 36.9 °C in humans). As the body loses heat from the surface, the extremities will be cooler. Figure 9.16 shows temperature gradients which exist in a human body.

The responses are brought about by **effectors** such as the sweat glands, hair erector muscles, sphincter muscles which cause vasoconstriction, and the muscles which shiver.

The coordination centre (the **thermoregulatory** centre) consists of two regions, the 'heat gain' centre and the 'heat loss' centre, and is situated in the **hypothalamus** of the brain (Figure 9.17). The negative feedback mechanisms which operate in thermoregulation are summarised in Figure 9.18.

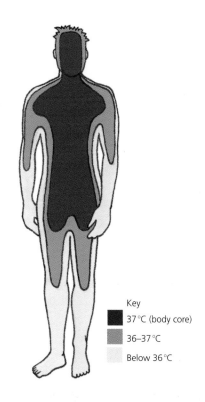

Key

37 °C (body core)

36–37 °C

Below 36 °C

*Figure 9.16 Temperature gradients in an unclothed human in an air temperature of 25 °C*

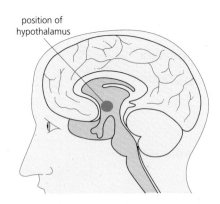

position of hypothalamus

**Figure 9.17** *Section through the human brain showing the position of the hypothalamus*

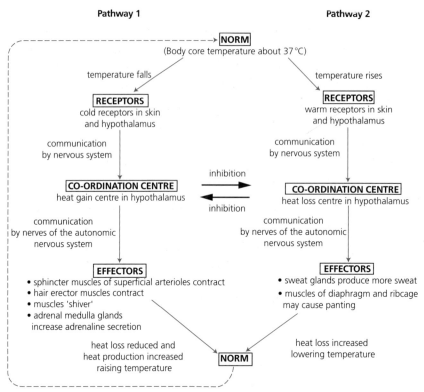

Pathway 1                                                    Pathway 2

**NORM**
(Body core temperature about 37 °C)

temperature falls                                    temperature rises

**RECEPTORS**
cold receptors in skin and hypothalamus

**RECEPTORS**
warm receptors in skin and hypothalamus

communication by nervous system

communication by nervous system

inhibition

**CO-ORDINATION CENTRE**
heat gain centre in hypothalamus

**CO-ORDINATION CENTRE**
heat loss centre in hypothalamus

inhibition

communication by nerves of the autonomic nervous system

communication by nerves of the autonomic nervous system

**EFFECTORS**
• sphincter muscles of superficial arterioles contract
• hair erector muscles contract
• muscles 'shiver'
• adrenal medulla glands increase adrenaline secretion

**EFFECTORS**
• sweat glands produce more sweat
• muscles of diaphragm and ribcage may cause panting

heat loss reduced and heat production increased raising temperature

**NORM**

heat loss increased lowering temperature

**Figure 9.18** *The negative feedback systems involved in thermoregulation in a mammal (compare with Figure 9.2)*

**NB:** There are two separate negative feedback mechanisms operating in thermoregulation. For example, if the body temperature rises, heat loss is increased by sweating. There is not just reliance on the fact that the hairs lie flat and that vasodilation occurs as the heat gain centre is inhibited.

**Pathway 1:** When the body temperature falls below the norm of 36.9 °C, cold receptors in the skin and hypothalamus are stimulated and transmit impulses along sensory nerve fibres to the heat gain centre of the hypothalamus. The heat gain centre transmits nerve impulses along nerves of the autonomic nervous system (see Section 7.11, page 259) to various structures with the following effects:

● the heat loss centre – inhibits its activity
● the sphincter muscles of superficial arterioles – contract, giving vasoconstriction
● the hair erector muscles – contract, pulling hairs erect
● various muscles – contract and relax frequently in shivering
● the adrenal medulla gland – stimulates the secretion of adrenaline which causes an increase in the metabolic rate.

As a result of reduced heat loss and more heat being produced the body temperature rises back towards the norm. Negative feedback has occurred.

**?**

6 Which of the responses in pathway 1 result in reduced heat loss, and which result in increased heat production? *(2 marks)*

**Pathway 2:** When the body temperature rises above the norm, warm receptors in the skin and hypothalamus transmit impulses along sensory nerve fibres to the heat loss centre of the hypothalamus. The heat loss centre is stimulated to transmit impulses to various structures with the following effects:

- the heat gain centre – inhibits its activity. This results in the relaxation of the muscles giving vasodilation, the flattening of hairs and the cessation of shivering. The metabolic rate also returns to normal
- the sweat glands – more sweat is produced and passed to the body surface
- the muscles of diaphragm and ribcage – stimulates panting.

As more heat is lost and less is produced, the body temperature falls towards the norm and negative feedback has occurred.

The above account is simplified and does not include priorities, such as the fact that the metabolic rate is only raised if other mechanisms do not increase the temperature sufficiently. Also, it does not include behavioural mechanisms such as curling up to reduce the surface area of the body and so reduce heat loss.

> **Remember**
>
> Thermoregulatory responses which are temporary and reversible must not be confused with permanent adaptations which allow mammals to survive in particular climates. For example, the thick fur of a polar bear is a permanent adaptation which reduces heat loss, not a thermoregulatory response.

**7** Cooling drinks? If the weather is very hot, the warm receptors in the skin of a sunbather are stimulated. The heat loss centre is switched on and sweating is increased, cooling the skin. If the sunbather drinks a pint of iced orange juice, this will cool the blood flowing through the stomach wall. As the cooled blood flows near the hypothalamus, the cold receptors are stimulated and the heat gain centre is switched on.

(a) What will happen to the rate of sweating in the skin as a result of the cold drink?

(b) What will then happen to the skin temperature?

(c) In the minutes following the iced drink, will the body lose more or less heat than before taking the drink?  *(3 marks)*

## 9.7  Control of digestive secretions in mammals

This section provides an example of a physiological process in a mammal's body which involves negative feedback systems and a combination of nervous and hormonal communication. It relates closely to the process of digestion in mammals (see *AS Biology*, Section 13.5, page 389).

Three of the main digestive juices which are secreted into the gut of a mammal are saliva, gastric juice and pancreatic juice. Figure 9.19 shows where these digestive juices are made and where they enter the alimentary canal. It would be a waste of energy and resources if these juices were secreted constantly. Also, the constant secretion of gastric juice and pancreatic juice would increase the risk of self digestion (see *AS Biology*, Section 13.5, page 389). Therefore, it is essential that their secretion is timed to have maximum effect following the intake of food and that the secretions are present as the food reaches the appropriate part of the gut.

*Figure 9.19* *The human alimentary canal showing the positions of the salivary glands, gastric pits and pancreas*

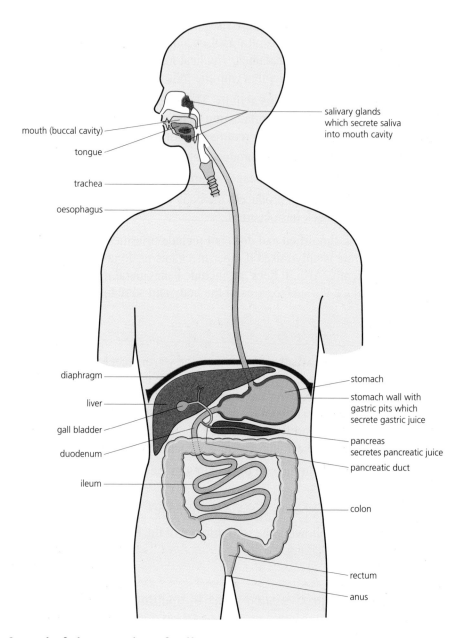

salivary glands which secrete saliva into mouth cavity

mouth (buccal cavity)

tongue

trachea

oesophagus

diaphragm

liver

gall bladder

duodenum

ileum

stomach

stomach wall with gastric pits which secrete gastric juice

pancreas secretes pancreatic juice

pancreatic duct

colon

rectum

anus

### Control of the secretion of saliva

Saliva is needed to act on the food in the mouth. The salivary glands are stimulated to secrete saliva by nervous communication. A mixture of simple and conditioned reflex actions occur (see Section 7.9, page 246).

● In the simple reflex actions, the stimuli are the contact of food with the tongue and the taste of the food. These are detected by the touch receptors and taste buds on the tongue. These reflexes ensure that saliva continues to be secreted as long as food is present in the mouth.

● In the conditioned reflex actions, the original stimuli have been replaced by associated ones. Here the sight and smell or even the thought of food can act as a stimulus and trigger the same response. This enables saliva to be already present when the food is eaten.

## Control of the secretion of gastric juice

It does not take long for food to reach the stomach. However, it remains in the stomach for up to 4 hours. It is clearly advantageous for some gastric juice to be secreted from the gastric pits in the stomach wall very soon after food enters the mouth so that it is present as the food enters the stomach. However, gastric juice must continue to be produced while the food remains in the stomach.

A response brought about involving nervous communication is rapid but short lived. On the other hand a response brought about when communication is by hormones is slow to start but lasts for a long time. Both communication systems are involved in the secretion of gastric juice. Stimulation occurs in three phases: the vagus phase, the local reflex phase and the hormonal phase.

- During the **nervous (vagus)** phase, the presence of food in the mouth and the act of swallowing act as stimuli. As the result of a simple reflex action (involving the vagus nerve), the gastric pits are stimulated to secrete gastric juice. This response takes place quickly and gastric juice is being secreted as the food reaches the stomach. The response lasts for about 1 hour.

- The **local reflex** phase takes place during the first 2 hours after the arrival of food in the stomach and bridges the gap until the final phase begins. The presence of food in the stomach distends the stomach, stimulating stretch receptors. A mixture of reflex actions, some of them just involving the nerve plexus in the stomach wall, trigger further secretion.

- The **hormonal** phase is slow to start and peaks 2 hours after the arrival of food. However, it lasts for about 4 hours. The chemical presence of food in the stomach stimulates certain cells in the stomach wall to secrete the hormone, **gastrin**. This circulates in the blood and eventually reaches the gastric pits, stimulating the secretion of gastric juice. Figure 9.20 shows the contributions of the three phases to total secretion.

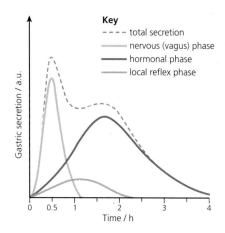

**Figure 9.20** *Graph showing the contributions of the three phases to total secretion of gastric juice*

## Control of the secretion of pancreatic juice

Pancreatic juice is released into the duodenum down the pancreatic duct as the acidified mixture of partially digested food, known as **chyme**, enters this region from the stomach. This process starts quickly and lasts for over 4 hours. Pancreatic juice consists of a mixture of enzymes and a solution of alkaline salts. These salts are essential for neutralising the chyme and providing an alkaline medium for the enzymes.

Once again a nervous response initiates the secretion quickly and a hormonal response, after a slower start, ensures that secretion lasts for a long time. When impulses are transmitted to the stomach wall along the vagus nerve during the nervous phase of gastric juice secretion, some impulses pass along other fibres in the nerve to the pancreas, stimulating the initial flow. This ensures that pancreatic juice is present as the first food from the stomach enters. However, most pancreatic juice is secreted in response to the hormones **secretin** and **cholecystokinin-pancreozymin (CCK-PZ)**.

- The acid nature of the chyme entering the duodenum stimulates cells in the intestine wall to secrete secretin. This circulates in the blood and eventually stimulates pancreatic cells to secrete the alkaline salts component of pancreatic juice.

*Figure 9.21 The actions of secretin and cholecystokinin-pancreozymin (CCK-PZ)*

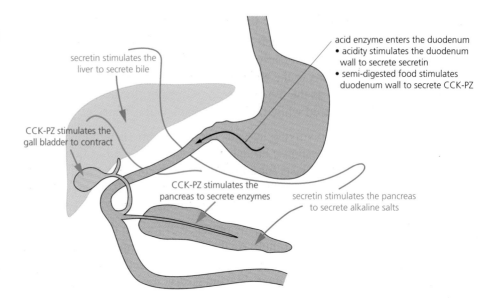

acid enzyme enters the duodenum
• acidity stimulates the duodenum wall to secrete secretin
• semi-digested food stimulates duodenum wall to secrete CCK-PZ

secretin stimulates the liver to secrete bile

CCK-PZ stimulates the gall bladder to contract

CCK-PZ stimulates the pancreas to secrete enzymes

secretin stimulates the pancreas to secrete alkaline salts

● Fats and proteins in chyme entering the duodenum stimulate cells in the intestine wall to secrete CCK-PZ which also reaches the pancreas via the blood. It stimulates pancreatic cells to secrete the enzyme components of pancreatic juice.

Secretin and CCK-PZ have other functions also. Figure 9.21 summarises the actions of these hormones.

**8** Explain how the production and action of secretin can be regarded as an example of a negative feedback system. *(1 mark)*

**9.8** **Comparing nervous and hormonal communication**

Figure 9.22 outlines the roles of nervous and hormonal means of communication in the control of the secretion of digestive juices.

### Remember

Nervous control gives rapid, short-lived responses.

Hormonal control gives responses which are slower to start but last longer.

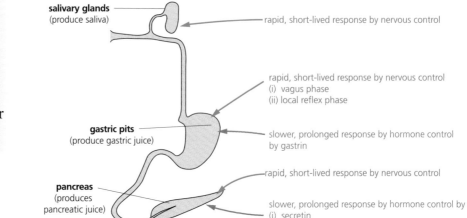

**salivary glands** (produce saliva)

rapid, short-lived response by nervous control

rapid, short-lived response by nervous control
(i) vagus phase
(ii) local reflex phase

**gastric pits** (produce gastric juice)

slower, prolonged response by hormone control by gastrin

rapid, short-lived response by nervous control

**pancreas** (produces pancreatic juice)

slower, prolonged response by hormone control by
(i) secretin
(ii) CCK-PZ

*Figure 9.22 Summary of control of digestive secretions*

Information is communicated from one part of the body to another either via the nervous system or by means of hormones (endocrine system). As has been seen, both methods are used to advantage in the control of digestive secretions. The functioning of the two systems is often linked, sometimes by means of the hypothalamus. For example, when the heat gain centre is stimulated via the nervous system, one of its actions is to stimulate the adrenal medulla gland to secrete adrenaline.

The nervous system and the nature of the nerve impulse have been discussed in Chapters 6 and 7, and various hormones and their actions have been referred to in this chapter. Therefore it is useful at this point to consider the differences between these two means of communication and these are summarised in Table 9.2.

*Table 9.2* *Differences between nervous and endocrine (hormonal) communication in mammals*

| Nervous system | Endocrine system |
|---|---|
| information is passed as **electrical** signals (nerve impulses) along nerve fibres and chemically only for short distances across synapses | information passed as **chemical** substances transported in the bloodstream |
| transmission is relatively rapid | transmission is relatively slow |
| the response is initiated rapidly | the response is initiated slowly |
| the response is short-lived | the response may carry on for a long time |
| the response is local as one nerve fibre only supplies one or a small group of effector cells | the response can be widespread as target cells in different parts of the body can respond to the same chemical |
| the response is reversible | often the response has a permanent effect (e.g. growth) |

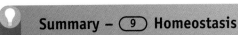

**Summary –** **9** **Homeostasis**

- Homeostasis is the maintenance of a constant internal environment. It enables organisms to maintain constant internal conditions independent of unfavourable, fluctuating external conditions. It allows them to be active in unfavourable environments.

- Homeostasis is brought about by negative feedback systems. These are systems in which a change from the norm (optimum state) is detected and triggers a mechanism which restores the norm.

- Negative feedback systems involve receptors which detect a change from the norm, a communication system (hormonal or nervous) and effectors which bring about a response which helps to restore the norm.

- Two separate mechanisms control changes in opposite directions from the norm. This gives more precise control and helps to keep fluctuation from the optimum to a minimum.

- The negative feedback systems controlling glucose concentration in the blood involve two hormones, glucagon and insulin. Glucagon is produced in response to falling glucose levels and its effects increase the amount of glucose in the blood. Insulin is produced when glucose levels rise and its effects reduce glucose levels. Interaction between these hormones maintains a steady concentration of glucose.

- Diabetes mellitus is a condition which occurs when a person cannot control the glucose concentration in the blood.

- There are two approaches to thermoregulation (control of body temperature) in animals – ectothermy and endothermy.

- Ectothermic animals derive most of their body heat from their surroundings and keep their body temperature fairly constant by their behaviour.

- Only mammals and birds are endothermic animals. They derive most of their body heat internally from respiration. They keep their body core temperature constant by controlling the rates at which they produce and lose heat energy.

- The negative feedback systems controlling temperature in endotherms involve nervous communication. There are warm and cold receptors in the skin and hypothalamus. These transmit impulses to the thermoregulatory (coordinating) centre in the hypothalamus region of the brain. This has a heat gain centre stimulated by cold receptors and a heat loss centre stimulated by warm receptors. These centres transmit nervous impulses to various effectors which bring about corrective changes.

- Body temperature can be raised by vasoconstriction of superficial arterioles, erection of hairs and curling up, all of which reduce heat loss, and by shivering and raising the metabolic rate which increase heat production.

- Body temperature can be lowered by vasodilation of superficial arterioles, not erecting hairs, spreading out the body and increasing sweating and panting. All of these increase the rate of heat loss.

- The secretion of digestive juices also involves nervous and hormonal communication and some negative feedback mechanisms.

- Saliva is secreted by the salivary glands as a response in simple reflex and conditioned reflex actions.

- Gastric juice is secreted by gastric pits in the stomach wall. Stimulation is in three phases. The vagus phase and local reflex phase involve nervous communication. In the third hormonal phase, prolonged secretion is stimulated by the hormone, gastrin.

- Pancreatic juice is secreted into the duodenum by the pancreas. Initial secretion is stimulated rapidly by nervous reflex but prolonged secretion is stimulated hormonally. Secretin stimulates the production of the alkaline salts component and cholecystokinin-pancreozymin stimulates production of the enzymes.

- The nervous system and hormones both provide means of communicating information around the body but there are many differences between these two methods.

## ? Answers

1 (a) Insulin reduces blood glucose levels by its effects; glucagon raises blood glucose levels by its effects, there is not just lack of insulin *(2)*.

   (b) This gives precise control with more rapid return to/less fluctuation from/the norm *(1)*.

2 The main target organ of both insulin and glucagon is the liver; these hormones reach the liver quickly in the blood *(2)*.

3 (a) Insulin would be digested in the gut and not enter the bloodstream *(1)*.

   (b) It enters the bloodstream gradually from the muscle or fat; if injected directly into the blood it could cause an overdose *(2)*.

   (c) Too much glucose would be removed from the blood; there would not be enough glucose for respiration of cells, especially brain cells; coma and possibly death would follow *(3)*.

4 The rat would need five times as much food as the iguana *(1)*.

5 The brown fat cells have smaller fat droplets, giving a larger surface area for enzymes to act upon; more mitochondria where later stages of respiration take place *(2)*.

6 Heat loss reduced by inhibiting heat loss centre (therefore sweating reduced), vasoconstriction, erecting hairs *(1)*. Heat production increased by shivering, increasing metabolic rate *(1)*.

7 (a) The rate of sweating would be reduced *(1)*.

   (b) The skin would become warmer *(1)*.

   (c) Less heat would be lost *(1)*.

8 Increased acidity in the duodenum triggers secretin which causes alkaline salts to be poured into the duodenum, reducing the acidity *(1)*.

**End of Chapter Questions**

NB: You will need to have covered Section 6.11, page 211 in order to answer part (c) (ii) of Question 6.

1   Describe (a) two similarities and (b) two differences between the negative feedback mechanisms involved in glucose regulation and thermoregulation.

*(Total 4 marks)*

N2.1  2   An experiment was carried out to investigate the relationship between the concentrations of glucose and insulin in the blood of healthy people. At the start of this experiment 34 volunteers each ingested a syrup containing 50 g of glucose. The concentrations of glucose and insulin were determined in blood samples at intervals over a period of 2 hours. The results shown in the graph below are mean values for the group of volunteers.

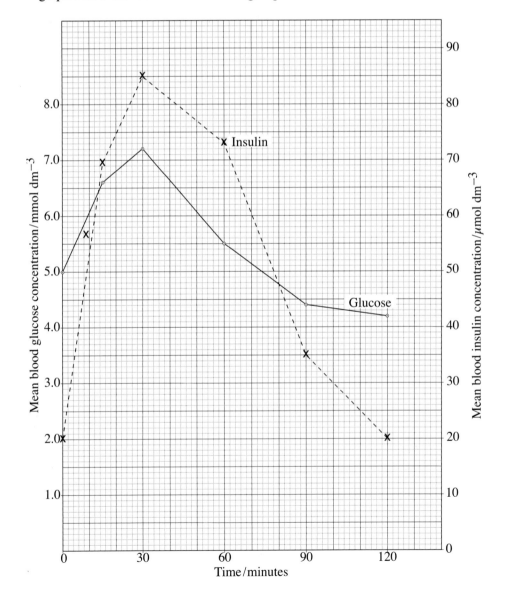

**(a)** From the graph, find the mean concentration of insulin 100 minutes after the start of the experiment. *(1 mark)*

**(b)** Describe and suggest an explanation for the changes in the concentration of glucose during the following time intervals.

   (i)  0 to 30 minutes                  *(2 marks)*

   (ii) 30 to 120 minutes.            *(2 marks)*

**(c)** Discuss the relationship between the concentrations of glucose and insulin as shown by this graph. *(3 marks)*

**(d)** At the start of a period of prolonged exercise, the blood glucose level begins to fall. Describe and explain how the level is controlled as the exercise continues. *(4 marks)*

*Edexcel 2000*                              *(Total 12 marks)*

**3 (a)** The graph shows changes in blood glucose concentration after a meal.      N2.1
                                                                          C2.3

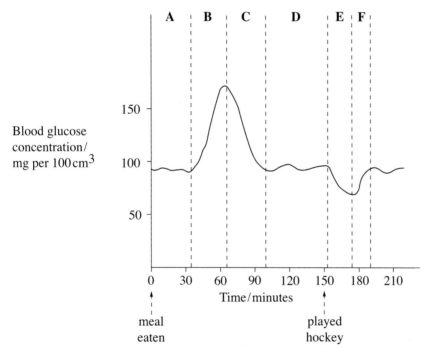

Explain the changes in blood glucose concentration in **each** of the stages **A** to **F** in the graph. *(8 marks)*

**(b)** Explain why it is an advantage for humans to have a constant body temperature. *(4 marks)*

*AQA (NEAB) 1998*                              *(Total 12 marks)*

**4**  The diagram shows some important features of homeostatic mechanisms in the body.

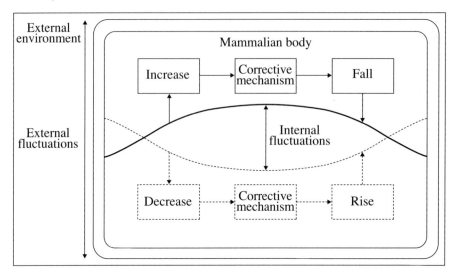

**(a)**  Use the information in the diagram to help explain the importance of a mammal maintaining a constant internal temperature.  *(4 marks)*

**(b)**  Explain the role of the hypothalamus and nervous system in the regulation of body temperature.  *(5 marks)*

N2.3

**(c)**  Explain why, in a normal healthy individual, the blood glucose level fluctuates very little.  *(6 marks)*

*AQA 2000*  *(Total 15 marks)*

N3.1
C2.2

**5**  In one form of diabetes, the pancreas is unable to make sufficient insulin. In an investigation, 20 people were divided into two groups. Group **A** contained 10 people with this form of diabetes, while Group **B** contained 10 people without diabetes (control group).

Blood samples were taken from each person at 30 minute intervals, and the amounts of glucose, insulin and glucagon measured. After 1 hour, each person ate a meal containing a large amount of carbohydrate. Mean concentrations were calculated for each substance at each sampling time.

The results are shown in the graphs below.

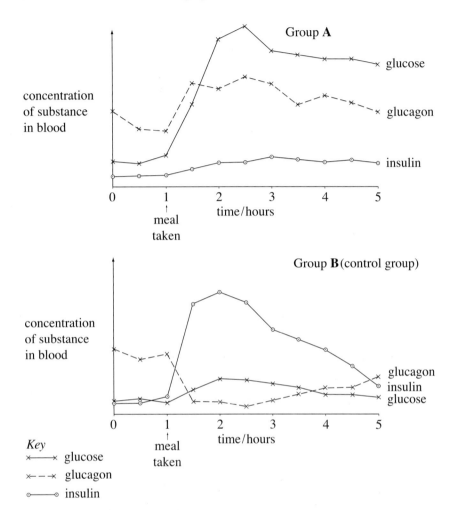

(a) (i) State **one** difference between Groups **A** and **B** in the way in which **glucagon** secretion responds to the intake of carbohydrate.  *(1 mark)*

(ii) State **two** differences between Groups **A** and **B** in the way in which **insulin** secretion responds to the intake of carbohydrate.   *(2 marks)*

(b) Explain the changes in blood glucose concentration in

(i) **Group A**                                                              *(3 marks)*

(ii) **Group B**.                                                            *(3 marks)*

(c) Suggest what would happen to the blood glucose concentration of people in Group A, if they ate no carbohydrate for another 24 hours. Explain your answer.                            *(3 marks)*

OCR 2000                                              *(Total 12 marks)*

**6** The graph shows how an injection of secretin affects the secretion of pancreatic juice by the pancreas.

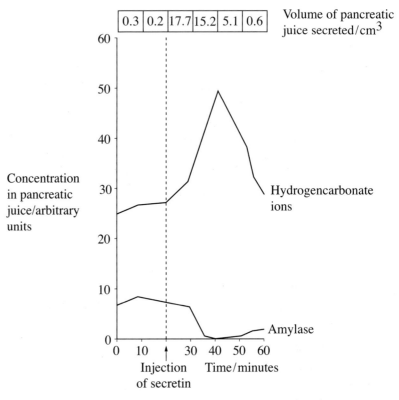

| 0.3 | 0.2 | 17.7 | 15.2 | 5.1 | 0.6 |

Volume of pancreatic juice secreted/cm$^3$

Concentration in pancreatic juice/arbitrary units

Hydrogencarbonate ions

Amylase

Injection of secretin     Time/minutes

**N2.1**

**(a)** **(i)** Use the graph to describe the effect of secretin on the pancreas.

**(ii)** Explain why the *concentration* of amylase in the pancreatic juice decreased shortly after the injection of secretin.     *(3 marks)*

**(b)** What other digestive secretion is stimulated by secretin?     *(1 mark)*

**(c)** Certain types of ulcer are thought to be made worse by the production of too much acid from the stomach. Doctors have used a number of different methods to treat these ulcers. Suggest how the following treatments might reduce the amount of acid secreted by the stomach.

**(i)** cutting the vagus nerve to the stomach     *(1 mark)*

**(ii)** giving the patient atropine, a drug which blocks the action of acetylcholine.     *(2 marks)*

*AQA (NEAB) 1997*     *(Total 7 marks)*

# 10 Excretion and water balance

The importance of maintaining a constant internal optimum state was explained in Chapter 9. Two further mechanisms which are involved in achieving this are the control of the level of nitrogenous waste products and the control of water balance (**osmoregulation**). In mammals, nitrogenous waste products are processed in the liver and passed out of the body (**excreted**) via the kidneys. The correct water potential of the blood plasma and body fluids is maintained homeostatically by the kidneys under the influence of antidiuretic hormone (ADH).

> **This chapter includes:**
> - what is excretion?
> - nitrogenous metabolic waste
> - the structure and function of mammalian kidneys
> - the role of the kidneys and ADH in water balance
> - control of the water budget in small desert mammals.

## 10.1 What is excretion?

The chemical reactions which take place in the body of an organism (**body metabolism**) produce some substances which are of no use to the body and which would poison the body if they were allowed to build up. In other words they are **toxic**. For example:

- **carbon dioxide** is produced during respiration
- various **nitrogen containing compounds** are produced during the breakdown of excess amino acids and nucleic acids
- bile pigments **bilirubin** and **biliverdin** are produced during the breakdown of haemoglobin.

Such products must be removed from the body so that their levels do not rise to the point where they become dangerous. The process of eliminating such substances from the body is called **excretion**.

In mammals, carbon dioxide is excreted from the **lungs** during breathing. The bile pigments are produced from haemoglobin in the **liver** and passed into the **bile**. They pass into the **duodenum** with the bile and are passed out of the body in the **faeces**. In vertebrates, the nitrogenous waste products are produced in the liver and excreted from the kidneys. The excretion of nitrogenous waste is discussed in detail in this chapter.

> **Definition**
>
> **Excretion** is removal of the waste products of metabolic processes from the body. These products would be toxic if they were allowed to accumulate.

> ❓
>
> 1 Why is it true to say that passing faeces out of the body involves excretion as well as egestion? *(2 marks)*

## 10.2 Nitrogenous metabolic waste

Proteins in the diet are digested, producing amino acids which are absorbed and transported in the blood. Also, any body proteins which are no longer required,

> **Remember**
>
> It is important to distinguish between excretion and **egestion** (sometimes called **defaecation**), which is the removal from the body of waste material, such as undigested food, which has not been part of body metabolism.

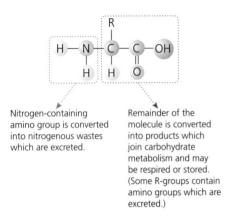

Nitrogen-containing amino group is converted into nitrogenous wastes which are excreted.

Remainder of the molecule is converted into products which join carbohydrate metabolism and may be respired or stored. (Some R-groups contain amino groups which are excreted.)

**Figure 10.1** *Structural formula of an amino acid – excess amino acids are toxic and must be broken down*

are hydrolysed into amino acids which pass into the blood. If amino acids which are in excess of the body's requirements were allowed to accumulate they would have toxic effects. This means that they cannot be stored and have to be broken down.

Amino acids contain nitrogen atoms in amino groups as shown in Figure 10.1 (see *AS Biology*, Section 3.16, page 83). When an amino acid is broken down, two products are formed. One product does *not* contain nitrogen and is usually recycled and respired, as it is converted into molecules which take part in carbohydrate metabolism. The other product (nitrogenous waste) contains nitrogen. It is toxic and is excreted.

In mammals, the main nitrogenous waste product is **urea**. Breakdown of amino acids and urea production take place mainly in liver cells. This involves two stages, **deamination** and the **ornithine cycle**. These processes are outlined below.

**Urea production**

1 **Deamination** (removal of the amino group) – the amino acid is oxidised, producing a keto acid and ammonia:

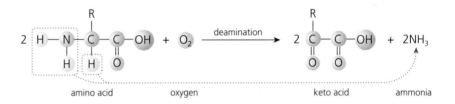

The keto acid is either fed into carbohydrate metabolism and respired or converted into lipid and stored. In this way much of the molecule is used. The ammonia produced is very toxic and is immediately converted into urea which is less toxic and can be tolerated in higher concentrations.

2 **The ornithine cycle** – this is a complex cyclical metabolic pathway in which ammonia and carbon dioxide are converted into urea. There is no need to memorise the details of the pathway. The resulting overall reaction is as follows:

$$2NH_3 \ + \ CO_2 \ \xrightarrow{\text{ornithine cycle}} \ CO(NH_2)_2 \ + \ H_2O$$

ammonia     carbon             urea      water
               dioxide

The urea passes out of the liver cells into the blood. The kidneys remove urea from the blood and it eventually passes out of the body in the urine.

Other animals may produce different nitrogenous waste products. Some of these are discussed in Box 10.1.

Extension

## Box 10.1 Alternative nitrogenous wastes

### Freshwater bony fish

Water is constantly entering the bodies of freshwater fish by osmosis, through the gills and lining of the pharynx, as the body fluids have a lower water potential than the river or lake water. The fish continually produce large amounts of dilute urine which eliminates this water. They break down their excess amino acids by deamination in the way described above, producing ammonia. However, they do not use valuable energy to convert this ammonia into urea. As there is such rapid production of very dilute urine it is possible to excrete the nitrogenous waste as ammonia. It is not toxic to the body as the concentrations are so low. Ammonia also diffuses directly into the surrounding water from the gills (Figure 10.2).

### Insects and birds

These two types of animals may be very different from each other but they face the same problem. They live active lives in the air, often flying for long distances. It is an advantage to them if they can conserve as much water as possible. Urea may not be as toxic as ammonia but it must still be passed out dissolved in sufficient water to prevent it reaching toxic levels. Insects and birds convert ammonia into **uric acid** and its salts. This is much less toxic and requires far less water for removal. Uric acid is not very soluble and can be precipitated out of solution and excreted in semi-solid form, or even stored harmlessly.

Insects extract nitrogenous wastes from their blood by specialised organs which convert them into uric acid. This is then passed into the rectum and mixed with the faeces. Water is reabsorbed from the mixture before it is passed out of the body.

In birds, the urine produced by the kidneys contains uric acid in saturated solution. It passes down tubes called **ureters** to a small cavity, the **cloaca**, on the underside of the bird. Here water is reabsorbed into the blood leaving the uric acid as a white paste. This mixes with the dark-coloured faeces and is expelled as the familiar 'bird droppings' (Figure 10.3). Uric acid produced by embryo birds developing inside eggs is stored in a special sac which is left behind when the bird hatches.

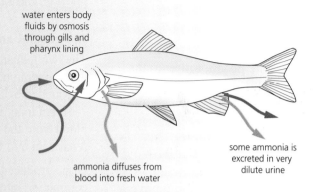

*Figure 10.2 A bony fish which lives in fresh water is constantly taking in water by osmosis. It can excrete its nitrogenous waste as ammonia as concentrations are too low to be toxic*

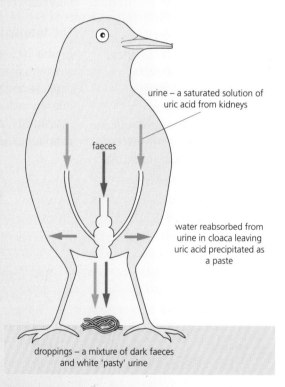

*Figure 10.3 Uric acid has low toxicity and solubility. It is excreted as a semi-solid paste enabling birds to conserve water*

**10.3** **The structure of the mammalian kidney**

### The general anatomy of the human renal (urinary) system

The two human kidneys, like those of other mammals, are situated at the back of the abdominal cavity at about the level of the waist. Each one is usually surrounded by a protective layer of fat. They are bean-shaped and about 7–10 cm long. Figure 10.4 shows how each kidney is supplied with oxygenated blood from the aorta via a renal artery. Blood drains away from each kidney in a renal vein into the vena cava.

Urine produced by each kidney passes down a tube called a **ureter** by muscular contraction into a muscular bag called the **bladder**. This sac stores urine and can extend to hold about 700 cm$^3$. When the sphincter muscle at the base of the bladder relaxes, urine passes down the **urethra** and out of the body. The elimination of urine from the body is called **urination** or **micturition**. Originally this is an autonomic reflex action (see Section 7.11, page 259). The stretch receptors in the bladder wall trigger a response in which bladder wall muscles contract and the sphincter relaxes. However, humans learn to exert voluntary control over this action early in life.

### The internal structure of a kidney

Figure 10.5 shows a photograph and a diagram of a longitudinal section through a mammalian kidney. The kidney is enclosed within a protective **fibrous capsule** and shows distinct regions. The outer region, the **cortex**, has a more uneven texture than the inner **medulla**. The medulla has zones called 'pyramids' which surround the central cavity, the **pelvis**. This cavity is continuous with that of the ureter.

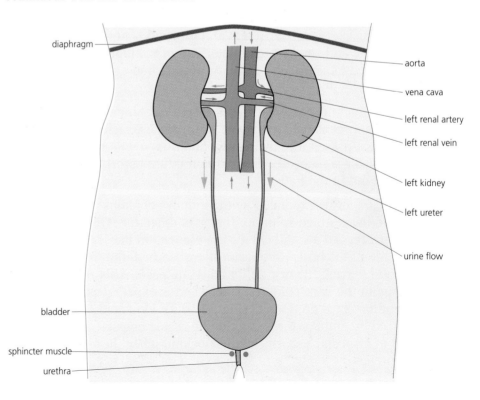

*Figure 10.4 The anatomy of the human renal (urinary) system and its blood supply*

Each kidney contains over one million microscopic tubules known as **nephrons**, each of which has a rich blood supply. These and a small amount of connective tissue make up the kidney. The positioning of nephrons relative to the regions of the kidney is shown in Figure 10.5b. The overall structure of a nephron and its blood supply is shown in Figure 10.6.

## 10.4 The ultrastructure and the functioning of a nephron

The role of the kidneys in excretion and osmoregulation is carried out by individual nephrons. So it is by studying how a nephron works that the overall function of the kidneys can be understood. In outline, each nephron operates by allowing blood to filter into it under pressure (**ultrafiltration**). The filtrate which enters the nephron from the blood contains any molecules under a certain size (with a relative molar mass – RMM – of under 68 000) whether they are useful or not. The cells in the nephron walls reabsorb molecules required by the body back into the blood, mainly by **active transport** (see *AS Biology*, Section 2.6, page 48). This is called **selective reabsorption**. They also **secrete** some unwanted substances into the filtrate. Some water re-enters the blood by osmosis. The solution which is left in the nephrons is **urine** and is passed out of the body. This is summarised in Figure 10.7. Each region of the nephron is adapted to carry out a part of this process and must be studied separately.

*(a)*

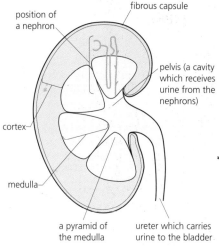

*(b)*

***Figure 10.5*** *(a) Longitudinal section through the kidney of a human, (b) LS of a mammalian kidney*

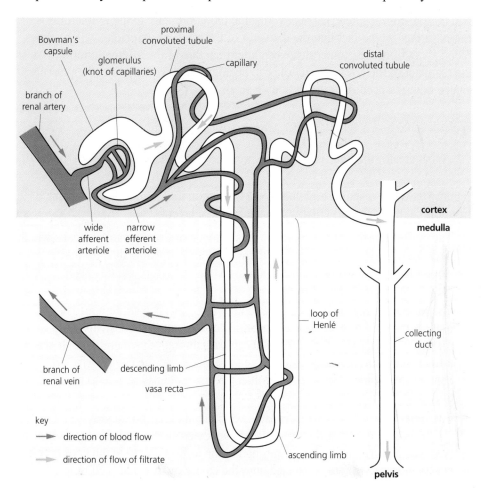

***Figure 10.6*** *The structure of a nephron and its blood supply*

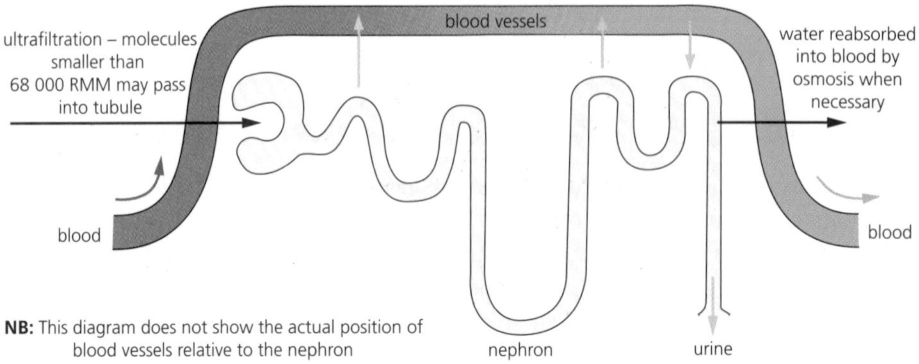

*Figure 10.7 A diagrammatic summary of the function of a nephron*

- *selective active reabsorption* of useful substances into blood
- water reabsorbed by *osmosis*

- further *selective reabsorption* into blood
- active *secretion* of unwanted substances into nephron
- water reabsorbed by osmosis into blood when necessary

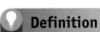

ultrafiltration – molecules smaller than 68 000 RMM may pass into tubule

blood vessels

water reabsorbed into blood by osmosis when necessary

blood

blood

**NB:** This diagram does not show the actual position of blood vessels relative to the nephron

nephron

urine

**Definition**

The structure consisting of a Bowman's/renal capsule and the glomerulus which it encloses is known as a **Malpighian body**.

### Ultrafiltration – The renal/Bowman's capsule and the glomerulus

Each nephron, which can be up to 14 mm long, has a cup-shaped blind end (**Bowman's capsule** or **renal capsule**) in the cortex of the kidney (Figure 10.8a). There is a network of capillaries (the **glomerulus**) in the depression of the capsule. Each glomerulus is supplied with blood via an **afferent arteriole** which is a branch of the renal artery. The blood is at high pressure as the renal arteries are short (and still not far from the heart). This hydrostatic blood pressure in the glomerulus is raised further by the fact that the diameter of the **efferent arteriole**, which carries blood away from the glomerulus, is smaller than that of the afferent arteriole. As the diameter of the efferent arteriole can be adjusted by constriction, the pressure can be adjusted according to the body's needs.

**2** What creates the hydrostatic blood pressure in the glomerulus? *(1 mark)*

Figure 10.8b and d shows how the blood flowing in the glomerulus is separated from the cavity of the Bowman's capsule by three layers:

- The **endothelium** of the capillary – this is made of thin (squamous) cells which have pores between them making the wall more permeable than normal capillaries. All constituents of blood plasma can pass through this layer. Blood cells cannot usually pass through.
- The **basement membrane** of the endothelium – this is a continuous layer of organic material to which the endothelial cells are attached. Usually, only molecules of RMM less than 68 000 can pass through this membrane. It acts as a filter (**dialysing membrane**) between the blood and the cavity. All constituents of blood plasma other than plasma proteins are able to pass through.
- The **inner wall** of Bowman's capsule which is made of **podocytes** – Figure 10.8b and d shows that the structure of podocytes allows any substances which have passed through the basement membrane to flow freely in the gaps between the 'branches', without resistance, into the cavity of Bowman's capsule.

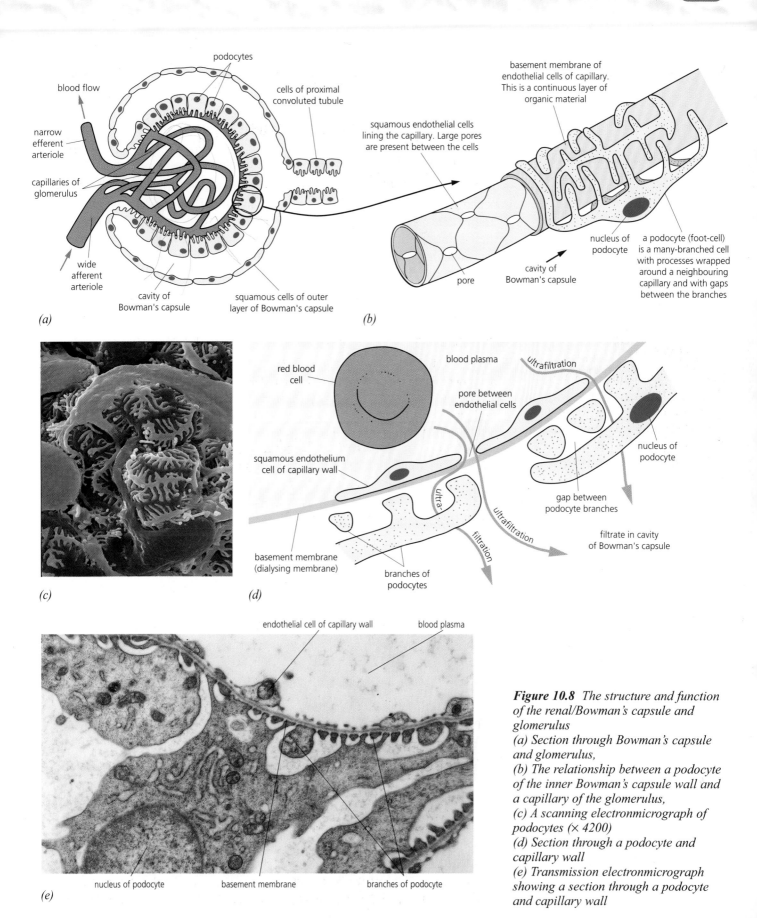

(a)

blood flow

narrow efferent arteriole

capillaries of glomerulus

wide afferent arteriole

podocytes

cells of proximal convoluted tubule

cavity of Bowman's capsule

squamous cells of outer layer of Bowman's capsule

(b)

basement membrane of endothelial cells of capillary. This is a continuous layer of organic material

squamous endothelial cells lining the capillary. Large pores are present between the cells

nucleus of podocyte

a podocyte (foot-cell) is a many-branched cell with processes wrapped around a neighbouring capillary and with gaps between the branches

cavity of Bowman's capsule

pore

(c)

(d)

red blood cell

blood plasma

ultrafiltration

pore between endothelial cells

nucleus of podocyte

squamous endothelium cell of capillary wall

gap between podocyte branches

filtrate in cavity of Bowman's capsule

ultra-

filtration

ultrafiltration

basement membrane (dialysing membrane)

branches of podocytes

(e)

endothelial cell of capillary wall

blood plasma

nucleus of podocyte

basement membrane

branches of podocyte

*Figure 10.8* *The structure and function of the renal/Bowman's capsule and glomerulus*
*(a) Section through Bowman's capsule and glomerulus,*
*(b) The relationship between a podocyte of the inner Bowman's capsule wall and a capillary of the glomerulus,*
*(c) A scanning electronmicrograph of podocytes (× 4200)*
*(d) Section through a podocyte and capillary wall*
*(e) Transmission electronmicrograph showing a section through a podocyte and capillary wall*

As the hydrostatic pressure of the liquid (blood) in the glomerulus is higher than that in Bowman's capsule, the plasma is filtered under pressure into the capsule (**ultrafiltration**). About 20% of the substances with molecules smaller than a RMM of 68 000 enter Bowman's capsule; the remaining 80% flow on with the rest of the blood into the efferent arteriole. Figure 10.9 shows a summary of this process involving the main components of the blood.

As can be seen in Figure 10.9, the filtrate (**glomerular filtrate**) which enters the nephron at the Bowman's capsule consists mainly of **inorganic ions**, **glucose**, **amino acids** and **urea**, all dissolved in **water**. The concentration of each of these solutes is the same as it is in the blood plasma. However, as the plasma also contains plasma proteins in solution, the total concentration of all solutes is greater in the plasma. The plasma therefore has a lower (more negative) solute potential than the glomerular filtrate (see *AS Biology*, Section 2.7, page 49). The glomerular filtrate is **hypotonic** to the blood plasma. Table 10.1 reminds you of these terms.

**Figure 10.9** *Summary of the process of ultrafiltration*

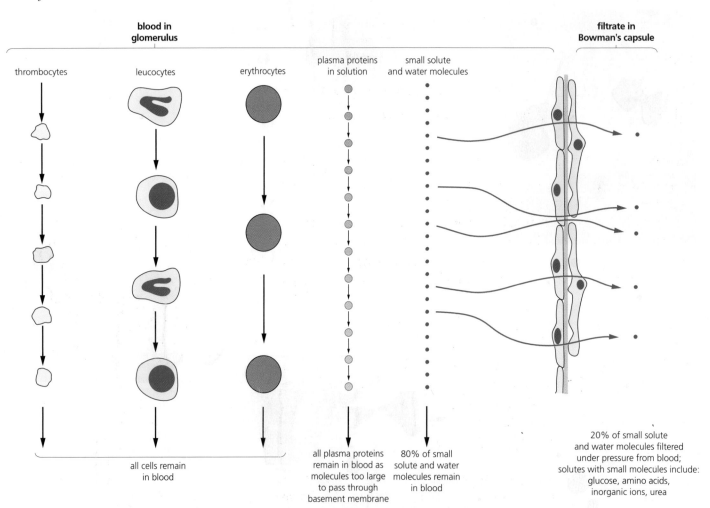

The concentration of any one filtered solute is the same in the filtrate as in the blood plasma. However, as blood plasma also contains plasma proteins in solution, its overall solute concentration is greater than that of the glomerular filtrate. In other words, the blood plasma has a lower solute potential than the filtrate (the filtrate is hypotonic to the blood plasma).

## Remember

| Solution with greater concentration of solute molecules | Solution with lower concentration of solute molecules |
| --- | --- |
| lower concentration of water molecules | higher concentration of water molecules |
| lower solute potential ($\psi_s$) | higher solute potential ($\psi_s$) |
| lower water potential ($\psi$) | higher water potential ($\psi$) |
| hypertonic | hypotonic |

*Table 10.1  A reminder of terms*

- Solutions with the same solute concentrations are **isotonic** with each other.
- Pure water has a water potential = 0 kPa (at standard temperature and pressure).
- Therefore the $\psi$ and $\psi_s$ of solutions have negative values.
- A 'higher' $\psi$ or $\psi_s$ can be referred to as a 'less negative' $\psi$ or $\psi_s$.
- A 'lower' $\psi$ or $\psi_s$ can be referred to as a 'more negative' $\psi$ or $\psi_s$.

The forces acting during the ultrafiltration process are:

- The difference between the hydrostatic pressure of blood and the hydrostatic pressure of the filtrate causes a net flow of liquid into the capsule.
- The difference between the solute potential of the filtrate and plasma causes a net flow of water to move by osmosis out of the capsule into the blood.

As the difference in hydrostatic pressures is greater than the difference in solute potentials, there is an overall net flow of filtrate into the capsule. Figure 10.10 summarises this point.

3  Explain how the structure of (a) the endothelial wall of the glomerular capillaries, (b) the basement membrane and (c) the podocytes helps them to carry out their roles efficiently in ultrafiltration.  *(3 marks)*

4  Compare the components of the glomerular filtrate with that of tissue fluid (see *AS Biology*, Section 10.12, page 300).  *(1 mark)*

In humans, the blood flow to the kidneys is so high that all the blood flows through the kidneys about once in every 5 minutes. Efficient ultrafiltration produces about 125 cm³ of glomerular filtrate every minute (180 dm³ per day!). As a typical adult has about 5–6 dm³ of blood, it follows that if all the glomerular filtrate were allowed to pass out of the body, dehydration would take place very quickly. The rest of the nephron is concerned with the reabsorption

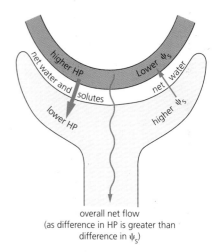

overall net flow
(as difference in HP is greater than difference in $\psi_s$)

key
HP = hydrostatic pressure
$\psi_s$ = solute potential

*Figure 10.10 Summary of the forces acting during ultrafiltration*

of essential solutes and the return of about 99% of the water, so that eventually about 1 cm³ of urine is produced per minute.

### Selective reabsorption – the proximal (first) convoluted tubule

The proximal convoluted tubule, which is the longest part of the nephron, is situated in the cortex. It is surrounded by many capillaries which are in close contact with the tubule walls. Figure 10.11a gives a diagrammatic representation of a longitudinal section through a portion of wall with an adjacent blood capillary. It also summarises the processes by which essential solutes and water are reabsorbed back into the blood. About 80% of the glomerular filtrate is reabsorbed here.

The cubical epithelial cells which line the tubule walls have numerous microvilli on their free surfaces which increase the surface area of wall exposed to the filtrate. In humans, the total surface area of the proximal tubule cells is about 50 m². The lining cells are in close contact with each other and fluid cannot flow between them. However, infoldings of the cell surface membrane create many intercellular and subcellular spaces outside the point of contact and next to the basement membrane of the blood capillary. The cells have many mitochondria near these infoldings.

Figure 10.11a shows how all the glucose and amino acids pass back into the blood by **active transport** across the infolded membrane into the subcellular spaces. This creates concentration gradients, so these solutes diffuse from the filtrate into the cells and from the subcellular spaces into the blood. The solutes are carried away in the blood, so maintaining the concentration gradients.

Some mineral ions are actively transported in a similar way to glucose and amino acids and are also reabsorbed into the bloodstream. For example, sodium ions are actively transported across the infolded membrane. They are followed by chloride ions.

*Figure 10.11 (a) LS through a portion of the wall of the proximal convoluted tubule and adjacent capillary, (b) Transmission electronmicrograph of proximal convoluted tubule cells*

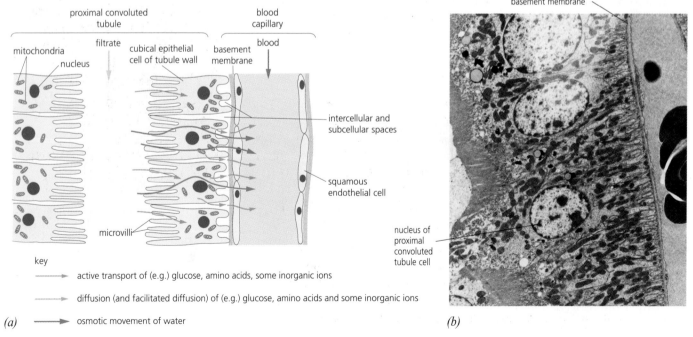

key

⟶ active transport of (e.g.) glucose, amino acids, some inorganic ions

⟶ diffusion (and facilitated diffusion) of (e.g.) glucose, amino acids and some inorganic ions

*(a)* ⟶ osmotic movement of water

*(b)*

The glomerular filtrate is hypotonic to the blood plasma when entering the tubule and the removal of so many solutes makes the filtrate even more dilute. Water therefore moves by osmosis from the filtrate into the blood and is carried away. By the time the filtrate has reached the end of the proximal tubule, a balance has been reached and the filtrate is isotonic with the plasma (i.e. it has the same overall concentration of solutes). As water leaves the filtrate, the urea becomes more concentrated than it is in the plasma and some diffuses back into the blood across the cells. Almost half the urea is unavoidably reabsorbed in this way.

So, as the filtrate enters the Loop of Henlé it is normally isotonic to the blood, and contains urea (and traces of other excretory products) and some mineral ions. If a person is suffering from diabetes mellitus (see Section 9.4, page 299), the concentration of glucose in the filtrate may be higher than usual. It may therefore not all be reabsorbed. This is why glucose may appear in the urine of a sufferer.

### The loop of Henlé

The loop of Henlé is a part of the nephron in the shape of a long hairpin bend as shown in Figure 10.12. The **descending limb**, which has thin walls *permeable* to water, penetrates deep into the medulla. The **ascending limb**, which has thicker walls relatively *impermeable* to water, returns to the cortex. The loop is surrounded by blood capillaries. One part of this capillary network, the **vasa recta**, also has the shape of a hairpin bend (see Figure 10.6, page 325).

The role of the loop of Henlé is simply to make the tissues and tissue fluid of the medulla between the nephrons (**interstitial tissues**) very concentrated with solutes. In other words, it makes their water potential lower than that of body fluids such as the blood (hypertonic to the blood). As the filtrate eventually flows down the last part of the nephron, the **collecting duct**, it passes through these medulla tissues. If the collecting duct walls are permeable to water, then water passes out of the collecting duct into these tissues by osmosis and is carried away in the blood. The urine left in the collecting duct will have a similar concentration to that in the medulla tissues and be hypertonic to the blood. Such a mechanism allows animals to conserve water when necessary.

### The working of the loop of Henlé

The cells in the walls of the thick part of the ascending limb actively transport chloride ions out of the filtrate into the surrounding tissues. Sodium ions follow passively. The surrounding tissues are therefore more concentrated with salt than the filtrate in the tubule but water cannot follow by osmosis as the walls of the ascending limb are *impermeable* to water. The longer the loop, the more sodium chloride can be transported and the more concentrated the medulla tissues can become. The deeper part of the medulla near the pelvis becomes most concentrated and so has the lowest water potential. There is a water potential gradient between the deep part of the medulla and the part near the cortex.

Filtrate entering the loop of Henlé from the proximal convoluted tubule is isotonic to the blood. It is carried through tissues of increasing solute concentration (lower water potential). The walls of the descending limb are *permeable* to water and water passes out of the filtrate into the surrounding tissues by osmosis. This water does not remain in the tissues and dilute them

### Remember

Significant features which ensure efficient reabsorption from the proximal convoluted tubule are:

- the length of the tubule and presence of microvilli give a large surface area for diffusion and facilitated diffusion
- the infoldings of the membranes next to the subcellular spaces give a large surface area which can accommodate many carriers for active transport
- the abundance of mitochondria near the infolded membranes produce ATP which releases energy needed for active transport
- the thin squamous endothelial cells of the capillary create a short distance for diffusion from the subcellular spaces to the blood
- the continuous flow of filtrate and blood maintains the concentration gradients between them.

**Figure 10.12** *The working of the loop of Henlé and collecting duct*

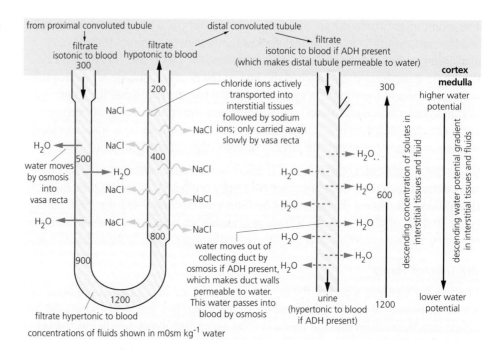

from proximal convoluted tubule

distal convoluted tubule

filtrate isotonic to blood 300

filtrate hypotonic to blood

filtrate isotonic to blood if ADH present (which makes distal tubule permeable to water)

cortex
medulla
higher water potential

200

NaCl

chloride ions actively transported into interstitial tissues followed by sodium ions; only carried away slowly by vasa recta

300

H₂O
water moves by osmosis into vasa recta

NaCl

500

NaCl

400

NaCl

H₂O

NaCl

H₂O

600

H₂O
NaCl

NaCl

H₂O

descending concentration of solutes in interstitial tissues and fluid

descending water potential gradient in interstitial tissues and fluids

H₂O

NaCl

800

NaCl

H₂O

900

water moves out of collecting duct by osmosis if ADH present, which makes duct walls permeable to water. This water passes into blood by osmosis

H₂O

H₂O

1200

filtrate hypertonic to blood

urine (hypertonic to blood if ADH present)

1200

lower water potential

concentrations of fluids shown in m0sm kg⁻¹ water

because it passes into the vasa recta and is carried away in the blood. This occurs because the blood in the ascending vasa recta is flowing from deeper, more concentrated regions of the medulla and so has a water potential lower than the filtrate in the adjacent part of the descending limb.

Owing to loss of water, the filtrate becomes more concentrated as it reaches the hairpin bend and has a water potential balancing the surrounding tissues. It is hypertonic to the blood. As the filtrate begins to flow up the ascending limb, the active removal of sodium chloride, as described above, leaves the filtrate hypotonic to the blood as it enters the distal convoluted tubule in the cortex.

**The loop of Henlé as a 'countercurrent multiplier'**

The mode of action of the loop of Henlé is often referred to as a **countercurrent multiplier** system. The filtrate flows in opposite directions in the two limbs of the loop of Henlé. This is referred to as a **countercurrent** type of flow. This countercurrent system can occur over a longer distance if the Loop of Henlé is long. If the pumping out of sodium chloride from the ascending limb and the withdrawal of water from the descending limb can occur over a longer distance

| Part of nephron | Permeability |
| --- | --- |
| Bowman's capsule | permeable |
| proximal convoluted tubule | permeable |
| descending limb of loop of Henlé | permeable |
| ascending limb of loop of Henlé | impermeable |
| distal convoluted tubule | permeable only if ADH present |
| collecting duct | permeable only if ADH present |

**Table 10.2** *Permeability to water of different parts of the nephron*

then higher concentrations of sodium chloride can be built up in the interstitial tissues of the medulla (and the water potential is lower). So the effect is **multiplied** if the loop is longer.

Mammals which need to conserve water excrete as little water in their urine as possible and produce highly concentrated urine (see Section 10.6, page 338). In order to do this, the medulla tissues must have a very low water potential so that much water passes out of the fluid in the collecting duct and is reabsorbed into the blood. This leaves urine which is hypertonic to the blood to flow on to the bladder and out of the body. For this reason desert mammals such as the kangaroo rat have relatively thick medullas in their kidneys, accommodating nephrons with long loops of Henlé. In contrast, mammals which live with an abundance of available fresh water, such as beavers, have kidneys with relatively thin medullas and short loops of Henlé (Figure 10.13).

**Definition**

A **countercurrent multiplier** system is one in which the effects brought about when liquids flow in opposing directions in close proximity, increase with the length of the system.

**5** Why is the small amount of urine produced by the kangaroo rat darker in colour than the copious amounts of urine produced by a beaver? *(1 mark)*

### The distal convoluted tubule

The cells of the wall of the distal convoluted tubule are similar to those of the proximal convoluted tubule. They have microvilli, many mitochondria and carry out active transport in a similar way. However, whereas the proximal tubule reabsorbs most of the filtrate, the distal tubule reabsorbs varying quantities of inorganic ions according to the body's needs. It can also secrete substances into the filtrate. For example, it controls the pH of the blood by secreting hydrogen ions into the filtrate when the blood is too acid; and if it is too alkaline, it secretes hydrogencarbonate ions into the filtrate.

The walls of the distal convoluted tubule are permeable to water if **antidiuretic hormone** (**ADH**) is present. Otherwise they are impermeable to water. If they are permeable, water from the hypotonic filtrate passes into the blood so that the filtrate is isotonic as it enters the collecting duct. If they are not permeable, hypotonic filtrate enters the duct.

### The collecting duct

The distal convoluted tubule transports the filtrate to the collecting duct. Several nephrons share one collecting duct. Here, final modification is made to the filtrate if necessary as it is transported through the medulla. The filtrate is then emptied, as urine, into the pelvis of the kidney.

The collecting duct walls, like those of the distal convoluted tubule, are permeable to water only if ADH is present. If the walls are permeable, then water from the filtrate passes into the medulla and is reabsorbed, leaving hypertonic urine. If they are not permeable, water remains in the filtrate and hypotonic urine is released. The role of ADH and the kidneys in maintaining water balance is discussed below.

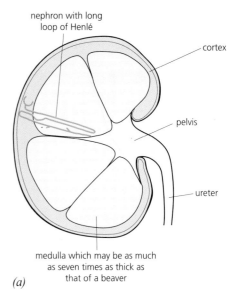
*(a)*

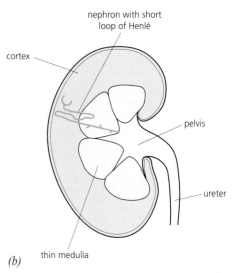
*(b)*

**Figure 10.13** *Longitudinal sections through the kidneys of (a) a kangaroo rat and (b) a beaver (not drawn to scale)*

**Remember**

The term **diuresis**, meaning the production of dilute urine, leads to two further terms being used. A **diuretic** is a substance which promotes the production of urine with a high water content. An **antidiuretic** leads to the production of a more concentrated urine. (Hence the name **antidiuretic hormone (ADH)**.)

**6** People with untreated diabetes mellitus may have glucose present in the urine (see Section 9.4, page 299). They also pass larger volumes of urine than non-diabetic people. Explain how the presence of glucose can lead to the production of large amounts of urine. *(2 marks)*

### 10.5 Osmoregulation – regulation of blood water potential

It is necessary to maintain a steady optimal water potential of the blood and tissue fluids so that the body cells are bathed in solutions which are isotonic to their cytoplasm. This prevents excessive amounts of water entering or leaving cells osmotically which would damage the cells.

If an excessive amount of water (or other beverage) is consumed, large amounts of water are absorbed into the blood, raising its water potential. On the other hand, if much water is lost by sweating, e.g. when playing sport on a hot day, the blood's water potential is lowered. It is also lowered following absorption of many inorganic ions after a salty meal.

The optimal water potential is controlled homeostatically by negative feedback systems similar to those discussed in the previous chapter (see Figure 9.2, page 294). Communication in the systems controlling water potential is mainly hormonal. Figure 10.14 shows how the antidiuretic hormone (ADH) is produced steadily by neurosecretory cells in the hypothalamus and is stored in the posterior pituitary gland.

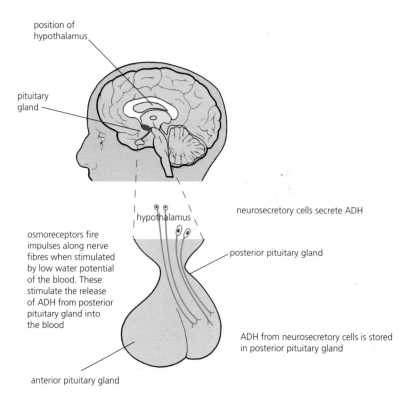

*Figure 10.14* *The mechanism of ADH release*

The receptors (**osmoreceptors**) which detect a change in water potential are also situated in the hypothalamus. If the water potential of the blood flowing near these osmoreceptors becomes lowered, nerve impulses are transmitted to the posterior pituitary gland and it is stimulated to release ADH into the blood.

ADH circulates in the blood. Its target cells are the cells of the walls of the distal convoluted tubules and collecting ducts. The hormone molecules fit into receptor sites in their cell surface membranes and they become permeable to water. Water is reabsorbed as described above, hypertonic urine is excreted and the water potential of the blood rises back towards the norm. (The osmoreceptors also stimulate a sensation of thirst. Although the water potential has been controlled, sweating will have led to a reduction in the volume of body fluids which is restored by drinking.)

Following drinking, the water potential of the blood rises and the osmoreceptors are no longer stimulated. The release of ADH from the posterior pituitary is greatly reduced. As any ADH which was in the receptor sites in the tubule walls does not remain active for long, the walls of the distal convoluted tubules and collecting ducts become much more impermeable to water. The water remains in the urine. Hypotonic urine is produced and the water potential of the blood becomes lowered as more water is lost relative to the solutes.

Figure 10.15 gives a summary of the negative feedback mechanisms involved in this process. Compare this with Figure 9.2 on page 294.

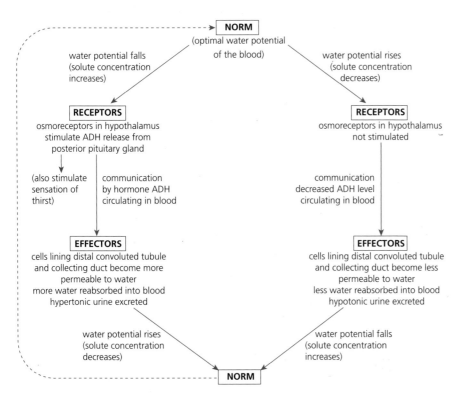

**NB:** Unlike the negative feedback systems illustrated in 9.6 and 9.18, this mechanism does not have separate systems for returning to the norm from different directions.

*Figure 10.15 The negative feedback systems involved in the regulation of the water potential of the blood (compare with Figure 9.2 on page 294)*

## Box 10.2 Diabetes insipidus (water diabetes)

This is a rare condition which must be clearly distinguished from diabetes mellitus which was discussed in Section 9.4 (page 299). A person suffering from **diabetes insipidus** produces large amounts of very dilute urine which is very hypotonic to the blood. This leads to a persistent thirst, risk of dehydration and an upset of the balance of ions in the body.

The condition can arise because the person cannot make or release enough ADH. This results in the lack of permeability of the walls of the distal tubules and collecting ducts. It follows that very little water can be reabsorbed in these parts of the nephrons and so remains in the urine.

Alternatively, the disorder may be present in people who release enough ADH, but whose cells cannot react to it. It is possible that they lack the appropriate receptor in the cell membranes of the distal tubule and collecting duct cells. This receptor is a protein. If a person does not inherit the gene which carries the code for this protein, the receptor will not be present. This rare form of diabetes insipidus is inherited. Most of the inherited cases are caused by a mutated X-linked gene and are therefore found more commonly in males than females (see Section 3.7, page 76). Sometimes the disease arises if the mechanism of responding to ADH is damaged by the use of certain prescription drugs.

## Box 10.3 Artificial kidney machines

Although a person can live normally with only one healthy kidney, if both are damaged by disease or injury this can be fatal. Kidney transplants are becoming more and more successful but a patient may have to wait for a long time before a suitable kidney is found. In the meantime, filtering of the blood can be carried out by **dialysis** using artificial kidney machines (Figure 10.16). Dialysis involves separating the small molecules from larger ones by allowing them to diffuse through a partially permeable membrane.

Figure 10.16 shows the relatively simple principles upon which a kidney machine works.

- Blood is pumped from the radial artery in the arm (or a 'shunt' vessel created between an artery and a vein). Heparin is added to prevent the blood clotting. The blood passes through the machine and an air bubble trap, and is returned to a vein in the arm. The blood circulates many times through the machine during a treatment session which may take up to 10 hours.

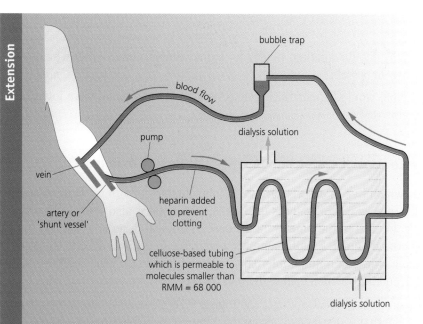

*Figure 10.16* *An artificial kidney machine*

- In the machine the blood flows through many tubes to give a large surface area. The tube walls are made of a material which allows only molecules of an RMM below 68 000 to diffuse through. It is these walls which act as a dialysing membrane.

- A prepared solution (the dialysis solution) flows around these tubes in the opposite direction to produce a countercurrent system and maximise exchange of materials between the blood and the dialysis solution (see *AS Biology*, Section 9.14, page 263).

- The dialysis solution contains glucose, amino acids and inorganic ions in solution in water. The concentrations of the various solutes and overall water potential are of the optimal level found in blood plasma. This solution is constantly flowing so that any materials diffusing into the fluid from the blood are carried away.

- As the blood flows past the dialysis solution small molecules (including water) diffuse down concentration gradients across the membrane between the two fluids. When a balance is reached the blood has the optimal level of the solutes, as found in the dialysis solution, and the correct water potential of the blood is maintained. As the dialysis fluid contains no nitrogenous wastes, such as urea, these diffuse out of the blood and are carried away in the fluid.

- Between treatments, nitrogenous wastes build up and inorganic ion concentrations and water balance are not well controlled. It is necessary, therefore, to have a carefully controlled diet during the few days between each treatment session.

**?**

**7** (a) Why should a patient treated by kidney dialysis have a diet which is low in protein? *(2 marks)*

(b) What would happen during dialysis if a patient's blood were low in glucose? *(1 mark)*

### 10.6 Control of water budget in small desert mammals

Animals which live in deserts survive in inhospitable conditions. Deserts are, by definition, places where water is in short supply. Food is also sparse. Deserts may be cold most of the time or, in the case of tropical deserts, extremely hot by day and cold by night. Small mammals which live in tropical deserts still need to maintain constant, optimal internal conditions by homeostasis and have many special adaptations which enable them to do so.

The kangaroo rat (*Dipodomys* sp.; see Figure 9.1b, page 294), which maintains a correct internal water balance despite having no available drinking water, provides a good example of such specialised mechanisms. Like other mammals, to maintain a constant water content in the body it must balance its water gain with its water loss. As water gain is low, it is essential that its water loss is kept to a minimum.

Table 10.3 shows the ways in which mammals, in general, gain and lose water. The kangaroo rat has no access to water and does not drink at all. The available food is mainly seeds which have one of the lowest water contents of all living material. Unlike most mammals, the main source of water is metabolic water. This can be seen in Table 10.4 which demonstrates water balance in a kangaroo rat fed only barley seeds for a month under controlled conditions.

The kangaroo rat keeps water loss to minimum by behavioural and physiological mechanisms.

| Water gain | Water loss |
|---|---|
| • **drinking** (for most mammals this is the major source of water) <br> • **eating food which contains water** (most food contains a high proportion of water, for example a lettuce leaf is 95% water) <br> • **producing metabolic water** (some metabolic processes, such as respiration, produce molecules of water during the chemical reactions) | • the passing of **urine** <br> • the passing out of **faeces** <br> • evaporation of water during **sweating**, and **breathing out** |

*Table 10.3 Methods of water gain and water loss in mammals*

| Water gain/g | | Water loss/g | |
|---|---|---|---|
| drinking | 0 | in urine | 13.5 |
| in food | 6 | in faeces | 2.5 |
| metabolic | 54 | breathed out | 44 |
| | | in sweat | 0 |
| Total | 60 | Total | 60 |

*Table 10.4* *Water balance in a kangaroo rat (fed 100 g of barley seeds in one month, and kept at 25 °C and 20% humidity)*

### Behavioural mechanisms

These are instinctive behaviour patterns. The rat is nocturnal. During the heat of the day, it rests in its burrow where the air is cool and humid. Evaporation from the lungs is slower in the cooler temperatures and any water vapour breathed out makes the atmosphere humid (raises its water potential). As the water potential gradient between the lung tissues and the air in the lungs is reduced, less water is lost by evaporation. In other words, the air breathed in is almost as damp as the air breathed out. The rat only leaves its burrow to search for food at night when it is cold and there is no need to cool by sweating.

### Physiological mechanisms

Figure 10.13a on page 333 shows that the deep medulla of the kidney of a kangaroo rat contains nephrons with long loops of Henlé. As explained in Section 10.4, the long loops produce very high salt concentrations in the medulla, allowing the production of highly concentrated urine. Large quantities of ADH are also produced which favour the production of concentrated urine.

The rat conserves water by producing urine which is 17 times as concentrated as its blood. (This is four times more concentrated than that which can be produced by humans.)

8  Explain how (a) high salt concentration in the medulla and (b) high levels of ADH allow the production of concentrated urine in a kangaroo rat.

*(2 marks)*

The nasal passages of the rat are specially adapted and have a large surface area. They become cooled by evaporation of water as the animal breathes in but most of this water condenses on the cool surfaces again as it breathes out. Nevertheless, loss of water during breathing is the major way in which water is lost. The colon of the rat is very efficient at reabsorbing water from the faeces and very dry faeces are produced, minimising water loss.

### Summary – 10 Excretion and water balance

- Excretion is the removal of the waste products of body metabolism which would be toxic if allowed to accumulate. Such waste products include carbon dioxide, bile pigments and nitrogenous waste products.

- Nitrogenous waste products include the breakdown products of excess amino acids. Amino acids are first deaminated, producing ammonia and keto acids. The keto acids are used in carbohydrate metabolism but the highly toxic ammonia must be removed.

- Mammals convert ammonia into the less toxic urea. This urea is produced in the liver and then transported to the kidneys from where it is excreted in the urine.

- The mammalian urinary (renal) system comprises two kidneys each with a ureter leading to a urinary bladder from which urine is evacuated through a urethra.

- A kidney shows distinct layers, the outer cortex and the inner medulla which surrounds a central pelvic cavity. The kidney is composed of about a million small tubules known as nephrons.

- Each nephron is supplied with blood vessels which form a knot of capillaries, the glomerulus, enclosed by the cup-shaped Bowman's capsule, and which surround all other parts of the nephron.

- Each nephron produces urine as follows:
  – Ultrafiltration takes place in the Malpighian body (Bowman's capsule and glomerulus). Small molecules are filtered from the blood plasma into the cavity of the Bowman's capsule under pressure. They pass through the thin porous capillary walls, the filtering basement membrane and the podocytes of the cavity wall which are specialised to allow free flow of substances between their branches. The glomerular filtrate contains glucose, amino acids, inorganic ions and urea, all dissolved in water. It is hypotonic to the blood as the solution contains no plasma proteins.
  – In the proximal convoluted tubule, glucose, amino acids and some inorganic ions are actively transported back into the blood during selective reabsorption. Water follows by osmosis and the filtrate becomes isotonic to the blood.
  – The loop of Henlé creates hypertonic conditions in the interstitial tissues and fluids of the medulla, giving a decreasing water potential gradient extending from the medulla's boundary with the cortex to the inner medulla. This enables water to be withdrawn from the collecting ducts if they are permeable to water. Urine passing out of the collecting ducts is then hypertonic to the blood. As it does this, the loop of Henlé operates as a countercurrent multiplier. Its ascending limb actively pumps chloride ions (followed by sodium ions) into the tissues. Water is withdrawn from the descending limb and is carried away in the blood in the vasa recta. The longer the loop, the greater the salt concentration built up in the tissues.
  – Before the filtrate passes down the collecting duct, it travels through the distal convoluted tubule where further substances are actively passed into and out of the filtrate according to the body's needs. Water passes into the blood by osmosis if the distal tubule walls are permeable to water.

- Antidiuretic hormone (ADH) makes the walls of the distal convoluted tubules and collecting ducts permeable to water when it fits into receptor sites in the membranes of cells in the walls.

- Osmoregulation (the maintenance of the optimal water potential of body fluids) involves the kidneys and ADH. If blood water potential falls, ADH is released into the blood. ADH enables water to be reabsorbed through the walls of the distal convoluted tubules and collecting ducts. This raises the water potential of the blood and leads to the production of hypertonic urine. If the water potential of the blood rises, then ADH release is reduced and the impermeability of the tubule walls prevents the reabsorption of water. More water passes out in the dilute urine and the water potential of the blood falls.

- Small desert mammals such as the kangaroo rat do not drink water and only obtain a little water in their food. They produce most of their water by metabolic processes such as respiration. As water is in short supply they have mechanisms which enable them to conserve water.

## Answers

**1** Faeces contain undigested food which is egested; bile pigments which are excreted *(2)*.

**2** Contraction of the left ventricle wall in the heart *(1)*.

**3 (a)** Thin squamous cells with pores *(1)*.

**(b)** Only permeable to molecules smaller than RMM 68 000 *(1)*.

**(c)** Branches allow free flow of substances between them into the cavity of the capsule *(1)*.

**4** They are the same *(1)*.

**5** The solutes, which provide the colour, are more concentrated in the urine of the kangaroo rat *(1)*.

**6** The presence of glucose in the filtrate in the collecting duct lowers its water potential; therefore less water is reabsorbed as the water potential gradient between the filtrate and the medulla tissues is reduced *(2)*.

**7 (a)** Proteins are digested into amino acids which may be in excess of requirements; these are converted to urea which may build up to toxic levels between treatments *(2)*.

**(b)** Glucose would diffuse into the blood from the dialysis fluid until the optimum level was reached *(1)*.

**8 (a)** More water passes back into the blood through collecting duct walls as there is a steeper water potential gradient between the filtrate and the medulla *(1)*.

**(b)** More ADH molecules in receptor sites make the collecting duct walls more permeable and more water can pass through and be reabsorbed *(1)*.

### End of Chapter Questions

**1** The diagram represents a nephron of a mammalian kidney. Letters A–F indicate regions of the nephron. Letters V–Z indicate points along the nephron.

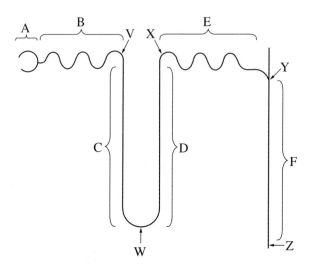

(a) Using the appropriate letters from the diagram, answer the following:

(i) Which region(s) are very long in a mammal which needs to conserve water? *(1 mark)*

(ii) In which regions would you expect to find large numbers of mitochondria in the cells of the walls? Explain your answer. *(2 marks)*

(iii) In which region(s) of the tubule is glucose normally absent from the filtrate? *(1 mark)*

(iv) At which points along the nephron might the filtrate be hypertonic to the blood? *(1 mark)*

(b) Explain why, although sodium ions are reabsorbed in B, the concentration of sodium ions in the filtrate is the same at point V as it was in A. *(1 mark)*

(c) Why is the concentration of sodium ions lower at point X than at point V? *(1 mark)*

(d) If ADH is present, why is the concentration of sodium ions greater at point Z than at point Y? *(1 mark)*

(e) What effect would a decrease in ADH release have on the water potential of the blood plasma? *(1 mark)*

(f) Haemoglobin molecules have an RMM less than 68 000. Why is haemoglobin not normally found in urine? What does it indicate if it is found in solution in urine? *(2 marks)*

*(Total 11 marks)*

**2** The drawing has been made from an electronmicrograph of part of the glomerulus and renal capsule (Bowman's capsule) of a nephron.

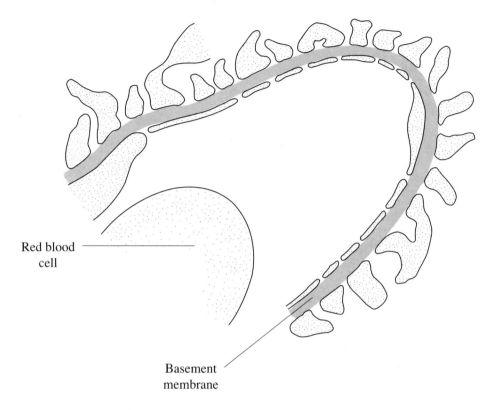

Red blood cell

Basement membrane

(a) Add an arrow to the drawing to show the route by which filtered substances pass into the renal capsule.
*(1 mark)*

(b) Explain how the arrangement of structures shown in the drawing allows effective filtration.
*(2 marks)*

(c) The table contains data concerned with the filtration of some substances in the renal capsule.　　　⌨️○ **N2.1**

| Substance | Radius of molecule /nm | Concentration of substance in renal capsule / Concentration of substance in plasma |
|-----------|------------------------|-----------------------------------------------------------------------------------|
| Urea | 0.16 | 1.0 |
| Glucose | 0.36 | 1.0 |
| Albumin | 3.55 | 0.01 |

Explain how this information supports the hypothesis that filtration is linked to molecular size.
*(2 marks)*

*AEB 1993*

*(Total 5 marks)*

3  Diuretics are substances which increase urine production by the kidneys.

(a) Complete the flow chart to show how ethanol acts as a diuretic.

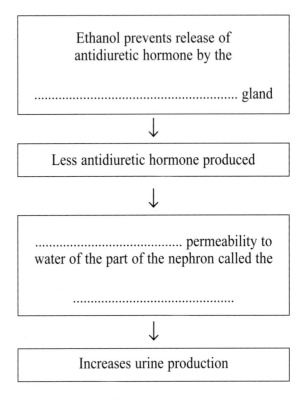

Ethanol prevents release of
antidiuretic hormone by the

.............................................................. gland

↓

Less antidiuretic hormone produced

↓

.......................................... permeability to
water of the part of the nephron called the

.............................................

↓

Increases urine production

*(2 marks)*

(b) (i) Explain why eating large amounts of glucose can lead to an increase
in urine production.

(ii) Some diuretics used in medicine inhibit the reabsorption of chloride
ions in the loop of Henlé. Explain how this would lead to an increase
in urine production.  *(3 marks)*

*AEB 1997*

*(Total 5 marks)*

4  Diabetes insipidus is a condition in which large volumes of dilute urine are
produced.

The volume of urine produced each day by a patient with diabetes insipidus
was recorded. On certain days the patient received an injection of pituitary
extract. The results are shown in the table on the next page. The asterisks (*)
indicate the days on which pituitary extract was injected.

N2.1

(a) Describe the effect of the injection of pituitary extract on the volume of
urine produced each day.  *(2 marks)*

(b) Name the substance responsible for this effect.  *(1 mark)*

(c) Describe how this effect is brought about.  *(2 marks)*

| Day | Volume of urine / dm³ | Day | Volume of urine / dm³ |
|-----|------------------------|------|------------------------|
| 1 | 5.4 | 9 | 6.3 |
| 2 | 6.0 | 10 | 4.9 |
| 3 | 5.8 | 11 | 5.1 |
| 4 | 5.0 | 12 | 5.5 |
| 5 | 7.0 | 13 * | 1.5 |
| 6 | 5.7 | 14 * | 2.1 |
| 7 * | 1.9 | 15 * | 2.4 |
| 8 | 2.2 | 16 * | 1.7 |

*London 1997*                                  *(Total 5 marks)*

**5**   The diagram below shows the simplified structure of a kidney tubule (nephron).

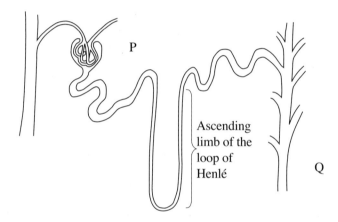

**(a)**   The table below shows the quantities of water and urea passing through P and Q in a period of 24 hours. The table also shows the quantities and percentages reabsorbed during the same period.   **N2.2**

Complete the table by writing the correct figures in the boxes labelled (i), (ii), (iii) and (iv).

| Substances | Quantity passing through P | Quantity passing through Q | Quantity reabsorbed | Percentage reabsorbed |
|------------|---------------------------|---------------------------|---------------------|-----------------------|
| Water | 180 dm³ | 1.5 dm³ | 178.5 dm³ | (i) |
| Urea | 53 g | 25 g | (ii) | (iii) |
| Glucose | 180 g | (iv) | 180 g | 100 |

*(4 marks)*

**(b)**   Describe how the ascending limb of the loop of Henlé is involved in adjusting the concentration of the filtrate as it passes through the medulla in the kidney.

*(4 marks)*

*Edexcel 2000*                                  *(Total 8 marks)*

C2.2   6   Read the following passage.

A few deaths have been due to drinking too much water. This occurred either because people thought that water was an antidote to *Ecstasy* (it is only an antidote to dehydration) or because the drug induced repetitive behaviour. Under the influence of *Ecstasy* people have been known to drink 20 litres of fluid and smoke 100 cigarettes within three hours. Drinking pure water to replace the fluid lost in sweating is dangerous because it does not put back the sodium ions and so dilutes the blood. This makes the cells of the body swell up, which is particularly lethal in the brain which can expand and be crushed against the skull. The centres in the brain that regulate breathing and the heart can be irreversibly damaged, and the person dies. *Ecstasy* increases the risk of brain damage by triggering the release of antidiuretic hormone (ADH).

*(Reproduced by permission of The Guardian)*

(a)   Describe the mechanisms which lead to copious sweating by a person dancing in a night club. *(5 marks)*

(b)   Explain in terms of water potential why the drinking of large amounts of pure water after copious sweating may lead to swelling of cells in the brain. *(5 marks)*

(c)   Explain how '*Ecstasy* increases the risk of brain damage by triggering the release of antidiuretic hormone (ADH)'. *(4 marks)*

*AQA B 2000*                                                   *(Total 14 marks)*

N2.1   7   The graph shows the rate of glucose absorption in and excretion from a mammalian kidney in relation to the glucose concentration in the plasma.

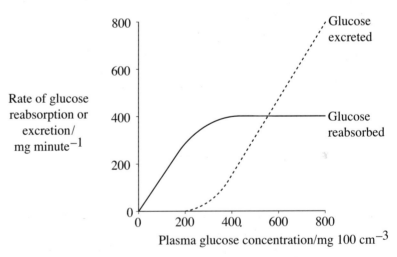

(a)   Draw a line on the graph to show the rate of filtration of glucose in the renal capsule. *(1 mark)*

(b)   In which part of the nephron is glucose reabsorbed? *(1 mark)*

(c)   Explain the shape of the glucose reabsorption curve. *(3 marks)*

*AQA A 2000*                                                   *(Total 5 marks)*

# 11 Control and coordination in plants

Plants, like animals, must respond to conditions and changes in the internal and the external environments if they are to maintain optimal conditions and survive. For example, it is essential that sufficient light energy is absorbed by plant leaves for photosynthesis. A plant in dim light responds by modifying the growth of its stems and leaves in a way which results in as much leaf surface being exposed to as much light as possible (Figure 11.1).

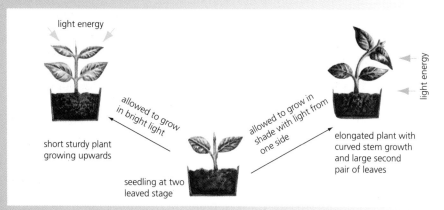

**Figure 11.1** *Plant responses involve modification of growth*

As in animals, any environmental stimulus is detected and the information conveyed to an area of the body which responds. In animals, this information is conveyed by chemical means (hormones) or by the nervous system. Responses (involving movement or secretion of chemicals) are brought about quite rapidly. In plants, information is conveyed chemically and responses usually involve modification of growth. These responses are brought about much more slowly.

**This chapter includes:**
- overview of responses in plants
- how plants grow
- tropic responses (tropisms)
- auxins and their role in growth, tropisms and other responses
- other plant growth regulators – gibberellins, cytokinins, abscisic acid and ethene
- synergism and antagonism
- phytochromes and the detection of light.

## 11.1 Plant responses – an overview

### Stimuli and detection

A change in an environmental factor (outside or inside the plant) which brings about a response is called a **stimulus**. If plants are to photosynthesise and reproduce, the availability of light, water, inorganic ions and a suitable temperature are essential and all of these factors can act as stimuli. Plants can respond to **light intensity**, the **direction of light** and the length of daylight (**photoperiod**) in a 24-hour period. Plants can also respond to gravity in such a way that their shoots are exposed to light and their roots reach water and inorganic ions while also supporting the plant. Climbing plants respond to touch which enables them to twine around other structures or plants until they reach the light.

Some of the ways in which plants detect stimuli are not fully understood, although the region of the plant which acts as the receptor or detector is often quite clear. The role of **phytochromes** in the detection of light and that of **statoliths** in gravity detection are discussed later in the chapter.

## Communication and control

Sometimes the response takes place at some distance from the area of detection. In this case a chemical is produced where the stimulus is detected and is transported or diffuses to the effector region where the response takes place. Such a chemical is acting in a similar way to an animal hormone. Chemicals like these used to be called plant hormones. However, in many cases they bring about responses in the area where they are produced and control the response themselves. The response usually involves influencing growth in some way. For these reasons these chemicals are now referred to as **plant growth regulators** or **plant growth substances**.

## Response (effect)

This usually involves a modification in the pattern of growth or development. For example, the direction of growth of a stem may be affected or a tip of a shoot may develop into a flower bud instead of continuing in vegetative (non-flowering) growth.

### 11.2 How plants grow

Two main types of plants produce flowers, **monocotyledonous** and **dicotyledonous** types. They have different patterns of growth. Figure 11.2 shows the general form of such plants when they are not in flower. Unless specifically stated, the following descriptions apply to plants of the dicotyledonous type. The terms which are used for the various parts of the plants are also shown in Figure 11.2.

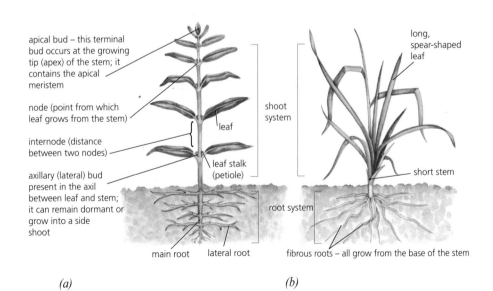

*Figure 11.2* *General pattern of the vegetative (non-flowering) parts of the two main types of flowering plant (a) dicotyledonous, e.g. wallflower, rose, (b) monocotyledonous, e.g. grass, wheat*

As most plant responses affect plant growth, it is necessary to understand the outlines of the ways in which plants grow before these responses can be discussed. The definition of growth sounds like a complicated way of describing the obvious but it is important not to regard increase in size caused by the temporary uptake of water or storage of fat (in animals) as growth.

Animals grow throughout their bodies but plants have special zones where new cells are produced and zones where cells increase in size. Regions where cell division takes place in plants are called **meristems**. These consist of closely packed, small, undifferentiated cells with thin cell walls and no large cell vacuoles (Figure 11.3). These cells can divide by mitosis (see *AS Biology*, Section 6.4, page 158).

Meristems at the tips (apices) of all roots and shoots are called **primary** (or **apical**) **meristems** and these provide cells for the growth in length of the roots and shoots. Each side shoot, even inside a bud, has a primary meristem at its tip. The cells which are produced, grow and then differentiate into cells of the various plant tissues, such as parenchyma, xylem and phloem, in the regions behind the meristems. The meristem of a root also provides cells for a root cap which covers the meristem, protecting the tip as the root pushes into the soil. Cells in the outer layers of the cap are worn away and are replaced by new cells from the meristem. Figure 11.4 shows the regions of root tips and shoot tips in which these events take place.

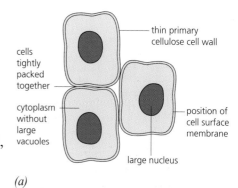

*(a)*

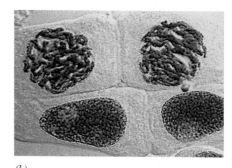

*(b)*

**Figure 11.3** *Meristem cells (meristematic cells)*
*(a) The main characteristics of meristem cells,*
*(b) Meristem cells seen under a light microscope (they can also be seen undergoing mitosis)*

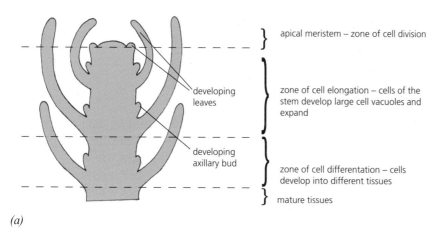

*(a)*

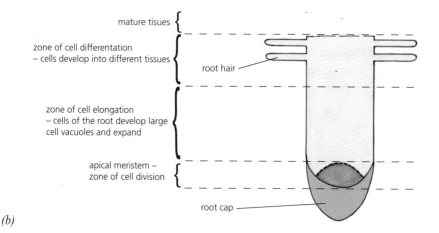

*(b)*

**Figure 11.4** *Longitudinal sections through*
*(a) a shoot tip and*
*(b) a root tip, simplified, to show the main regions involved in growth and development*

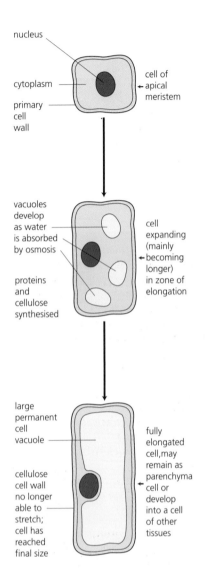

*Figure 11.5* *Stages involved in cell enlargement*

*Table 11.1* *Examples of tropic responses*

The cells elongate and enlarge in the region just behind the meristem which is known as the **zone of cell elongation**. Figure 11.5 shows the various stages involved in this cell enlargement.

The cells produced in the meristem have very thin **primary cell walls** with fibrils of cellulose scattered irregularly in a gel of hemicelluloses and other polysaccharides (see *AS Biology*, Section 1.16, page 15). They develop small vacuoles, absorbing water by osmosis, and eventually large permanent cell vacuoles are formed. As they absorb water they increase in size. This is possible as long as the cell walls are flexible. However, as each cell develops, further layers of more densely packed cellulose fibres are deposited in the cell wall. The layers are orientated in different directions, reducing the wall's flexibility. Eventually the cell cannot expand any more. The cell expansion which has taken place in this zone of elongation behind the meristem has been mainly caused by the uptake of water but true growth also takes place as further organic material is synthesised. For example, more proteins and other macromolecules are produced in the cytoplasm as well as cellulose molecules being added to the cell walls.

The mature cells then differentiate into cells of the various tissues in the region behind (known as the **zone of cell differentiation**). However, some cells do not develop large vacuoles, remain undifferentiated and retain the power of cell division. These are found between the xylem and phloem. They provide the cells for growth in the diameter of the roots and shoots. These cells make up the **secondary meristems** (or **cambium**) which become active as the plant matures and provide further cells for the xylem and phloem tissues.

### 11.3 Tropic responses (tropisms)

The direction in which shoots and roots grow is determined by their response to external stimuli. It is advantageous if shoots grow towards the direction of light rays as leaves, held at approximately right angles to the stem, are then able to absorb as much light energy as possible. Similarly it is advantageous if roots grow downwards into the soil as they are then able to absorb water and inorganic ions and provide anchorage for the plant. The type of response where the direction of growth is determined by the direction of the stimulus is called a **tropic response** or **tropism**. Table 11.1 summarises the two different types of tropism discussed in this chapter.

| Tropism | Stimulus | Response |
|---|---|---|
| phototropism | direction of light – exposure to light from one side (unilateral light) | shoots bend towards the light – they are positively phototropic (roots do not respond) |
| geotropism | force of gravity | shoots grow upwards against the force of gravity (negatively geotropic) |
| | | roots grow downwards with the force of gravity (positively geotropic). |

Simple observations can reveal these responses. A shoot placed in a position with shade on one side and exposed to sunlight on the other grows in a curve towards the sun's rays and then continues to grow towards the direction of the light. Similarly, anyone who has planted seeds in the dark earth is sure that the shoot will emerge upwards from the ground while the root will grow downwards. However, controlled experiments such as those shown in Figure 11.6 are necessary to verify these observations.

> **Definition**
>
> A **tropism** is a plant response in which the direction of growth is determined by the direction of a stimulus.

1  Study Figure 11.6 and answer the following:

(a) Why is it necessary to carry out experiments with controls before being able to verify that a shoot or root is responding to a particular stimulus?

(b) In Experiment 1, is A or B the control? Explain your answer.

(c) Several seedlings are used in each of the experiments. Why is this necessary?                    *(5 marks)*

*Figure 11.6* *Experiments demonstrating tropic responses (a) Experiment 1 – demonstrating positive phototropism in cress seedlings, (b) Experiment 2 – demonstrating positive geotropism in roots of germinating broad bean seeds*

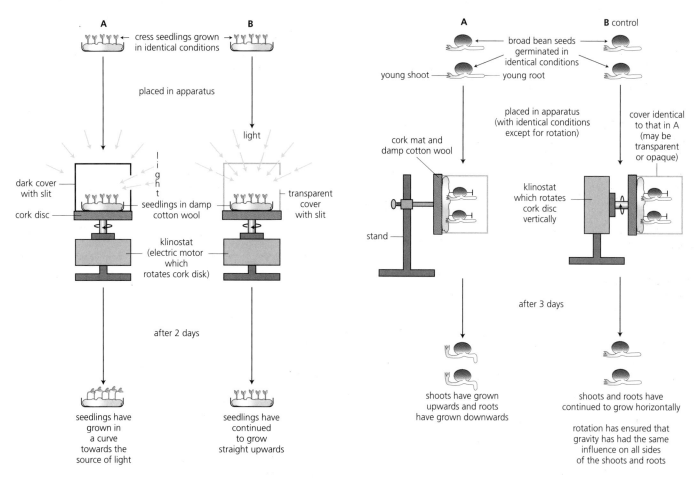

(a)                                          (b)

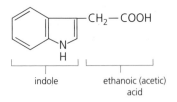

*Figure 11.7  The structure of indole-acetic acid (IAA)*

**11.4  Auxins and their effects**

### The nature of auxins

Auxins are a group of chemicals which act as plant growth regulators. Their name is derived from the Greek word *auxein* – to grow. Although they affect many aspects of plant growth and development, a chemical is considered to be an auxin if, amongst other things, it can promote the elongation of stem cells. The first auxin to be isolated was **indole-acetic acid** (**IAA**). Its chemical structure is shown in Figure 11.7. It is easily synthesised and is used commercially and in experimental work.

The highest concentrations of auxins are produced in shoot apical meristems, growing leaves and fruits. They move away from their point of origin, probably by diffusion from cell to cell, and can be transported over longer distances from shoots to roots mainly in the phloem. They are eventually broken down by enzymes.

**?**

2  The diffusion of auxins through stems is said to be 'polar'. Study Figure 11.8 and explain what is meant by 'polar diffusion'.                      *(2 marks)*

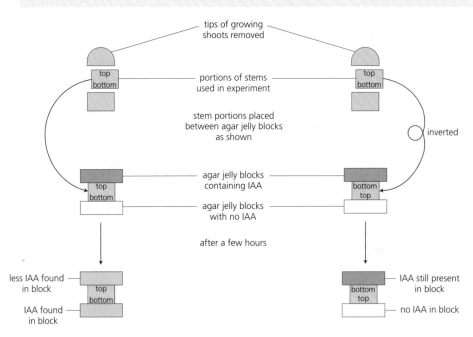

*Figure 11.8  Experiment demonstrating polar diffusion of IAA*

### The role of auxins in the growth in length of shoots

Experimental evidence gathered over many years has led to the following theory about the processes involved in the growth in length of stems. A few of the experiments involved are described in Figure 11.9. Many of these experiments have been carried out on **coleoptiles** rather than naked stems, but as coleoptiles react in a similar way to stems, the results have been used in helping to develop the ideas listed below. Figure 11.10 shows a coleoptile.

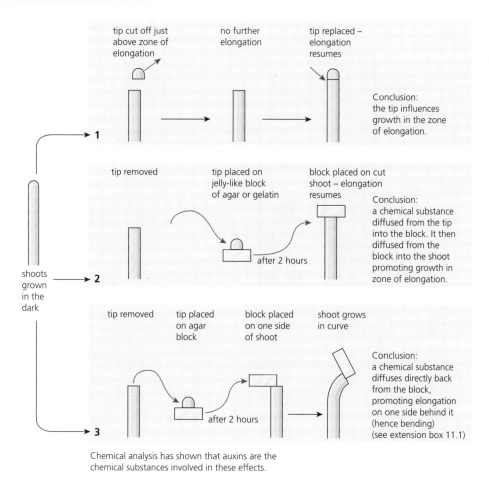

**1**

tip cut off just above zone of elongation → no further elongation → tip replaced – elongation resumes

Conclusion: the tip influences growth in the zone of elongation.

shoots grown in the dark

**2**

tip removed → tip placed on jelly-like block of agar or gelatin (after 2 hours) → block placed on cut shoot – elongation resumes

Conclusion: a chemical substance diffused from the tip into the block. It then diffused from the block into the shoot promoting growth in zone of elongation.

**3**

tip removed → tip placed on agar block → block placed on one side of shoot (after 2 hours) → shoot grows in curve

Conclusion: a chemical substance diffuses directly back from the block, promoting elongation on one side behind it (hence bending) (see extension box 11.1)

Chemical analysis has shown that auxins are the chemical substances involved in these effects.

*Figure 11.9 Experiments which provide evidence that auxins are involved in stem growth*

Although details vary, shoots of dicotyledonous plants emerge above ground in a curved way protecting the stem tip. They straighten later. For example:

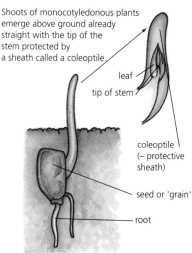

tip of stem surrounded by leaves
seed
root

*(a)*

Shoots of monocotyledonous plants emerge above ground already straight with the tip of the stem protected by a sheath called a coleoptile.

leaf
tip of stem
coleoptile (– protective sheath)
seed or 'grain'
root

As growth and tropic responses are easier to measure if the shoot is straight, many experiments have been carried out using coleoptiles.

*(b)*

*Figure 11.10 Emerging shoots of (a) dicotyledonous and (b) monocotyledonous plants*

## Growth in length in stems:

● Auxins are synthesised in the cells of shoot apical meristems.
● These auxins diffuse away from the tip towards the zone of elongation.
● The auxins bind to specific receptor sites on the cell surface membranes of the small, newly formed cells.
● This activates some molecules in the cell surface membranes to pump hydrogen ions (by active transport) from the cytoplasm into the primary cell walls, lowering their pH.
● The low pH provides the optimum conditions for enzymes to act, which break bonds between adjacent cellulose microfibrils. The microfibrils can slide freely past each other and this keeps the walls flexible.
● The cells absorb water by osmosis and the flexible cell walls allow the cells to expand as the extra water exerts increased hydrostatic pressure against them.
● As the cells mature they become situated further from the tip owing to the formation of new cells. Eventually the auxins in the receptor sites are destroyed by enzymes and the pH of the cell walls rises. Bonds form more readily between the increased number of cellulose microfibrils and the cell walls become more rigid and cannot expand any more.

*Figure 11.11 Summary of the role of auxins in shoot growth*

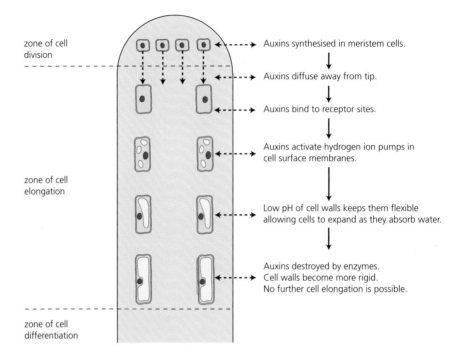

Figure 11.11 illustrates these ideas. It must be remembered that other processes are going on at the same time. For example protein synthesis (possibly also stimulated by the presence of auxin) is taking place as more proteins are required for cell membranes and metabolism in the cytoplasm. Also cellulose is being added to the growing cell walls.

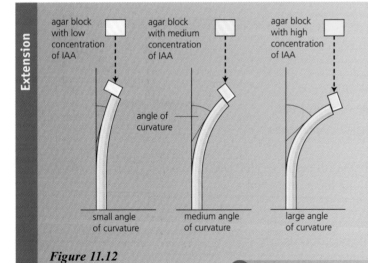

*Figure 11.12*

**Extension**

### Box 11.1 Bioassay

Bioassay is a technique in which the concentration of a particular substance is determined by measuring the extent of its biological effect. In 1928, a Dutch scientist named Went found that when an agar block containing auxins was placed on one side of a shoot (as in Figure 11.9, Experiment 3), the angle of curvature produced depended upon the amount of auxins in the block. This idea is shown in Figure 11.12. The results of such experiments could be used to *compare* the relative concentrations of auxins in different blocks.

**?**

3 (a) List *three* conditions which must be the same in the various experiments shown in Figure 11.12 in order for the comparisons to be valid.

(b) How is it possible to modify such experiments to *determine* the absolute concentration of auxins in a particular agar block? *(5 marks)*

## The role of auxins in tropic responses

Tropic responses have been studied for over a century by many scientists, with some important early experimental results being obtained by Charles Darwin and his son.

## Phototropism in shoots

The role of auxins in the positive phototropic response of shoots is explained below. Figure 11.13 describes some of the experiments which have led to the following ideas.

*Figure 11.13 Experiments which provide evidence of the role of auxins in the phototropic response in shoots*

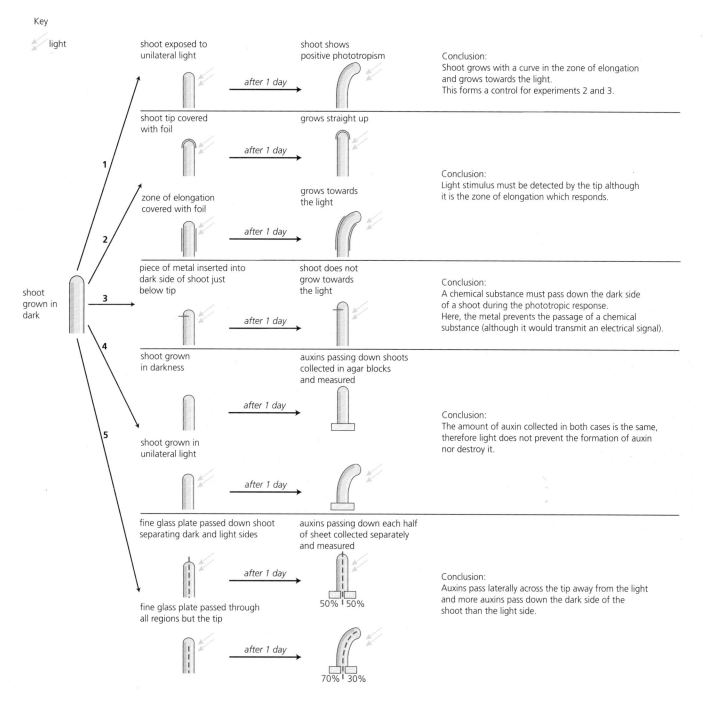

The main stages in the response are thought to be as follows:

● Auxins produced in the tip move across the tip away from the side exposed to the light towards the dark side. (In this way the tip detects the direction of the light and acts as a **receptor** region.)

● Therefore a higher concentration of auxins diffuses down the dark side of the shoot than the light side.

● More auxin molecules bind to receptor sites of the cells on the dark side of the zone of elongation than on the light side.

● The cell walls on the dark side of the shoot remain more flexible for longer than the cell walls on the light side.

● The cells on the dark side of the shoot elongate more than the cells on the light side, which leads to curved growth, and the shoot bends towards the direction of the light. (The response of the curved growth takes place in the zone of elongation and so it can be considered as the **effector** region in this response.)

Figure 11.14 summarises this theory.

*Figure 11.14 Summary of the role of auxins in the phototropic response of shoots*

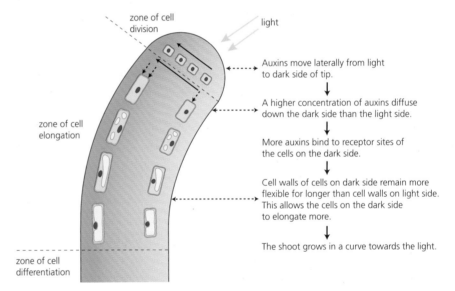

membrane of amyloplast    starch grain

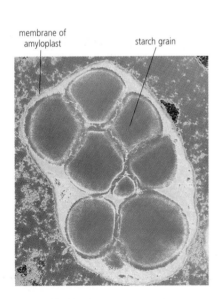

*Figure 11.15 Transmission electronmicrograph of an amyloplast in a root cap cell (×18 000). Amyloplasts are heavy and sink through the cytoplasm under the influence of gravity*

### Geotropism

The receptor region in this response in roots is thought to be the **root cap**. If the root cap is removed, a horizontally placed root does not respond to gravity and continues to grow horizontally. Many of the root cap cells contain large **amyloplasts** (see Figure 11.15) in their cytoplasm. Amyloplasts are membrane-bound organelles which contain starch grains. They can be found in many plant cells where starch is stored. However, the large, heavy amyloplasts found in root cap cells can be seen to sink through the cytoplasm to the bottoms of the cells which contain them under the influence of gravity.

It is thought that they act as **statoliths** (gravity detectors) and they may press against certain areas of endoplasmic reticulum or cell surface membrane as they sink and stimulate a particular reaction. Although this mechanism is not understood, it is known that if starch grains are removed chemically from its

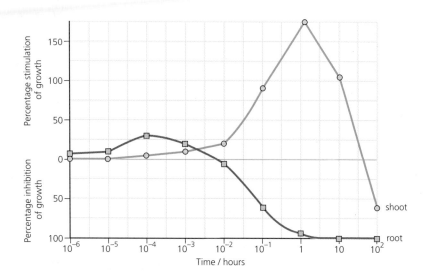

root cap cells, a horizontally placed root is unable to respond to gravity until the starch grains reform the next day. (Some doubt about the role of amyloplasts has arisen as it has been shown that some genetically engineered plants without amyloplasts still show the geotropic response.)

The role of auxins in this response is now very doubtful. For many years it has been known that the concentrations of auxins in roots are much lower than the concentrations in shoots. They are carried from the shoot to the root in the phloem. The high concentrations of auxins found in the stems actually *inhibit* elongation of cells in roots (Figure 11.16). The low concentrations found in roots have no effect on elongation of stem cells. These facts have been used to develop a very tidy theory to explain negative geotropism in shoots and positive geotropism in roots. This theory is explained in Figure 11.17.

However, this theory is very unlikely to explain the response, especially in roots, as:

● It has been demonstrated experimentally that auxins are able to move only from the 'shoot end' towards the root cap in a portion of root and not from the root cap backwards as suggested in the theory.

● In an intact root, it is known that auxins are carried into the root in the phloem from the shoot. There is no evidence that auxins are synthesised in the root tip or cap.

● Concentrations of auxins in roots are so low that it has not been possible to show whether there is a significantly higher concentration of auxin on the underside than on the upper side.

● On the other hand, much evidence points to other growth regulators having a role. **Abscisic acid** (see Section 11.7, page 364), a growth inhibitor, is known to be present in root caps. **Gibberellins** (see Section 11.5, page 360), growth promoters, have also been shown to be present in higher concentrations on the more rapidly growing sides of horizontally placed roots and shoots.

Clearly, there is much to be discovered about the roles of growth regulators in the geotropic response. While auxins may play a part in the shoot's response, they seem unlikely to have a significant role in the reaction of roots.

germinating seed placed so that the young root and shoot are horizontal

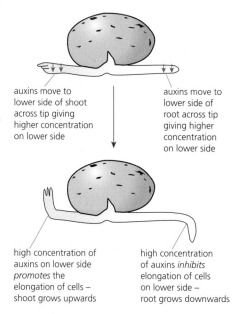

auxins move to lower side of shoot across tip giving higher concentration on lower side

auxins move to lower side of root across tip giving higher concentration on lower side

high concentration of auxins on lower side *promotes* the elongation of cells – shoot grows upwards

high concentration of auxins *inhibits* elongation of cells on lower side – root grows downwards

**Figure 11.17** *An old theory which suggested that auxins play a major part in the geotropic response. It is considered now that these ideas do not fit the facts (see text)*

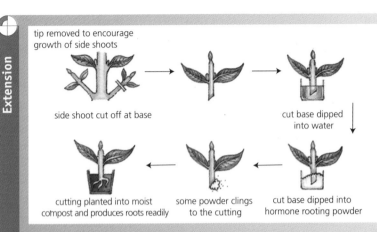

tip removed to encourage
growth of side shoots

side shoot cut off at base

cut base dipped
into water

cut base dipped into
hormone rooting powder

some powder clings
to the cutting

cutting planted into moist
compost and produces roots readily

*Figure 11.18 Taking cuttings*

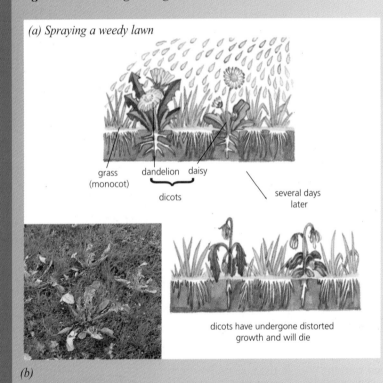

(a) Spraying a weedy lawn

grass
(monocot)    dandelion  daisy

dicots

several days
later

dicots have undergone distorted
growth and will die

(b)

*Figure 11.19 Use of
selective weedkiller on
a lawn*

*Figure 11.20 Seedless grapes*

### Box 11.2  Commercial uses of auxins

An understanding of some of the functions of auxins in plants has led to their exploitation commercially. It is possible to manufacture several synthetic forms of auxins on a large scale and different types are used in different ways:

#### In hormone rooting powders

As some auxins promote the growth of roots from the cut ends of stems, the treatment shown in Figure 11.18 helps cuttings to grow roots more quickly.

#### In selective weedkillers

If plants are sprayed with certain forms of synthetic auxins, they can react differently. At high concentrations, monocotyledonous plants (monocots) are unharmed, while dicotyledonous plants (dicots) can be killed. The dicots are more sensitive than monocots to the effects of auxins and absorb more auxins through their broader leaves. Their growth is disrupted, their stems elongate and collapse and the repeated division of cells can cause growths which block xylem vessels and phloem sieve tubes (see *AS Biology*, Section 15.9, page 447).

As grasses and cereals are monocots and the weeds found amongst them are mainly dicots, spraying with appropriate solutions kills the weeds while leaving the cereal crops or grasses of a lawn largely unharmed (Figure 11.19).

#### In fruit setting

Spraying young developing fruits with auxins can prevent them falling off prematurely and so let them grow on the plant (the fruit are allowed to '**set**'). If certain flowers are sprayed with a mixture of auxins, gibberellins and cytokinins they can form fruits without being fertilised (parthenocarpy). Such fruits are seedless. There is a large market for such fruits, for example seedless grapes (Figure 11.20).

### In micropropagation

If clusters of undifferentiated plant cells are grown in mixtures containing the correct proportions of auxins and cytokinins (see Section 11.6, page 363) the cells differentiate and eventually form new plants. Such techniques allow many plants to be produced rapidly in a small space. The plants produced are genetically identical to the plant providing the original cells (see *AS Biology*, Figure 6.10, page 166).

**Figure 11.21** *Plantlets produced by micropropagation*

### The role of auxins in apical dominance

The tip of the main stem is known as the **apex** (plural: apices). As seen in Figure 11.2 on page 348, the apex with its apical meristem usually forms part of an apical bud which will grow to give a new length of stem. The figure also shows that lateral buds occur in the axils of leaves at nodes along the stem. Each lateral bud is capable of growing into a side shoot and has a meristem in its central stem tip, but this is dormant.

While the apical bud is present it is unusual for a lateral bud to grow into a side shoot (the apex is dominant). However, if the apical bud is removed, then the side shoots start to grow. This is the principle used by gardeners when they cut out the tips of main stems to produce bushier plants (Figure 11.22).

Auxins produced by the apical meristem in the apical bud are transported down the stem. It is known that it is the effect of these auxins which suppress the growth of lateral buds. If an apical bud is removed but is replaced by an agar block containing auxins which can move down the stem, the lateral buds do not grow (Figure 11.23). It is not known how auxins exert their effects on the lateral buds as their concentrations are often quite low at the levels of the buds. In some way the reduced level of auxins may allow the lateral buds to inactivate the abscisic acid (growth inhibitor) which they contain. Also it has been suggested that the presence of auxins causes the formation of ethene which inhibits side shoot growth.

*(a)*

*(b)*

**Figure 11.22** *A fuchsia plant. Customers often prefer a bushy plant (b) rather than a single-stemmed plant (a). To produce such a bushy plant, gardeners pinch out the apical bud after two sets of leaves have formed. Side shoots then grow from the lateral buds. Later, the apical buds of these side shoots are removed to increase bushiness*

**?**

**4** Study Figure 11.23 and explain why it is necessary to carry out experiment D. *(1 mark)*

*Figure 11.23 Experiment which demonstrates the role of auxins in suppressing the growth of side shoots*

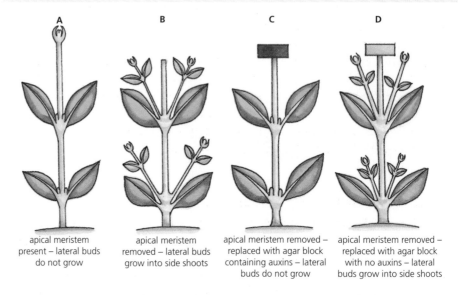

apical meristem present – lateral buds do not grow

apical meristem removed – lateral buds grow into side shoots

apical meristem removed – replaced with agar block containing auxins – lateral buds do not grow

apical meristem removed – replaced with agar block with no auxins – lateral buds grow into side shoots

### Other effects of auxins

Auxins affect the growth and development of plants in many ways. These are often poorly understood and may involve interaction between auxins and other plant growth regulators. Some of these effects are as follows:

● Auxins stimulate **adventitious** roots to grow. These are roots which grow directly from stem tissue rather than as branches from other roots.

● Auxins are produced naturally in the ovary after an ovule has been fertilised and promote the growth of the fruit, although they delay its ripening and prevent it falling off the plant before it is ripe.

● Auxins, with gibberellins, promote the division of cambium cells in secondary meristems.

**11.5** **Gibberellins and their effects**

### The nature of gibberellins

Gibberellins (gibberellic acids) are a group of compounds with similar chemical structure (Figure 11.24). They have been found in plants of all types (flowering and non-flowering) as well as in algae, fungi and some bacteria. Over 90 different gibberellins have been isolated. Many of these act as growth regulators, although some may be inactive forms. In flowering plants, it is thought that gibberellins are synthesised in young tissues of the shoots, in developing seeds, and possibly small amounts are produced in leaves. They can diffuse away from their regions of production. Gibberellins have many different effects upon growth, depending upon the species of plant and the type of gibberellin present. A few of these effects are discussed below.

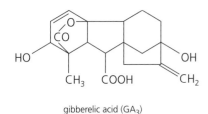

gibberelic acid (GA₃)

*Figure 11.24 Structure of gibberellic acid*

**Box 11.3 The discovery of gibberellins**

Japanese farmers have known for over a century that some rice seedlings grow abnormally long, early in the season. They called these seedlings 'foolish seedlings' (see Figure 11.25). In the 1930s it was established that this condition occurred in plants which were infected by the fungus then known as *Gibberella fujikuroi*. In 1935, a mixture of compounds was extracted from this fungus which was shown to stimulate growth when applied to rice roots. It was called gibberellin A and thought to be the agent which caused this disease when secreted into the plant by the fungus. Since then many similar gibberellins have been found to occur naturally in plants themselves as well as the fungus. It seems that the foolish seedlings were receiving an 'overdose'.

*Figure 11.25* 'Foolish seedlings'

## The role of gibberellins in stem elongation

Gibberellins, like auxins, promote growth by promoting cell elongation. However, unlike auxins, if gibberellins are applied to a dwarf variety of a plant, the plant will grow to be of normal size. It is thought that dwarf varieties, such as dwarf beans, possess a mutated gene which prevents them from synthesising gibberellins. Introducing the gibberellins into the plant allows them to grow normally (Figure 11.26). Gibberellins can also stimulate cell division and have an effect on cell differentiation.

**Did You Know**

Dwarf varieties of plants are very popular with gardeners as they produce flowers or fruit on a smaller plant which does not need staking. Plants are treated commercially with substances which prevent them forming gibberellins so that they develop into dwarf plants.

*Figure 11.26* Runner beans (left) and dwarf beans (right)

## The role of gibberellins in the germination of seeds

The role of gibberellins in initiating the **germination** of cereal grains has been studied in some detail. A cereal grain is a fruit containing one seed. Its structure is shown in Figure 11.27. Like all seeds, a cereal seed contains an **embryo** and stored food reserves. Here the food reserves are stored in a mass of tissue called an **endosperm**. The main reserves are starch which occupies the more central regions of the endosperm, and proteins which are stored mainly in an outer layer of the endosperm, the **aleurone layer**. Before any seed can start to grow, its food reserves must be hydrolysed into soluble forms such as glucose and amino acids so that they can be transported to the growing parts of the embryo.

*Figure 11.27  Longitudinal section through a barley grain*

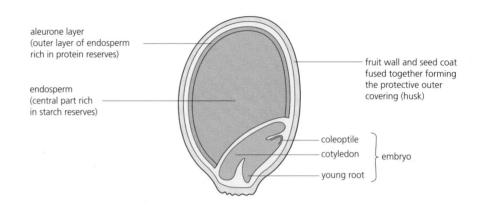

**Remember**

Germination is the growth of an embryo into a young plant using the food reserves present in the seed. When the plant starts to produce its own food by photosynthesis, germination is over.

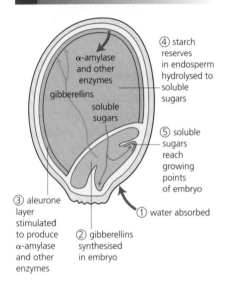

*Figure 11.28  The role of gibberellins in the onset of germination in a barley grain*

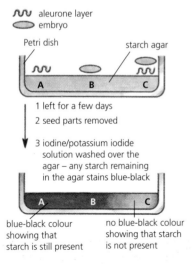

*Figure 11.29*

Gibberellins have a vital part to play in the germination process which begins as follows:

- The seed absorbs water.
- Gibberellins are produced by the embryo and diffuse across the endosperm to the aleurone layer.
- Gibberellins switch on the genes in the cells of the aleurone layer which causes them to use some of their protein reserves and secrete several enzymes including α-amylase. These enzymes diffuse into the other parts of the endosperm. In particular α-amylase converts starch reserves to maltose. Maltase converts maltose to glucose. The glucose diffuses into the embryo and is used for respiration and for the formation of macromolecules such as cellulose.

These processes are illustrated in Figure 11.28.

**5** Figure 11.29 shows an experiment using a Petri dish containing inert agar jelly impregnated with starch. The results show that the starch was hydrolysed only in trial C. Study Figure 11.29 and then, using your knowledge of the germination process, suggest

(a) why the starch was not hydrolysed in trial A

(b) why the starch was not hydrolysed in trial B.  *(2 marks)*

## Other effects of gibberellins

Different gibberellins have many different physiological effects on the development of plants, working with or against the action of other growth regulators (see Section 11.9, page 366):

- Unlike auxins, gibberellins inhibit the growth of adventitious roots.
- Like auxins, they stimulate the growth of fruits and are used in sprays to encourage the growth of seedless fruits such as grapes. They also stimulate the stalk of each grape to lengthen which separates the grapes on a bunch, discouraging fungal growth.

## 11.6 Cytokinins and their effects

### The nature of cytokinins

Cytokinins are compounds which stimulate cell division (*cytokinesis = cell division*). They have a chemical structure similar to adenine (a nitrogenous base present in DNA and ATP). The first cytokinin to be extracted and chemically identified was obtained from maize (*Zea mays*) in 1963 and so was called zeatin. Its chemical structure and the similarity to adenine is shown in Figure 11.30. Over 200 natural and synthetic cytokinins are now known and they have been found in mosses, fungi and bacteria as well as in flowering plants. They are believed to be produced in root tips of flowering plants and translocated in the xylem to the shoot system, becoming most concentrated in areas of cell division such as meristems, young leaves and developing fruits and seeds.

### Effects of cytokinins

As stated above, their main role is in the stimulation of cell division which they do while interacting with other substances such as auxins and gibberellins. The balance between the various growth regulators is very significant and knowledge of this is exploited in micropropagation techniques (see Box 11.2, page 359). For example, if a mass of undifferentiated plant tissue is cultured in a solution containing a high cytokinin to auxin ratio, shoot development is promoted. If the proportions are then changed to a lower cytokinin to auxin ratio, root development is favoured and the whole plantlet eventually develops.

Cytokinins stimulate the growth of lateral buds. In this way they have the opposite effect to auxins.

It is believed that as a leaf grows older and becomes yellow as its chlorophyll pigments break down (as part of the process known as **senescence**), its levels of cytokinins fall. It is thought that the cytokinins may be translocated in the phloem from these leaves to younger developing leaves. Evidence for this idea is provided by the observation that a leaf which has become detached from a plant soon yellows and senesces. However, if such a leaf is treated with a spot of a cytokinin known as kinetin in one area, that area alone remains green and nutrients move into this green 'island' (Figure 11.31).

*(a)*

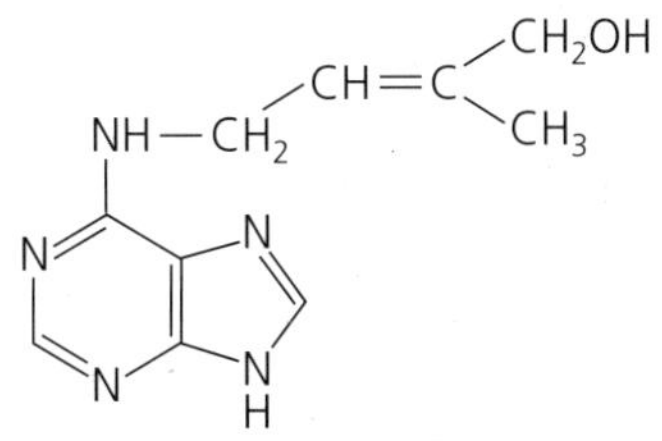

*(b)*

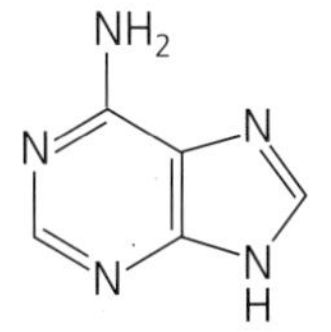

**Figure 11.30** *The structure of (a) a zeatin molecule (a cytokinin) and (b) an adenine molecule*

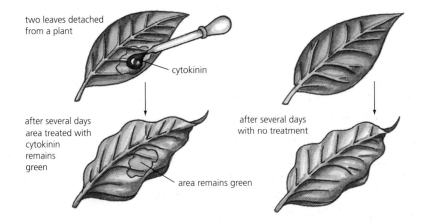

two leaves detached from a plant

cytokinin

after several days area treated with cytokinin remains green

area remains green

after several days with no treatment

**Figure 11.31** *Experiment to show that cytokinins can delay senescence in leaves*

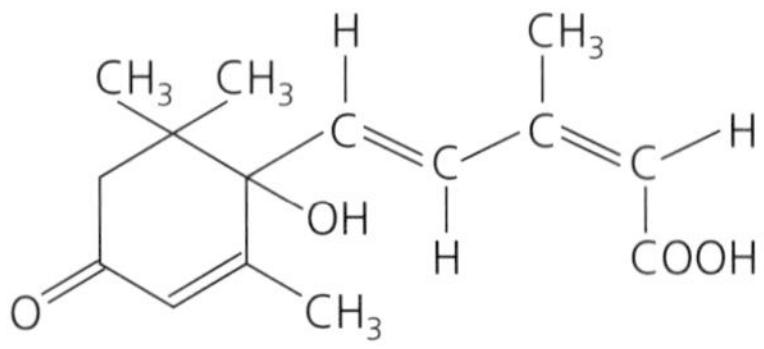

Figure 11.32 *The structure of abscisic acid*

## 11.7 Abscisic acid and its effects

### The nature of abscisic acid

Abscisic acid is a single compound with the chemical structure shown in Figure 11.32. It gained its name because one of its earliest known effects was thought to be the promotion of the **abscission** (falling off) of fruits in cotton plants. It is thought to be produced at least partly in chloroplasts or other plastids. It is therefore mainly made in the leaves and can be translocated in xylem and phloem, and it can diffuse through parenchyma cells. In general it is a growth inhibitor.

### Effects of abscisic acid

Abscisic acid inhibits shoot growth, although it does not have such an inhibitory effect on root growth. In doing this it opposes the actions of all the above growth regulators. It is particularly effective in times when the plant is stressed, for example by being short of water. It is also thought to play a part in the geotropic response of roots.

In times of stress, during a water shortage when the leaves begin to wilt, the level of abscisic acid in the leaves rises markedly. It is thought to promote the closing of the stomata.

**?**

**6** Why might it be an advantage to a plant to
   (a) slow its shoot growth in times of stress
   (b) close its stomata when water is in short supply?          *(2 marks)*

If abscisic acid is present in a seed which has dropped from a plant in the autumn, it can prevent the seed from beginning to germinate (partly by inhibiting the effects of gibberellins). The seed is described as **dormant**. It is not until the abscisic acid has been destroyed that the seed can begin to germinate. Abscisic acid is destroyed by exposure to the cold of winter.

Despite its name, the role of abscisic acid in promoting leaf fall (abscission) is now in doubt although it may play a part in abscission of fruits. It keeps buds in a dormant state in overwintering plants during the cold months of winter.

**?**

**7** What is the advantage to a seed of a period of dormancy?     *(2 marks)*

## Box 11.4  How leaves and fruits fall off plants

The leaves of deciduous trees are lost in the autumn in temperate zones. Some tropical plants shed their leaves before a dry season, evergreen plants lose a few leaves all of the time and fruits and unfertilised flowers are always finally shed by plants. This shedding of parts of plants is called **abscission** and is a necessary plant survival strategy.

8  Why are the following an advantage to a plant?
   (a)  the shedding of leaves in the autumn          (2 marks)
   (b)  the shedding of fruits.                        (1 mark)

When part of a plant is shed, it is essential that the transport systems and vulnerable tissues are sealed to prevent nutrients leaking from the phloem, and to prevent the entry of pathogens. The main stages in the abscission process are shown in Figure 11.33.

As described, auxin is thought to delay abscission. The simple experiment shown in Figure 11.34 supports this idea. The levels of auxin production in leaves and ripening fruits goes down as time for abscission draws near. However, levels of ethene, which promotes abscission, rise as leaves senesce and fruits ripen. Abscisic acid levels have also been shown to rise in some fruits before they drop. However, it is not certain that abscisic acid (despite its name) is important in causing leaves to fall. Although applying abscisic acid artificially in large quantities does increase leaf drop, this could be because it stimulates ethene production. Abscisic acid levels are known to rise in some plants before leaf fall in the autumn but this may be because of its role in keeping the buds dormant so that they do not grow until the next spring.

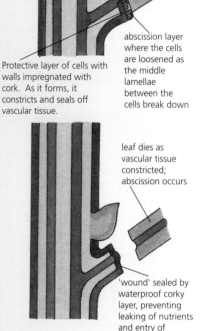

lateral bud

vascular tissue

leaf stalk

abscission layer where the cells are loosened as the middle lamellae between the cells break down

Protective layer of cells with walls impregnated with cork.  As it forms, it constricts and seals off vascular tissue.

leaf dies as vascular tissue constricted; abscission occurs

'wound' sealed by waterproof corky layer, preventing leaking of nutrients and entry of pathogens

**Figure 11.33**  *Leaf fall in a woody plant*

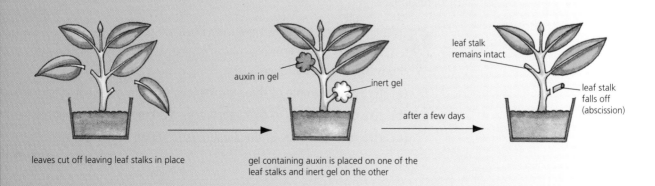

leaves cut off leaving leaf stalks in place

auxin in gel

inert gel

gel containing auxin is placed on one of the leaf stalks and inert gel on the other

after a few days

leaf stalk remains intact

leaf stalk falls off (abscission)

**Figure 11.34**  *Experiment showing that auxin delays abscission*

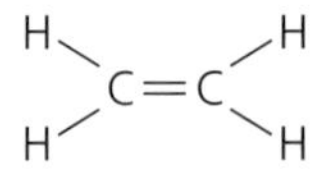

*Figure 11.35* *The structure of ethene*

## 11.8 Ethene and its effects

### The nature of ethene

Ethene is a gas with a simple organic molecule (Figure 11.35) which is made in most plant organs. A water-soluble substance which is involved in the production of ethene has been shown to be transported in plants, but ethene itself does not diffuse freely in air spaces. It tends to act where it is produced and be given off from plant surfaces.

### Effects of ethene

Ethene is well known for speeding up the ripening process in fruits and increasing their respiration rates. It was shown in the 1930s that ripe fruits such as bananas and apples emit ethene. The increased amount of ethene in the atmosphere then speeds up the ripening of nearby fruits. This has commercial advantages and disadvantages. Fruits are usually imported in an unripe state to prevent damage. They can then be ripened quickly for sale by placing them in an atmosphere with raised ethene levels. However, there is a danger that a load of imported fruit will arrive overripe and damaged because a few fruit in the pack were already ripening and emitting ethene during transport.

Ethene is a growth inhibitor and, like abscisic acid, it can inhibit shoot and root growth especially during times of stress. It may also be involved in the geotropic response in roots. It also promotes leaf and fruit fall (abscission).

**Did You Know**

Ancient Egyptians used to expose figs to various gases to ripen them, and the ancient Chinese used to burn incense in closed rooms to stimulate the ripening of pears. It was not until the 20th century that the active component of the gases used was shown to be ethene. Even now fumes from a leaky oil-fired boiler can cause fruit to ripen too quickly in a kitchen!

## 11.9 Synergism and antagonism

Unlike animal hormones which usually have individual effects, plant growth regulators often work in conjunction with each other. When two or more substances amplify each other's effects, producing a greater response than either substance alone, they are said to be working **synergistically**. On the other hand if two substances have the opposite effects and oppose the action of each other, they are working **antagonistically**. The response which is brought about depends upon the balance between the two substances present. This interaction between different growth regulators helps the plant to exert a fine control in its responses.

**?**

**9** Study Table 11.2 and then

(a) give *one* example of three substances working together synergistically

(b) give *one* example of two substances working together synergistically

(c) give *two* examples of the antagonistic action of two substances.

*(4 marks)*

| Growth regulator | Stem growth | Growth of adventitious roots | Apical dominance | Fruit growth | Fruit ripening | Fruit abscission |
|---|---|---|---|---|---|---|
| auxins | promotes | promotes | promotes | promotes | inhibits | inhibits |
| gibberellins | promotes | inhibits | promotes | promotes | – | – |
| cytokinins | promotes | – | opposes as promotes growth of lateral buds | promotes | – | – |
| abscisic acid | inhibits | – | – | – | – | promotes |
| ethene | inhibits | – | – | – | promotes | promotes |

*Table 11.2* *Some examples of synergism and antagonism between growth regulators; a dash indicates no marked effect known*

### 11.10 The detection of light by phytochrome pigments

Light, the visible part of the sun's electromagnetic spectrum, is all important to plants. It is their vital source of energy. Most growth responses of the shoot system directly or indirectly enable the plant to trap as much light energy as possible by photosynthesis. The phototropic response in shoots (see Section 11.3, page 350) is one example of such a response.

It is not surprising therefore that light acts as a stimulus for many responses. Although it is not known to be involved in phototropism, plants possess a chemical photoreceptor which acts as the light detection system in many responses. This photoreceptor is a blue-green pigment called **phytochrome**. It is a conjugated protein (see *AS Biology*, Section 3.17, page 86) found in leaves, some seeds and growing points. It is present in such small quantities that it does not affect the colour of the plant itself. It occurs in two forms known as $P_R$ and $P_{FR}$.

- $P_R$ is chemically stable unless it is exposed to light. It readily absorbs red light (wavelengths around 660 nm) and is converted into the other form of phytochrome, $P_{FR}$. This conversion is very rapid, taking only seconds in light of high intensity.
- $P_{FR}$ is chemically unstable and slowly changes into $P_R$ in the dark. If exposed to light right on the border of the visible spectrum known as far-red light (wavelengths around 730 nm), it absorbs this light and is converted into $P_R$ much more rapidly.

Sunlight contains far more red light than far-red light and so effectively what happens in nature is that on exposure to sunlight, $P_R$ is converted to $P_{FR}$ which builds up, and in the dark, $P_{FR}$ slowly breaks down to $P_R$ and so levels of $P_R$ rise. These ideas are summarised in the Remember box.

$P_R$ and $P_{FR}$ have different physiological effects on the growth and development of plants. In general, $P_{FR}$ is metabolically active and $P_R$ is not. The amount of $P_{FR}$ acts as a daylight length detector system and particular metabolic processes which lead to a response are either triggered or not by high levels of $P_{FR}$.

**Remember**

red light (660 nm)

$P_R \xrightarrow{\hspace{1.5cm}} P_{FR}$

far-red light (730 nm)

slow conversion in darkness as unstable

During prolonged hours of darkness, $P_R$ levels build up.

In daylight (which contains much more red light than far-red), $P_{FR}$ levels build up.

### The role of phytochrome in the germination of some seeds

Although most seeds germinate below the ground in the dark, some very small seeds, such as lettuce seeds, do not contain enough food reserves to sustain them until they reach the light if they germinate too far below ground level. Such seeds only germinate if they are exposed to light, even if only briefly. This ensures that they only germinate when they are near the surface of the soil.

Such seeds possess phytochrome pigment. Exposure to sunlight stimulates $P_R$ to convert into $P_{FR}$ and this initiates germination. Figure 11.36 shows examples of experiments carried out on lettuce seeds which led to the discovery of phytochromes. The lettuce seeds were allowed to absorb water in the dark, and then given 5-minute periods of illumination with red and far-red light in various sequences as shown. They were then left in the dark and the percentage of seeds which germinated in each case was calculated.

It was discovered that whenever the last exposure was to red light, then $P_{FR}$ was left in the seed. Although this would eventually disappear in the dark, it remained long enough to initiate germination.

*Figure 11.36* *Early experimental results interpreted in terms of phytochromes*

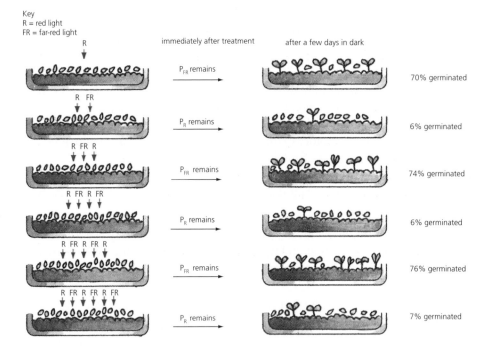

Key
R = red light
FR = far-red light

### The role of phytochrome in etiolation

Dicotyledonous plants which are grown in the dark become very elongated with long internodes on their frail stems. The leaves remain small, unexpanded and yellow as chlorophyll pigments do not form. Such a plant is **etiolated**. If it is not eventually exposed to light, it will die. Figure 11.37 shows how plants grown in the light and in the dark differ in their growth patterns. Etiolation is a survival mechanism. All the food reserves in the plant are used to make the plant as tall as possible, increasing the chances of it reaching a light area. There is no advantage in developing green leaves in the dark!

(a)

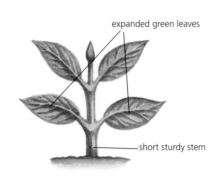

expanded green leaves

short sturdy stem

(b)

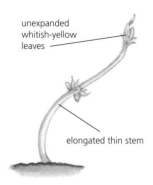

unexpanded whitish-yellow leaves

elongated thin stem

In the dark, $P_R$ is present but there is no $P_{FR}$, whereas in the light $P_{FR}$ builds up. This suggests that $P_{FR}$:

- inhibits the lengthening of internodes
- promotes leaf expansion
- promotes the development of chlorophyll.

**Figure 11.37** *Etiolation (a) Seedlings of broad bean (i) showing normal growth (ii) showing etiolation (b) Growth habits of a plant grown in the light and in the dark*

### The role of phytochrome in the timing of flowering

In the autumn when the nights become longer and the days shorter, there is a slow fall in the level of $P_{FR}$ in plants over the season. The levels remain low throughout the winter and early spring. In some plants (short-day plants) which produce flowers in these seasons, $P_{FR}$ inhibits flowering. As summer approaches, the days become longer and the nights shorter. $P_{FR}$ levels build up as the nights are too short for it all to convert into back into $P_R$. In plants which flower in midsummer (long-day plants), $P_{FR}$ stimulates flowering. This process of flowering at different times of the year in response to the length of daylight in a 24-hour period is called the **photoperiodic** response.

Table 11.3 summarises the effects of red light (and therefore $P_{FR}$ build-up) on aspects of plant development. In all the examples given, a long period of darkness or exposure to far-red light has the reverse effect as $P_{FR}$ is removed by conversion to $P_R$.

**Table 11.3** *Effects of red light on plant development*

| Response | Exposure to red light ($P_{FR}$ build-up) | Long dark period or exposure to far-red light ($P_{FR}$ removed) |
|---|---|---|
| germination of lettuce seeds | promotion | inhibition |
| lengthening of internode | inhibition | promotion |
| expansion of leaves | promotion | inhibition |
| development of chlorophyll | promotion | inhibition |
| flowering in 'short-day' plants | inhibition | promotion |
| flowering in 'long-day' plants | promotion | inhibition |

**Summary –** ⑪ **Control and coordination in plants**

● Plant responses are initiated by detection of a stimulus. Chemical substances – growth regulators – produced by the plant, control the responses. They all involve modification of plant growth or development. The substances may or may not move from the site of detection before bringing about a response.

● Plant growth involves the production of new cells by mitotic cell division at meristems. Primary meristems are present at shoot tips and root tips and provide cells for lengthening the stems and roots. The cells then elongate in the region behind the tip and absorb water until the cell walls harden and allow no more expansion. The cells then differentiate into the different tissues.

● The cell elongation is promoted by growth regulators known as auxins. These are produced in the tip and diffuse back to the zone of elongation. They maintain the flexibility of cell walls, so allowing cell expansion. Eventually the auxins are destroyed by enzymes and the cell walls harden.

● A tropism is a plant response in which the direction of growth is a response to the direction of a stimulus. An example of this is shown in the way shoots grow towards a source of light shining from one side (positive phototropism).

● The phototropic response in shoots is controlled by auxins. Auxins, produced in a shoot tip, move away from the light side. As a result more move down the dark side of the shoot than the light side, promoting more elongation on the dark side and a resultant curvature of the shoot towards the light.

● Roots are positively geotropic (grow downwards with the force of gravity). It is not certain which growth regulators are involved in this response but amyloplasts in root cap cells detect the direction of the force of gravity.

● Other effects of auxins are listed in Table 11.4.

● Other plant growth regulators include gibberellins, cytokinins, abscisic acid and ethene. Some of their effects are also listed in Table 11.4.

● Growth regulators rarely work alone. Sometimes they work together, complementing each other's actions, i.e. synergistically. For example, gibberellins promote cell elongation in the presence of auxins. Sometimes the action of one substance opposes the action of another. For example abscisic acid opposes the action of auxins when it inhibits the lengthening of stems in times of stress. This action is antagonistic. This interaction between substances allows fine control of growth responses.

● Plants have phytochrome pigments which act as a light detection system. A stable but inactive form of phytochrome ($P_R$) absorbs red light in the sun's spectrum and is converted into an active but unstable form ($P_{FR}$) which can trigger or inhibit chemical processes which lead to various responses. In the dark this active form slowly reconverts into $P_R$ again. (It converts very quickly if exposed to far-red light but there is very little of this in sunlight.) $P_{FR}$ can stimulate germination in some small seeds, has a role in the etiolation of plants grown in the dark, and in the timing of flowering in the photoperiodic response.

| Growth regulator | Effect |
|---|---|
| auxins | • stimulate cell elongation, especially in shoots<br>• active in phototropic response in shoots<br>• inhibit the growth of lateral buds into shoots (apical dominance)<br>• stimulate the growth of adventitious roots from the bases of stems<br>• stimulate the growth and development of fruits<br>• delay fruit ripening and leaf senescence<br>• inhibit the fall (abscission) of leaves and fruits<br>• stimulate cell division in cambium |
| gibberellins | • stimulate cell division and cell elongation in the presence of auxins<br>• promote normal growth in dwarf varieties<br>• stimulate germination in seeds by stimulating enzyme production<br>• inhibit the growth of adventitious roots<br>• stimulate the growth and development of fruits |
| cytokinins | • stimulate cell division while interacting with auxins and gibberellins<br>• oppose apical dominance by promoting growth of lateral buds<br>• inhibit senescence in leaves |
| abscisic acid | • inhibits shoot growth in times of stress<br>• may play a part in the geotropic response in roots<br>• promotes dormancy in seeds and buds in winter<br>• stimulates the closing of stomata in water shortage<br>• stimulates abscission of fruits |
| ethene | • stimulates ripening of fruits<br>• stimulates leaf and fruit abscission<br>• inhibits shoot and root growth in times of stress |

*Table 11.4 Summary of some of the effects of plant growth regulators*

# 11 Control and coordination in plants

## ? Answers

1  (a) A control has all the same conditions except the one stimulus being tested; if the result differs from that of the control, that stimulus brought about the result *(2)*.

   (b) B; identical conditions to A except for stimulus of unilateral light whose effect is being studied *(2)*.

   (c) To eliminate the significance of any abnormal result *(1)*.

2  Diffusion only takes place in one direction; in a direction away from the tip *(2)*.

3  (a) Same species of plant; same age/size of shoot; same temperature; block left for same length of time; same light/atmospheric conditions *(max. 3)*.

   (b) Use a block with a known concentration of IAA and measure angle of curvature; compare angle produced by another block with this standard *(2)*.

4  To show that the agar block itself was not responsible for suppressing the growth of side shoots *(1)*.

5  (a) The aleurone layer is not stimulated to produce enzymes *(1)*.

   (b) The embryo may produce gibberellins but no aleurone layer is present to produce enzymes *(1)*.

6  (a) Maintenance of food reserves for vital processes *(1)*.

   (b) Conserve water by slowing transpiration through stomata *(1)*.

7  If it germinated immediately upon being shed in the autumn, the seedling would be vulnerable to frost damage and low light levels in the winter; dormancy delays germination until the spring *(2)*.

8  (a) Light levels and temperature too low in winter for efficient photosynthesis; advantageous to minimise water loss (ground may be frozen) and not to have to maintain leaves *(2)*.

   (b) either shed as part of seed dispersal or useless after seeds have been dispersed *(1)*

9  (a) Auxins, gibberellins and cytokinins in promoting stem growth/fruit growth *(1)*.

   (b) Abscisic acid and ethene in inhibiting stem growth/auxins and gibberellins in promoting apical dominance/abscisic acid and ethene in promoting fruit abscission *(1)*.

   (c) Abscisic acid/ethene opposing auxins/gibberellins/cytokinins in stem growth; gibberellins opposing auxins in adventitious root growth; cytokinins opposing auxins/gibberellins in apical dominance; ethene opposing auxins in fruit ripening; auxins opposing abscisic acid/ethene in fruit abscission *(max. 2)*.

## End of Chapter Questions

1 Distinguish between each of the following pairs of terms:

   (a) cell growth and cell differentiation

   (b) a meristematic cell and a parenchyma cell

   (c) amyloplast and chloroplast

   (d) fruit ripening and fruit abscission

   (e) a dormant seed and a seed in a packet in a garden centre

   (f) synergism and antagonism

   (g) a photoreceptor substance (e.g. phytochrome) and a growth regulator substance

   (h) etiolation and normal growth.

   *(Total 8 marks)*

2 During germination, the leaves that emerge from cereal grains are protected by coleoptiles. These coleoptiles produce indole-3-acetic acid (IAA) which can be collected by placing cut coleoptiles on to blocks of agar.  N2.1

   (a) An experiment was carried out to compare the amount of IAA produced by maize coleoptile tips in light and in dark conditions. Tips were removed from coleoptiles of maize seedlings and placed on blocks of agar. Some of these were left in the light and some in the dark.

   (b) Diagram 1 below shows how the experiment was set up and the relative amounts of IAA collected in the agar blocks at the end of the experiment.

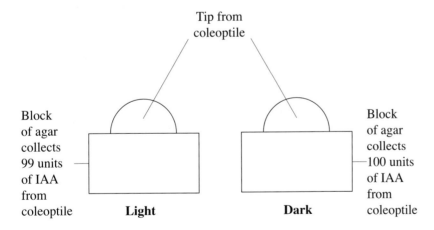

   What do the results of this experiment suggest about the production of IAA? *(2 marks)*

**(b)** Diagram 2 below shows a second experiment carried out to investigate the movement of IAA through maize coleoptiles.

Short lengths of coleoptiles were cut from maize seedlings and further cuts were made in their sides to remove a small section from each coleoptile. Agar blocks containing IAA were placed on the upper cut surfaces of these coleoptiles (block A). IAA that had travelled through the coleoptiles was collected in agar blocks that had been placed on their lower cut surfaces (block B) and in the cut sides (Block C).

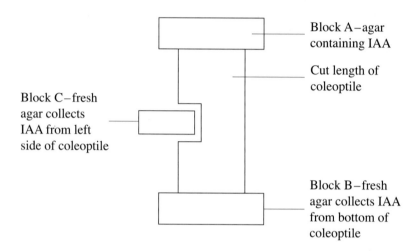

Block A—agar containing IAA

Cut length of coleoptile

Block C—fresh agar collects IAA from left side of coleoptile

Block B—fresh agar collects IAA from bottom of coleoptile

The table below shows the relative amounts of IAA collected at the side and lower surface of a group of coleoptiles that have been in dark conditions and a second group that have been illuminated from the right.

| Agar blocks | Relative amounts of IAA collected | |
|---|---|---|
| | Dark | Light from right side |
| Block A | 100 | 100 |
| Block B | 73 | 65 |
| Block C | 27 | 35 |

(i) Describe what the results of this investigation show about the movement of IAA through coleoptiles. *(4 marks)*

(ii) IAA stimulates elongation of cells in the stem and coleoptile of maize. Explain how the movement of IAA helps the maize plant to carry out photosynthesis efficiently. *(2 marks)*

**(c)** IAA is one of a group of plant growth substances called auxins. Describe *two* ways in which knowledge of the action of auxins is used commercially to affect plant growth or development. *(4 marks)*

*Edexcel 1997*        *(Total 12 marks)*

**3** An investigation was carried out into the effect of different concentrations of indoleacetic acid (IAA) on the growth of coleoptiles (shoots) from 5 day old barley seedlings.

The tips of 70 coleoptiles were cut off, and the remainder of each coleoptile was trimmed to a length of 10 mm. Ten of these trimmed coleoptiles were placed in each of six solutions of IAA, and then incubated in the dark at 20°C for 72 hours. A further ten trimmed coleoptiles were incubated for the same length of time in distilled water.

The coleoptiles were then measured, and the mean increases in their lengths were plotted against the concentration of IAA used. The whole experiment was repeated at a temperature of 30°C. The results are shown in the graph below.

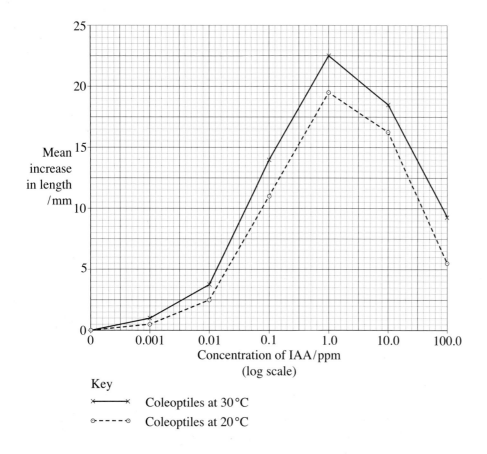

Key

×——————× Coleoptiles at 30 °C

∘- - - -∘ Coleoptiles at 20 °C

**(a)** Describe the effect of different concentrations of IAA on the growth of the coleoptiles.                                                        *(3 marks)*     N3.1

**(b)** Suggest an explanation for the differences between the results obtained at 20 °C and 30 °C.                                                      *(2 marks)*     N3.1

(c) (i) Suggest why the tips were removed from the coleoptiles. *(2 marks)*

(ii) Suggest why the investigation was carried out in the dark. *(1 mark)*

(d) Explain *one* way in which auxins, such as IAA may be used in horticulture and agriculture. *(2 marks)*

*Edexcel 1999*

*(Total 10 marks)*

4 The statements below refer to some effects of two groups of plant growth substances, auxins and gibberellins.

Copy the table. If the statement is correct place a tick (✓) in the appropriate box and if the statement is incorrect place a cross (✗) in the appropriate box.

| Effect | Auxins | Gibberellins |
|---|---|---|
| Promote cell elongation | | |
| Promote root formation in cuttings and calluses | | |
| Promote fruit ripening | | |
| Inhibit lateral bud development | | |
| Promote the breaking of dormancy in seeds | | |

*Edexcel 2000*

*(Total 5 marks)*

5 One of the effects of gibberellin is to stimulate stem elongation in dwarf pea plants. This forms the basis for a technique (known as bioassay) for measuring the quantity of this plant growth regulator in tissues and organs. To measure the effect of a range of known quantities of gibberellic acid, 42 seeds from pure-bred dwarf pea plants were germinated, grown under identical conditions for two weeks, and then divided into seven batches (A to G). Each batch of seedlings was sprayed with a different dose of gibberellin dissolved in water at weekly intervals for five weeks. A further batch of six seedlings (H) from a pure-bred tall pea plant was grown under identical conditions but these were not sprayed. Two weeks after the final spraying, the stem lengths of all the plants were measured; the mean lengths of batches A to H are shown in the table on the next page.

| Batch | Weekly dose of gibberellin / μg | Mean length of stem / mm |
|-------|--------------------------------|--------------------------|
| A | 0.00 | 152 |
| B | 0.05 | 204 |
| C | 0.10 | 251 |
| D | 0.50 | 408 |
| E | 1.00 | 454 |
| F | 5.00 | 600 |
| G | 10.00 | 623 |
| H | not sprayed | 627 |

(a) With reference to the table, describe the effect of spraying gibberellin on pea seedlings. *(3 marks)*

N2.1
C2.2

(b) (i) Suggest a likely cause of the dwarfism of the pea plants used in this investigation. *(1 mark)*

(ii) State **one** reason for the use of seeds from pure-bred plants. *(1 mark)*

(c) Make **two** criticisms of the design of this investigation. *(2 marks)*

(d) Describe the role of gibberellin in the germination of cereals such as wheat and barley. *(3 marks)*

Growth retardants are substances which are used to limit the height of both cereals and ornamental species which are grown for their flowers.

(e) Describe **two** economic advantages of using growth retardants. *(2 marks)*

OCR 2000 *(Total 12 marks)*

# 12 Populations and communities

Some of the fundamental concepts of ecology were dealt with in Chapter 14 of *AS Biology*. This chapter extends those topics in four main ways. First of all, it looks at the factors which affect the size of a population of organisms. Secondly, it deals with the processes by which communities develop in a relatively uncolonised habitat, such as a bare rock surface or a ploughed field. Thirdly, it considers some of the ecological sampling methods which are used to study the distribution of organisms in a habitat. Lastly, it considers the meaning of the term conservation, and how conservation measures can be employed in farming practices. Throughout the chapter boxes are used to remind you of the meaning of some of the basic ecological terms.

## 12.1 Population growth

Populations of organisms are usually restricted in their size by **limiting factors** (see *AS Biology*, Box 14.2, page 420), such as the availability of food (for animals) or light (for plants and other photosynthesisers). These factors maintain the size of a population at a fairly constant level. If the limiting factors are removed, the population is able to grow very rapidly. This is clearly demonstrated if a small number of bacterial cells are inoculated into a nutrient broth, and the number of cells is monitored for several hours (Figure 12.1).

At first the number of bacteria may remain constant (A). This is called the **lag phase**, and is due to the bacteria having to adapt to their new food supply, for example by synthesising new enzymes to break it down. Soon afterwards a period of **exponential growth** takes place, called the **logarithmic** or **log phase**

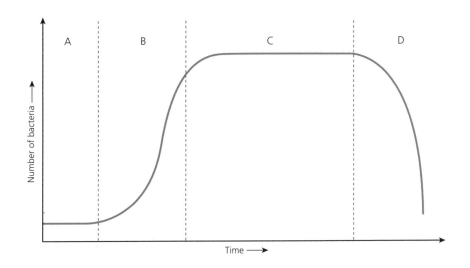

Figure 12.1 *Bacterial growth curve in a liquid culture*

### This chapter includes:
- factors which affect population growth
- carrying capacity and environmental resistance
- density-dependent and density-independent mortality
- intraspecific and interspecific competition
- predator–prey relationships
- estimating the size of a population
- colonisation and succession
- climax communities
- zonation
- measuring species diversity
- sampling methods
- the $\chi^2$ test of association
- conservation.

### Remember
A **population** is a group of organisms of the same species living in a particular habitat. For example, a group of rabbits living in a field, or a collection of woodlice under a rotting log can each be described as a population. The **habitat** is simply the place where they live (the field or the log). A **community** is the name given to all the organisms, of different species, living in a habitat, or one group of these, such as the plant community.

*Figure 12.2  Algal bloom on a freshwater lake*

*Figure 12.3  Changes in the population of sheep in Tasmania following their introduction in 1814. The red line shows the year-by-year numbers, the black line is the averaged population trend*

(B) when the cells are dividing into two in unit time intervals (this can be as frequent as every 20 minutes in some species). Eventually this exponential phase ends, as limiting factors begin to have an effect. These include exhaustion of the nutrients in the broth, or build-up of toxic waste products, such as carbon dioxide. The growth curve then enters the **stationary phase** (C) when the number of cells being produced equals the number which die. The curve on the graph so far is described as S-shaped or **sigmoid**. The stationary phase does not last indefinitely. Eventually the limiting factors result in a greater number of cells dying than the numbers produced by cell division, and the curve enters the **death phase** (D).

**1** Wine is made by fermentation of sugars by yeast cells. What is the factor which eventually limits the growth of the yeast cells in a vat of wine?

*(1 mark)*

Some natural populations can undergo periods of exponential growth. For example, if a freshwater lake is enriched with excess nutrients, such as nitrates or phosphates, algae may rapidly increase in numbers in a similar way. This enrichment is called **eutrophication**, and results in **algal blooms** (Figure 12.2).

If the nutrient enrichment is temporary, as may occur naturally following heavy seasonal rains, the numbers of algal cells will eventually decrease again, as the low nutrient levels become a limiting factor. A more complete account of eutrophication and its causes is given in *AS Biology*, Chapter 15.

Other populations may increase rapidly when conditions are favourable, reaching a maximum size when limiting factors maintain them at a constant level. This was the case following the introduction of sheep to the island of Tasmania in 1814 (Figure 12.3). The sheep were able to exploit their new habitat, so that over the next 40 years their numbers increased dramatically. However, from the middle of the 19th century, the numbers remained fairly constant, fluctuating around a plateau shown by the trend line.

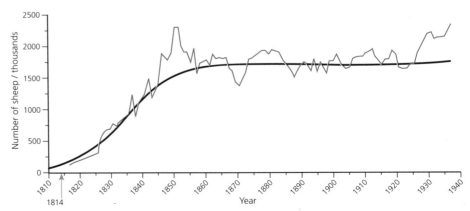

**2** Suggest two interactions with other organisms which may have helped stabilise the population size of the sheep. *(2 marks)*

### Carrying capacity and environmental resistance

The maximum population size that an environment can support is called its **carrying capacity**. The value of the carrying capacity will depend upon factors such as food supply, competition, predation, climate and disease. These factors, which determine the growth rate of a population, are known as the **environmental resistance**. Carrying capacity and environmental resistance can be illustrated graphically (Figure 12.4).

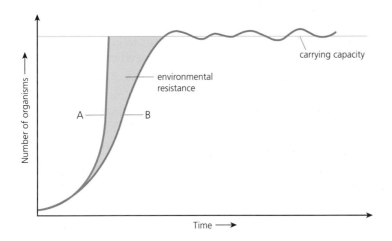

*Figure 12.4* Curve A shows the theoretical maximum growth rate of a population. Curve B shows the growth rate limited by various factors known collectively as the environmental resistance. The maximum population size the environment will support under these conditions is the carrying capacity

Some populations show a great fluctuation in numbers with time. A famous study of great tits, *Parus major* (Figure 12.5), carried out in Sweden from 1953 to 1964 found that the size of the population depended greatly on their food supply.

The numbers of birds were seen to vary enormously from year to year (Figure 12.6). These fluctuations were correlated with the annual crop of beech seeds,

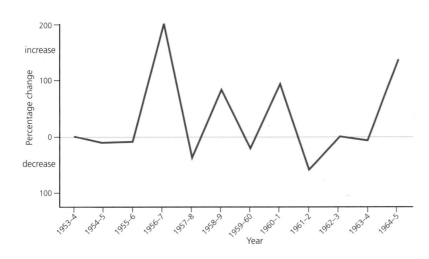

*Figure 12.5* Great tit (Parus major)

*Figure 12.6* Fluctuations in the numbers of great tits in Veluwe, Sweden from 1953 to 1965

*Figure 12.7* *Correlation in the numbers of breeding great tits and the size of the beech crop. The population increased in size in the year following each large beech crop*

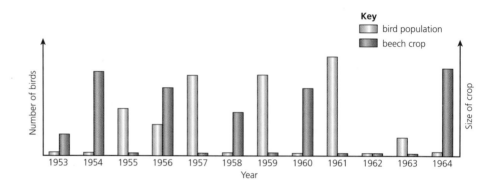

Key
bird population
beech crop

**Remember**

The term '**environment**' is a rather imprecise one. It simply means the surroundings of an organism, and is made up of an **abiotic** component (the non-living, chemical and physical surroundings) and a **biotic** component (biological surroundings, or other organisms). All the organisms together with their environment are called an **ecosystem**, which is the fundamental unit of ecology. We talk of a 'grassland' ecosystem or a 'freshwater lake' ecosystem. An ecosystem contains several different habitats supporting different communities of organisms. For example, a lake has communities occupying the bottom of the lake (**benthic** organisms) and free-swimming (**pelagic**) ones.

which form the staple diet of the birds in that part of Sweden (Figure 12.7). Juvenile birds are likely to die in winter, when food is scarce and temperatures low. Following a bumper crop of seeds (for example in 1960) more juveniles survived the winter, so that the tit population in 1961 was large. Similar high numbers were recorded in other years, following a good seed crop.

**?**

**3** Explain why some animal populations such as great tits show large changes in numbers, while others (such as the sheep of Tasmania) do not.

*(2 marks)*

When a population 'explosion' of the sort seen with the great tits takes place, the number of organisms in a particular area increases. In other words there is a high **population density**. With mobile animals such as birds, a period when food is in short supply will often lead to them leaving the area in search of other food sources. In other words **emigration** occurs. Population size is altered by a combination of four factors. These are:

● increased deaths (**mortality**) or **emigration** which both *decrease* numbers
● increased births (**natality**) or **immigration** which both *increase* numbers.

The size of a population may be limited by either an increase in mortality or a decrease in natality. Usually these terms are given as rates, expressed as a number per adult in the population. For example, a mortality rate of 5% means five deaths per 100 adults or 50 per 1000 adults.

**12.2** **Density-dependent and density-independent mortality**

If the death rate of a population changes with its density, this is called a **density-dependent mortality rate**. It is an important way that some populations are prevented from increasing in size when they reach a high density. The death rates of other populations are not affected by their density, and these are referred to as having a **density-independent mortality rate**. This is easiest to visualise as a graph (Figure 12.8).

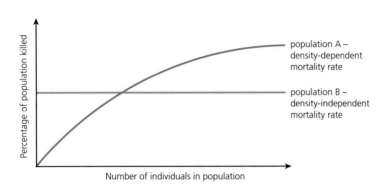

*Figure 12.8 Graphical representation of density-dependent and density-independent effects on mortality of a population*

The graph indicates that as the population size in an area increases, the percentage of individuals in population A which are killed also rises, i.e. the mortality rate depends on the population density. Whereas in population B, the percentage which are killed remains constant with increasing numbers, so it shows a density-independent mortality.

Births (natality) can also be density-dependent. However, ecologists who study population changes regard a low birth rate as being equivalent to a high mortality, and refer only to mortality rates in the life cycle.

The biologist H.N. Southern carried out a detailed study of population dynamics in the tawny owl (Figure 12.9) in Wytham Wood, near Oxford, from 1947 to 1959. The results illustrate how some mortality factors in this species are density-dependent, while others are density-independent.

The woods, covering an area of 525 hectares, contained over the years, between 17 and 32 pairs of owls, each occupying a well-guarded territory. Each pair held a territory for about five years, during which they could breed once a year, producing up to three eggs in each clutch. Hence there was potential for the population to greatly increase in numbers. The relatively stable population size was due to a number of factors which affected mortality, such as:

*Figure 12.9 A tawny owl*

- failure of a pair of birds to breed
- failure of breeding pairs to produce the maximum number of eggs
- failure of eggs to hatch
- death of chicks before they leave the nest
- failure of fledglings to obtain a territory in the wood (either through death or because they emigrate from the wood).

This last factor, competition for a territory, is density-dependent. If a large number of pairs of adult birds is occupying the territories in a wood, the probability that young birds will be able to establish a territory is low. If they do not have a territory, they will be unable to feed. They are therefore much more likely to die, or emigrate from the wood. Conversely, in a year when the adult birds are reduced in number (for example through deaths from exposure during a cold winter) the young birds will be able to occupy a territory and their mortality rate will be low. This factor therefore has a regulating effect on the population, tending to maintain its numbers at a relatively constant value, i.e. it contributes to the density-dependent mortality. Predation, disease and competition can all regulate a population's size by acting as density-dependent factors.

> **Remember**
>
> The **biotic** component of the ecosystem is the living organisms. Factors associated with these, such as primary production, predation or disease are **biotic factors**. The **abiotic** component of the ecosystem consists of the physical and chemical environment of the organisms, and **abiotic factors** are those such as temperature, oxygen concentration and soil type.

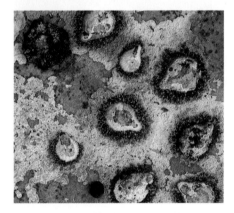

***Figure 12.10*** *The limpet,* Patella cochlear

> **Remember**
>
> **Intra**specific competition occurs between members of the **same** species. **Inter**specific competition occurs between members of **different** species.

***Figure 12.11*** *Effect of intraspecific competition on the maximum length (blue line) and biomass (black line) of the limpet* Patella cochlear

On the other hand, some factors affecting mortality are density-independent. For example, a very severe winter will result in the deaths of a certain proportion of owls, and this *proportion* will be unrelated to the total number present (Figure 12.8). Density-independent factors can be biotic *or* abiotic, although they are commonly 'catastrophic' abiotic events, such as droughts, floods or fires. Density-dependent factors are *always biotic*.

### 12.3 Intraspecific and interspecific competition

We noted in Section 12.1 (above) that populations are usually restricted in their size by limiting factors, such as availability of food, light or space. If any of these resources are in short supply, there will be **competition** between individual organisms for that resource. If the competition is between members of the *same* species, this is called **intraspecific** competition (as in the example of competition between tawny owls for territory). If the competition is between members of a *different* species, this is known as **interspecific** competition.

#### Intraspecific competition in limpets

The limpet *Patella cochlear* (Figure 12.10) shows an effect of intraspecific competition. This mollusc lives on the rocky shore, where it grazes on algae. When there is a low population density of the limpets, they are able to grow to a larger size (Figure 12.11). As the population density increases (shown on the graph as numbers per square metre) the maximum length of the animals decreases due to competition between individuals for food and space, so that above a certain density a near constant biomass of the population per square metre is maintained.

Intraspecific competition occurs between members of one species for a common resource. Therefore the greater the population density, the greater will be the degree of competition. Intraspecific competition is very important to evolution, since it exerts selection pressures which may result in changes to gene frequencies; in other words, natural selection in favour of beneficial characteristics (see Section 4.3, page 110).

Interspecific competition occurs when individuals of different species, living in the same area, require the same resources. However, it has been shown that two species which have very similar requirements (that is have the same **ecological niche**) will be unable to coexist.

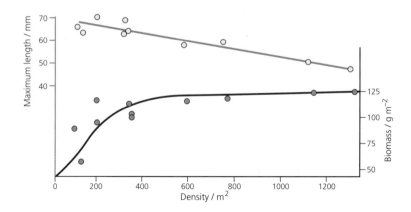

### Remember

#### The ecological niche

Any species of organism will have a particular set of biotic and abiotic environmental conditions in which it can survive and reproduce. This set of conditions is called the species' **ecological niche**. For example, a particular plant species might only grow in alkaline soil of pH 7–8.5, a temperature range of 10–20 °C, and low soil water content. These factors would define its ecological niche, and reflects the plant's specialism in the community. Similarly an animal's ecological niche will be in part determined by its trophic level and feeding method, such as herbivore, grazer, browser or filter-feeder. Other factors

may decide an animal's niche, such as the time when it is active (day or night), method of locomotion, and body shape and size. In simple terms, the niche is an organism's 'role' in a habitat.

In different habitats a niche may be filled by a different species. For example, the role of large terrestrial grazing herbivore is filled by deer in Europe, antelope in Africa and kangaroo in Australia. However, two different species cannot occupy the *same* niche in *one* habitat. One will always out-compete the less well-adapted species. This is known as the **competitive exclusion principle**.

## Interspecific competition in Paramecium

The competitive exclusion principle was proposed by the Russian ecologist G.F. Gause in 1934. Gause carried out experiments which demonstrated the principle, using two species of the single-celled ciliate *Paramecium* (*P. aurelia* and *P. caudatum*). He cultured these two organisms separately and in a mixed culture, where they were competing for the same food source. When they were cultured separately, their growth curves each followed the expected sigmoid shape (Figure 12.12a). But when they were cultured together, the faster growing *P. aurelia* out-competed *P. caudatum*, the numbers of which fell as time went on (Figure 12.12b).

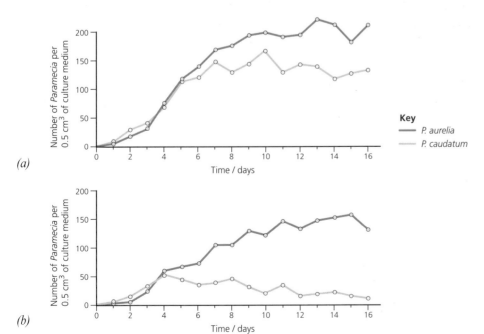

*Figure 12.12 Interspecific competition between two species of* Paramecium *cultured (a) separately and (b) together (after G.P Gause, 1934)*

Gause also investigated interspecific competition between *P. aurelia* and a third species of *Paramecium*, called *P. bursaria*. When these two species were cultured together the numbers of both increased and neither out-competed the other. Gause discovered this was because each tended to feed in a different part of the culture tubes. In other words they occupied a *different ecological niche*.

### Interspecific competition in barnacles

The *Paramecium* studies of Gause were carried out as laboratory experiments. It is much more difficult to demonstrate interspecific competition between organisms in their natural environment. However, such a study was carried out by the ecologist J.H. Connell in 1961, and it remains a classic demonstration of the competitive exclusion principle.

Connell investigated populations of two species of barnacle which are common on rocky shores around the coast of Britain. Barnacles are crustaceans which as adults remain attached to the rocks in one place, and are unable to move (they are **sessile**). One species, *Chthalamus stellatus*, lives higher up the shore, around the high tide mark. Another species, *Semibalanus balanoides*, occupies the middle and lower part of the shore (Figure 12.13). Both species colonise the shore through their larval stages, which swim in the plankton until they attach themselves to the rocks to complete their life cycle. The adult barnacles feed by filtering microscopic organisms from the water.

Connell removed all barnacles from experimental areas of rock and recorded the fate of larvae of both species after they had settled onto the surface. He found that both species were able to colonise the rock at any height up the shore, but in the lower intertidal zone, the larger *Semibalanus* out-competed the *Chthalamus* by smothering and crushing them as they grew. Only at the highest parts of the inter-tidal zone did *Chthalamus* survive, where *Semibalanus* was unable to live. This is because the upper shore is exposed to the air (and sun) for long periods of time, which results in desiccation of the barnacles. *Chthalamus* can withstand greater desiccation than *Semibalanus*. When young *Chthalamus* lower down the shore were kept free of *Semibalanus* competitors, they were perfectly able to grow on the rock.

*Figure 12.13 Interspecific competition between two species of barnacle on a rocky shore*
*(a) Chthalamus stellatus*
*(b) Semibalanus balanoides*

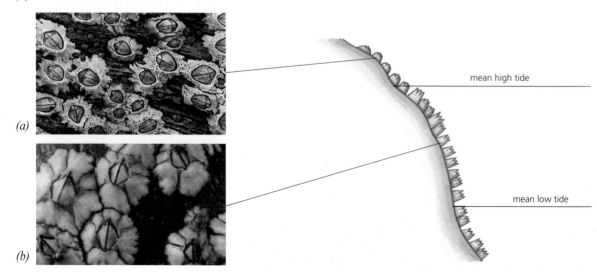

(a)

(b)

mean high tide

mean low tide

**4** Explain, in a sentence, the difference between intraspecific and interspecific competition. *(1 mark)*

### 12.4  Predator–prey relationships

A **predator** is a carnivore which feeds on another species, its **prey**. The numbers of predators that a habitat can support (its carrying capacity) will depend upon the population density of the prey. If there are many prey animals, more predators will survive and breed. Conversely, if the prey numbers fall, the number of predators will fall also. In turn this might be expected to lead to a rise in the numbers of prey again, maintaining an equilibrium of both predator and prey species (Figure 12.14).

The numbers of both species oscillate, but are slightly out of phase with each other. The numbers of the prey reach a peak just before the predator. This model of predator–prey interaction has been supported by a number of laboratory experiments with small invertebrates. For example, Figure 12.15 shows a predator–prey graph for two species of mite. The herbivorous mite *Eotetranychus sexmaculatus* is prey to a carnivorous species called *Typhlodromus occidentalis*. The experimental results fit the pattern predicted in Figure 12.14 well.

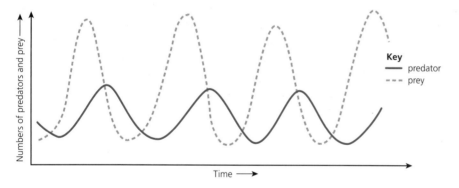

*Figure 12.14  Theoretical predator–prey cycles*

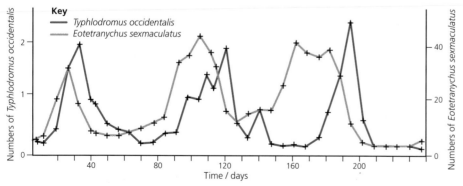

*Figure 12.15  Predator–prey graph for two species of mite,* Typhlodromus occidentalis *(predator) and* Eotetranychus sexmaculatus *(prey) (after M.K. Sands, 1978)*

Collecting predator–prey data for larger organisms in the wild is much more difficult. A famous example which is often cited in biology textbooks is that of the North American snowshoe hare and its predator, the lynx (Figure 12.16).

*Figure 12.16* *Predator–prey cycles for the lynx and snowshoe hare between 1845 and 1935*

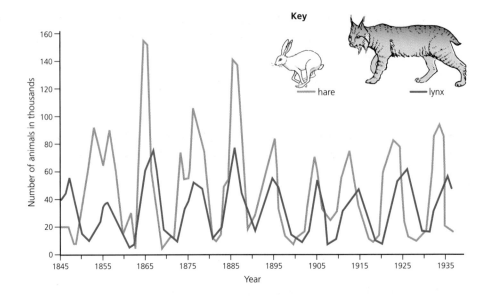

The data for this graph came from the records of the Hudson Bay Company of Canada. They recorded the numbers of pelts (skins) of both species that were obtained by trappers over the years shown. The cyclic fluctuations are often supposed to be due to predation by the lynx maintaining the numbers of hare, which can form up to 90% of the lynx's diet. However we now know that this apparent 'cause and effect' is not quite so straightforward. In parts of Canada in recent years, the lynx population has fallen to very low levels, yet the hare population continues to oscillate as before. It seems instead that it is the hare population which controls the population of the lynx. In turn the hare population is controlled by the plant species which are its food source. When the food plants of the hare are over-grazed, they produce toxic substances which prevent further grazing. These toxins remain in the plants for up to three years, during which the hare numbers fall and the plants recover. The plant populations are also affected by changing climatic conditions and fluctuating numbers of insect pests. The situation 'in the wild' is often much more complex than predicted by the simple predator–prey model, or even laboratory experiments.

## 12.5 Estimating the size of a population: the Lincoln Index

If a population of animals is large, it will be impossible to count all of them. One way we can obtain an estimate of their numbers is to use a method called the **mark-release-recapture** method, which generates an estimate of the population size called the **Lincoln Index**. This method is particularly suitable for use with small mobile animals living in a fairly well-defined habitat. This might be a population of woodlice under a log, or a population of amphipod shrimps living amongst some dead seaweed at the top of a beach.

A large number of the animals are first caught, counted and marked with a small dot of non-toxic paint, or 'Tippex'. These are then returned to their habitat and allowed to mix with the rest of the population. After a suitable length of time, a second sample of the animals is taken and counted. The number of marked animals in this sample is recorded. The ratio of the number of animals marked in the first sample to the total number in the population is assumed to be the same as the ratio of the number of marked animals recaptured to the total caught in the second sample, i.e.

$$\frac{\text{number marked in first sample}}{\text{total number in population}} = \frac{\text{number of marked animals recaptured}}{\text{total caught in second sample}}$$

Suppose a systematic search of a pile of seaweed at the strand line of a beach yielded 257 shrimps. Each was marked and released back into the seaweed. After several hours, a second search produced 193 shrimps, 31 of which were marked. We can use the above equation to calculate the population size (N):

$$\frac{257}{N} = \frac{31}{193}$$

Therefore:  $N = (257 \times 193) \div 31 = 1600$ (to the nearest whole shrimp).

The Lincoln Index method has a number of limitations:

- It can only be used for animals which inhabit a limited, well-defined habitat, as with the examples above.
- There must be no emigration or immigration from the habitat. For example, the shrimps must not burrow down into the sand between the two samples.
- Enough time must be allowed for the released animals to mingle with the rest of the population. For flying insects, this might be as little as three hours, but for slower animals such as snails, it could be a number of days.
- The method of capture must not alter the habitat so that the animals move away.
- The marking method must not harm the animals, reduce their chances of survival or affect their behaviour patterns.

You can see that the situations where the mark-release-recapture method is likely to be suitable are going to be limited. On top of this, if a reasonably accurate estimate is to be made, at least 10% of the released individuals from the first sample must be recaptured in the second sample.

**Remember**

The total population estimated by the mark-release-recapture method is given by:

$$\frac{\text{number marked and released} \times \text{total number recaptured}}{\text{number of marked animals recaptured}}$$

5 A student collected 112 woodlice from a pile of logs. She marked them with a small spot of paint on their lower surface and released them back into their habitat. The following day she collected 140 woodlice from the same habitat, of which 15 were marked.

(a) Calculate the size of the woodlouse population. *(1 mark)*

(b) Suggest why the woodlice were marked underneath, and not on their backs. *(1 mark)*

(c) Fifteen marked animals were recaptured. Was this enough for an accurate estimate of the population size? Explain your answer. *(2 marks)*

## 12.6 Colonisation and succession

If a field were to be ploughed up and then left abandoned for many years, the community of organisms living in it would go through a series of changes. At first small annual herbaceous plants would grow, perhaps from weed seeds in the soil, or carried by birds or other animals, or by the wind. These first **colonisers** would be able to exploit the abandoned soil, through their rapid rate of growth and reproduction, and lack of competition. They would be followed by other species, including larger perennial plants and small shrubs. These species take longer to establish themselves, but eventually would probably out-compete the smaller plants, for example by depriving them of light or water, so that the structure of the plant community would change. After many years, trees might grow from saplings, until the field contained a community dominated by mixed deciduous trees, with smaller shrubs and herbaceous plants. This change in the composition of a community with time is called **succession**.

The type of succession described above is known as a **secondary succession**. This is where the changes in the community structure take place in a location which has previously supported a community (in this case when it was farmland). Another example is the development of new communities following forest fires. Secondary successions are often rapid, due to the presence of soil and a 'seedbank' within the soil.

*Figure 12.17* *An Australian* Eucalyptus regnans *forest*

**Did You Know**

Although they are destructive events, natural forest fires can sometimes be essential to the maintenance of an ecosystem. Australia is home to about 700 varieties of eucalyptus trees, including vast forests dominated by the giant gum tree, *Eucalyptus regnans* (Figure 12.17).

This is the tallest flowering plant in the world, and produces millions of seeds every year. However, few of the seeds produce new trees. They have a very low rate of germination, and many seeds are carried underground by ants, to depths where they are unable to germinate. Those that do germinate are unlikely to survive, being killed by lack of light, drought, high temperatures, fungal diseases or being eaten by herbivores. The result is that whole forests consist of mature trees of a certain age, but with no younger saplings. Fire is the abiotic factor which allows regeneration of the gum tree forests. Every 10 to 20 years natural fires destroy vast areas of the forest. These can be very large. In 1994 a fire consumed 600 000 hectares of forest, threatening the outskirts of Sydney. The largest ever fire, in 1939, destroyed much of the State of Victoria. Devastating though they are to humans and other organisms, these fires eliminate competition, allowing the gum tree seeds to germinate and grow, explaining the uniform age of the trees in the forests. The eucalyptus species have leaves which are full of oil, and this actually *helps* them to burn.

A **primary succession** takes place in a previously uncolonised substrate, such as a newly formed pond, bare rock, or some sand dunes. For example, rocks eroded from the upper slopes of mountains, above the vegetation line, will fall under gravity down the sides of the mountain, forming scree deposits. The first organisms to colonise such an inhospitable substrate (**pioneer** species) are usually **lichens**, carried on the wind as spores (Figure 12.18). These are mutualistic associations of a fungus with either an alga or blue-green bacterium.

Lichens are well adapted for the role of coloniser. They are tolerant of desiccation and can grow without soil. The algal cells live inside the hyphae of the fungus, and can photosynthesise, making organic molecules. The fungus shelters the alga and provides it with carbon dioxide for photosynthesis, while the fungal hyphae can penetrate small cracks in the rocks, secreting acidic compounds which continue the surface weathering of the rock.

**Figure 12.18** *Lichens growing on a tiled roof*

**?**

**6** Explain the term 'mutualism'. (See *AS Biology*, Section 13.1, page 375.) *(1 mark)*

Death of the lichens, along with any plant debris deposited by the wind, provides organic materials for decomposition, preparing the way for other species to colonise the rock. Next in the succession will probably be **mosses**. These simple plants are also dispersed as spores by the wind. They have tiny root-like rhizoids which allow them to attach to surfaces, and do not need a deep soil as do more complex flowering plants. Mosses grow in clumps, trapping more dead material and water between the individual plants, and continuing the formation of a primitive soil. They are also capable of surviving a high degree of desiccation, reviving when rain eventually falls on them.

Larger plants such as **ferns** and **grasses** may colonise the rocks next, taking advantage of the thin soil which has collected in crevices in the rocks, and adding to its organic content when these plants die and decompose. Eventually this soil may build up enough for **herbaceous flowering plants** to become established, followed by **shrubs** and small **trees**. After a lengthy period of time, perhaps hundreds of years, what was a bare rock surface may become a diverse community dominated by deciduous trees such as oak. This stable community which eventually forms, when succession stops, is called a **climax community** and is in a dynamic equilibrium. A point that is very important to understand about succession is that it is brought about by changes which are *caused by the vegetation itself*. Each group of plant species modifies the habitat, making it easier for subsequent groups in the succession to establish themselves. The flow chart opposite shows stages in a primary succession.

The first organisms to colonise a habitat, such as lichens, are known as **pioneers**. The change in communities, from the pioneer community, through the succession to the final climax community, is called a **sere**, and each distinct community in the succession is known as a **seral stage**. There are different types of seres, depending on the habitat they occupy. The sere described above, involving colonisation and succession of bare rock, is called a **lithosere**. A sere starting in a freshwater habitat is a **hydrosere**, and one involving a dry habitat, such as sand dunes, is a **xerosere**.

| bare rock |
| :---: |

↓

| colonisation by lichens, weathering of rock and production of dead organic material |
| :---: |

↓

| growth of moss, further weathering, and the beginnings of soil formation |
| :---: |

↓

| growth of small plants such as grasses and ferns, further improving soil |
| :---: |

↓

| larger herbaceous plants can grow in the deeper and more nutrient-rich soil |
| :---: |

↓

| climax community dominated by shrubs and trees |
| :---: |

You might have noticed that so far, all the descriptions of succession have only mentioned plant communities. This is not to say that the animals in the communities do not undergo succession too. It is just that the plants are the more conspicuous and easily recognised members of a community, and are sometimes used to define the climax community. For example, other climax communities include temperate grassland and coniferous forest.

 **Did You Know**

**Surtsey: colonisation and succession on a new island**

On the 14th of November 1963, 20 miles off the south coast of Iceland, hot magma erupted from a crack in the ocean floor, 130 metres deep. The magma, interacting with the cold sea water, produced a continuous series of explosions, which ejected over a cubic kilometre of black ash and cinders, overnight forming a new volcanic island (Figure 12.19a). The island, about 2.7 km$^2$ in area, was named Surtsey, after a legendary Nordic fire-eating giant. Lesser eruptions continued for several years. By February 1964, an enclosed crater had formed, isolating the lava flows from the sea, so that the lava spread over the surface of the island from the middle, forming a solid rock surface, until the eruptions finally ceased in 1967.

From the moment that Surtsey was formed, biologists realised its unique potential as a natural laboratory for studying the processes of colonisation and succession. Visits to the island were restricted, in order to minimise the possibility of organisms being introduced as a result of human activity, and the biologists set about sampling the living organisms which arrived at the island, a research programme which has continued for nearly 40 years.

It was not long before the scientists discovered life on the island. Six months after the eruption, several species of bacteria and moulds were recorded, as well as insects and birds using the island as a resting place. Simple species of plants were soon established in the crater, such as lichens and mosses, their spores carried to the island by the wind and on the feet of birds. The first flowering plant to reach the island, a pioneer species called sea rocket, was recorded in 1966. Its seeds are dispersed by water and probably floated to Surtsey. Other plants soon followed, their seeds carried by birds, on water, or attached to floating objects such as driftwood. Ferns were able to colonise cracks in the rock surface.

All of these species allowed the process of succession to continue, by altering their environment, making it possible for other species of plants and animals to become established. Gulls and other seabirds began breeding on the island in 1986. They helped greatly to fertilise the soil that had formed, and carried in spores and seeds of lichens, mosses, ferns and flowering plants, so that by 1995, 52 different plant species had been recorded. The gulls' habit of building their nests around the edge of the colony helped to spread the area covered by vegetation (Figure 12.19b).

The process of succession is still continuing on Surtsey, although the island will not remain a terrestrial habitat forever. Already nearly half its original area has been eroded by the constant action of Atlantic waves and storms, and it is likely that within a few hundred years it will completely disappear.

(a)

(b)

*Figure 12.19* Surtsey island (a) shortly after its formation, (b) Surtsey today – the green areas are vegetation

### Climatic climax and plagioclimax

Climax communities are often determined by the climate of the environment in which they develop. When this happens, it is referred to as a **climatic climax**. Early ecologists believed that a particular climate would always lead to the establishment of a particular climatic climax community. Nowadays ecologists accept that at a local level, this is an over-simplification. There are many different climax communities which are governed by several factors as well as climate. These include soil (edaphic) conditions, local climate (microclimate), effects of grazing and trampling, and human activities. However, major climatic climax communities are apparent on a global scale. Where a major type of vegetation covers a large geographical area, this is called a **biome**. Biomes are mainly determined by two abiotic factors: **temperature** and **rainfall**. A combination of high temperatures and high rainfall is needed for a tropical rain forest biome, whereas low temperatures and low rainfall characterise a cool desert biome. There are a number of intermediate biomes between these two extremes (Figure 12.20).

Some plant communities are prevented from reaching a climax community as a result of human interference. Such actions as heavy grazing by domestic animals, ploughing, mowing or deliberately setting fire to vegetation, act as arresting factors in a succession. This can result in a stable plant community known as a **deflected** or **plagioclimax**. A good example of this in Britain is heather moorland on upland areas, which owes its existence to grazing and regular burning of areas of the heather, which arrests further succession. Without human interference, a climax community woodland dominated by birch and pine would result. Other familiar areas of land are deflected climaxes, such as lawns and grass verges, which are mown or sprayed to prevent weeds taking over.

Average annual rainfall

low

Average temperature

high

*Coniferous forest*

*Deciduous forest*

*Tropical rainforest*

*Tundra*

*Temperate grassland*

*Savannah*

*Cool desert*

*Temperate desert*

*Tropical desert*

***Figure 12.20*** *The principal terrestrial biomes result from different combinations of mean temperature and rainfall*

## 12.7 Zonation

**Zonation** occurs where there is an environmental gradient which affects the distribution of organisms. It results in bands or **zones** of organisms which are distributed along the gradient. This is easiest to understand by looking at particular examples of zonation, such as on a rocky seashore.

*Figure 12.21* *Zonation on a rocky shore in Pembrokeshire, West Wales*

7  You have seen an example of animals which show zonation on a rocky shore earlier in this chapter. What were they?          *(1 mark)*

A rocky shore shows distinct bands of marine algae (seaweeds) and animals (mainly barnacles and molluscs) which are clearly visible from a distance (Figure 12.21).

The zonation is directly or indirectly brought about by the action of tides. Tides cover the shore twice a day, and the maximum and minimum heights that they reach vary throughout the month, with the phases of the moon. Around the coast the tidal range differs at different locations, but may be several metres in height. An organism that lives near the top of the tidal range may be immersed under sea water for only minutes each day, whereas one near the bottom of the tidal range will be underwater for many hours. This variation in immersion time affects the organisms in a number of ways, such as:

● Exposure at low tide, which is greater on the upper and middle shore, produces an increased rate of **desiccation**.

● Seaweeds need to be immersed to be able to carry out **photosynthesis**, so species living on the lower shore can do so for longer periods. However, water absorbs some light wavelengths and reduces light intensity, so those species lower down the shore will receive less light when the tide is in.

● Emersion (being out of water) may result in **overheating** of organisms in the sun, which is more of a problem for those further up the shore. When covered by water their temperature is maintained at a constant level.

● Many animals, such as limpets or barnacles, can only feed when the tide is in. Time for **feeding** will be limited higher up the shore.

● **Grazing** by animals will affect the distribution of seaweeds, and **predation** by aquatic predators is greater on the lower shore.

● Other aspects of physiology, such as **gas exchange** via gills, requires the animal to be covered by water, or when the tide is out to be protected in an enclosing shell.

The net result of biotic and abiotic factors such as these is that different species of algae and invertebrate animals tend to be found in particular zones in the upper, middle and lower shore (Figure 12.22).

It is important to realise that these zones are not the same on every rocky shore. Even if two shores have the same tidal range, the zonation is affected by other factors, such as degree of exposure to wave action, and aspect (which direction the shore is facing).

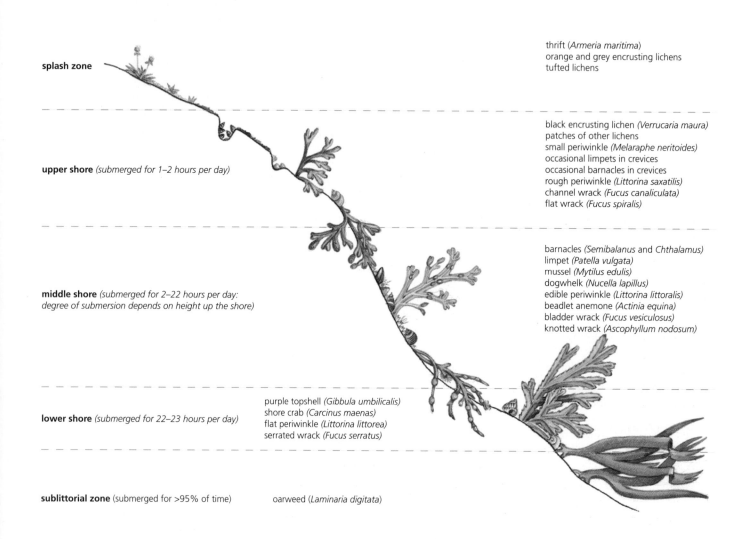

splash zone

thrift (*Armeria maritima*)
orange and grey encrusting lichens
tufted lichens

upper shore (*submerged for 1–2 hours per day*)

black encrusting lichen (*Verrucaria maura*)
patches of other lichens
small periwinkle (*Melaraphe neritoides*)
occasional limpets in crevices
occasional barnacles in crevices
rough periwinkle (*Littorina saxatilis*)
channel wrack (*Fucus canaliculata*)
flat wrack (*Fucus spiralis*)

middle shore (*submerged for 2–22 hours per day:
degree of submersion depends on height up the shore*)

barnacles (*Semibalanus* and *Chthalamus*)
limpet (*Patella vulgata*)
mussel (*Mytilus edulis*)
dogwhelk (*Nucella lapillus*)
edible periwinkle (*Littorina littoralis*)
beadlet anemone (*Actinia equina*)
bladder wrack (*Fucus vesiculosus*)
knotted wrack (*Ascophyllum nodosum*)

lower shore (*submerged for 22–23 hours per day*)

purple topshell (*Gibbula umbilicalis*)
shore crab (*Carcinus maenas*)
flat periwinkle (*Littorina littorea*)
serrated wrack (*Fucus serratus*)

sublittorial zone (*submerged for >95% of time*)

oarweed (*Laminaria digitata*)

*Figure 12.22 Zonation on a sheltered rocky shore (not drawn to scale). Different communities are found at different heights up the shore from the low water mark. The zonation results from the degree of submersion, or exposure (emersion) at each height*

## Case study: zonation and succession on Slapton shingle ridge

Slapton Ley is the largest natural freshwater lake in the South West of England, and is part of the Slapton Ley National Nature Reserve, managed by the Field Studies Council. The Ley is separated from the sea by a ridge composed of shingle. Geologically the ridge is a very recent feature, formed by onshore drift at the end of the last ice age. It only settled into its present position three to five thousand years ago. Before 1850, the ridge was regularly breached by rough seas, until a road was built along it, stabilising its structure (Figure 12.23).

The ridge shows zonation of plant communities from the sea side to the Ley, influenced by gradients of abiotic and biotic factors. These zones reflect the succession that presumably took place after the ridge stabilised in its present form, and still continues today.

**Figure 12.23** *Slapton Shingle ridge and the Ley*

On the side of the ridge nearest the sea, abiotic conditions are very harsh. It can be very windy, and the wind carries salt spray, which will 'burn' the leaves of most plants, by osmotic loss of water. The shingle retains very little water, and any that reaches plant roots may have a high dissolved salt content, and a higher than average pH. The shingle is exposed, and ground temperatures may be very hot on sunny days and very cold during the winter. Very few species of plants can grow under these conditions. However, near the top of the ridge a few species, such as the sea spurge, sand couch grass, and yellow horned poppy, have adaptations which allow them to survive (Figure 12.24a and b). These plants are known as the **pioneer** species. They show **xerophytic** adaptations such as:

● extensive root systems which collect any water that falls on the shingle
● rolled leaves, with stomata in pits which reduce transpiration
● a thick waxy cuticle to the leaves, restricting water loss
● a low-growing habit, reducing the likelihood of wind damage
● dead leaves which are retained, covering the living leaves and acting as a barrier to water loss.

Many species (e.g. sea spurge) are also **halophytes** – plants that can tolerate a salty water supply to the roots.

Pioneer species help to change the conditions on the ridge, which results in species which are less well adapted to the extreme conditions being able to survive. The roots of the pioneers stabilise the shingle, and when the plants die,

**Remember**

**Xerophytes** are plants adapted to survive conditions of low water availability, such as in deserts or growing on sand dunes or shingle (see *AS Biology*, Section 12.4, page 348). **Halophytes** are plants which can tolerate a very salty soil, such as that found on the banks of estuaries. Most plant species are mesophytes, growing under average conditions of water availability.

their organic remains decompose, adding to the soil humus, which retains valuable moisture. The decomposing humus produces acid compounds, which decrease the soil pH. Overall the soil structure is improved, enabling species such as sea mayweed, sea campion, sea bindweed, thrift, sea radish and Danish scurvy grass to grow. These plants form the **maritime turf community** (Figure 12.24c). They, too, show some xerophytic adaptations, and some, such as sea radish, are **biennials**, only living for two years. They store food reserves in their root systems in the first year, and expend the stored energy in flowering during the second year. In effect they 'stretch out' a one-year cycle into two years, enabling them to economise on resources.

The maritime turf species further improve the quality of the soil, which becomes deeper, more acidic and richer in inorganic nutrients with increasing distance from the sea. The average height of the plants increases too, providing shelter from the wind and drying effects of the sun, allowing other plant species to survive, as well as invertebrate animals such as insects, spiders and snails.

Half-way across the ridge, just before the road, the maritime turf species gradually merge with a type of vegetation which shows fewer xerophytic adaptations. This includes plants such as red clover, sea carrot, rest harrow, yarrow, ribwort plantain, black knapweed, hawkbit and bird's foot trefoil. These make up what is called the **hardy generalist** community (Figure 12.24c), reflecting the fact that although the abiotic conditions have improved from what they were nearer to the sea, they can still be harsh, so that only 'hardy' species can survive. Some of these are legumes, such as clover and bird's foot trefoil. They have root nodules containing symbiotic bacteria which fix nitrogen (see *AS Biology*, Section 13.9, page 405). In this way the nitrate content of the soil is increased when the plants die and decompose. Many of these plants are agricultural weeds adapted to thrive on disturbed ground, in this case from trampling and occasional storms.

On the side of the road furthest from the sea, the vegetation changes to a **meadow community** (Figure 12.24d). This is characterised by many species of grass, dandelion, vetches, hedge bedstraw, chervil, hogweed, ground ivy, goosegrass, dovesfoot cranesbill, toadflax and creeping cinquefoil. At this position on the ridge the abiotic stresses that the plants are exposed to have decreased substantially, and conditions are less hostile. The soil is deeper, with a

*Figure 12.24 Zonation and succession at Slapton*

(a) pioneer species: sea spurge and sand couch grass

(b) another pioneer: yellow horned poppy

(c) maritime turf and hardy generalist species

lower pH and more nutrients, and the wind speed is less. A greater number of less hardy species can survive, and the height of the vegetation has increased dramatically. In other words, there is much greater interspecific competition for factors such as light and growing space. Biotic, rather than abiotic factors have become more important in determining which species can survive.

At the back of the ridge near the Ley, the meadow community gives way to the **scrub and wood community** (Figure 12.24e). This contains a number of larger shrubs and small trees, such as gorse, blackthorn, apple, elder and sycamore, as well as smaller plants such as foxglove, brambles, nettles and ivy. These species are well adapted to out-compete those of the meadow community, but require higher nutrient levels for this. Many are tall, or fast growing such as brambles and gorse, or shade tolerant, such as wood sorrel.

The zonation we can see on the Slapton shingle ridge is likely to reflect the succession of plant communities which colonised the ridge over time. The first colonisers would have been pioneer species, which would have improved the abiotic conditions, allowing the growth of maritime turf species, followed by hardy generalists and meadow and scrub communities. In other words each zone represents a **seral stage** in the succession. The zones nearest to the sea are dominated by abiotic factors such as salinity, soil quality and wind speed, and their communities show a low species diversity. In the less hostile environments nearer to the Ley, the diversity of species increases, and the zones are dominated by the biotic factors of competition.

As well as showing an increase in species diversity, successions tend to show certain other trends through the seral stages. These include an increase in the:

- size of organisms
- biomass per unit area
- amount of dead organic matter in the soil
- number of heterotrophic organisms, such as animals
- specialisation of ecological niches, as interspecific competition increases.

The climax community on the shingle ridge is the scrub and wood community, although this could not really be described as a climatic climax. It is a good example of how the climax is determined by many local factors. In this case the microclimate is important, as well as edaphic factors – the soil never becomes

*(d) meadow species*

*(e) scrub and wood community*

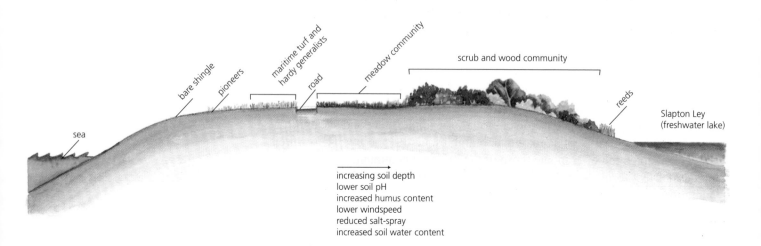

labels: maritime turf and hardy generalists, meadow community, scrub and wood community, bare shingle, pioneers, road, reeds, sea, Slapton Ley (freshwater lake)

increasing soil depth
lower soil pH
increased humus content
lower windspeed
reduced salt-spray
increased soil water content

*Figure 12.25 Section through Slapton shingle ridge showing locations of different plant communities (not drawn to scale)*

deep enough to support large trees. It is also an example of a plagioclimax, because the dynamic equilibrium is maintained by human interference. Regular cutting of the gorse and brambles, and trampling by humans on the paths along the ridge both affect the succession (Figure 12.25).

## 12.8 Species diversity

We have used the term 'diversity' above without really considering its full meaning. It is not merely the *number* of different species (this is called the species *richness*) but is also a measure of the *relative abundance* of each species. This is easier to understand by considering an example. Imagine three communities, each made up of a total of 60 organisms, drawn from combinations of six species, A to F (Table 12.1).

Community 1 has the highest diversity. It has the joint highest species richness (6) and each species has a similar relative abundance. Community 2 has the same species richness as community 1, but is dominated by one species (A) so that the diversity of this community is lower than in community 1. Community 3 has a lower diversity than community 1, due to its lower species richness.

> **Remember**
>
> You should be clear about the difference between succession and zonation. **Succession** refers to the change in the composition of a community with time. **Zonation** is where different communities are present at the same time, distributed as bands or zones along an environmental gradient.

| Species | Community 1 | Community 2 | Community 3 |
|---------|-------------|-------------|-------------|
| A | 10 | 50 | 20 |
| B | 9 | 1 | 20 |
| C | 11 | 3 | 20 |
| D | 10 | 2 | 0 |
| E | 8 | 2 | 0 |
| F | 12 | 2 | 0 |
| Total | 60 | 60 | 60 |

*Table 12.1 Species composition of three different communities*

There have been a number of ways devised to turn numbers such as these into a single measure of species diversity, called a **diversity index**. One such method is the **Simpson's index**. This is calculated from the formula:

$$\text{Simpson's index (d)} = \frac{N(N-1)}{\Sigma n(n-1)}$$

Where  N = the total number of all organisms

n = the numbers of individuals of one species

$\Sigma$ = the Greek letter sigma, meaning 'the sum of'.

The higher the value of d, the greater is the species diversity. Take for example community 1 in table 12.1. The values of $(n-1)$ and $n(n-1)$ are shown in table 12.2.

| Species | n (community 1) | n − 1 | n(n − 1) |
|---|---|---|---|
| A | 10 | 9 | 90 |
| B | 9 | 8 | 72 |
| C | 11 | 10 | 110 |
| D | 10 | 9 | 90 |
| E | 8 | 7 | 56 |
| F | 12 | 11 | 132 |
| Total | N = 60 | | $\Sigma n(n-1) = 550$ |

*Table 12.2  Data for calculation of Simpson's index for community 1*

So for community 1:

$$d = \frac{60 \times 59}{550}$$

$$= 6.44 \text{ (high diversity)}$$

By the same method, the Simpson's index for community 2 = 1.44 (low diversity).

**8** Calculate the Simpson's index for community 3 in Table 12.1. Show your working. *(3 marks)*

Simpson's index is easy to apply, and can be used for both animals and plants. The actual species do not have to be named, so long as individuals can be recognised as belonging to the same species. However, the method must only be used with data that consists of *numbers* of organisms (rather than, for example, percentage cover with plants). It is also important to realise that the index is *not a statistical test*. It gives no measure of the *significance* of the difference between the diversities of two communities, just a general indication of that difference.

Two communities with indices of 3.97 and 4.12 could be regarded as similar in diversity, two with indices of 2.66 and 4.99 as different. Compare this with statistical tests such as the Chi-squared test (see Section 3.11, page 94 and Section 12.10, page 407) which give values which allow us to tell the *mathematical probability* of obtaining differences between results by chance, for various types of data.

## 12.9 Sampling methods

Ecological investigations always involve the collection of data and its subsequent analysis. An ecologist will almost never have access to all the possible information about a particular habitat. For example, if measurements of the size of flat periwinkles (a mollusc living on seaweed) on a rocky shore are being taken, it would be impossible to measure all of the individual animals. The ecologist would be forced to measure a smaller sub-set of the population, which could then be used to draw conclusions about the whole population. This sub-set is called a **sample**. It is crucial that the measurements made or counts taken in the sample avoid **bias**. The ecologist above might find it easier to spot periwinkles which had yellow shells than those with brown shells. If the yellow periwinkles were also bigger than the brown ones, this would introduce a bias in the sample, leading to a bias in the conclusions about periwinkle size.

### Random sampling

The area from which a sample is taken is known as a **quadrat**. The size of the quadrat chosen depends on the size and distribution of the animal being sampled. For plants growing on a school playing field, a quadrat size of 0.5 × 0.5 m, or 0.25 m², might be suitable, whereas lichens growing on a tree, or barnacles on a rocky shore could be sampled with a 10 × 10 cm quadrat. Sampling elephants on the plains of Africa needs 'quadrats' with sides several kilometres long (they are virtual sampling squares drawn on a map, and sampling is carried out from an aircraft!).

To avoid bias in the placing of the quadrat in a particular area, the location must chosen at random, without any selection by the investigator. One way that this can be done is to have a numbered grid across the sampling area, and to position the quadrat at coordinates on this grid (Figure 12.26).

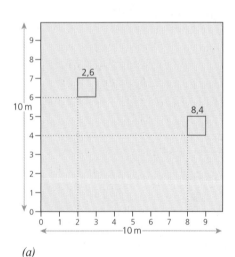

*(a)*

**Figure 12.26** *(a) A 10 m² grid with 1 m² quadrats positioned at coordinates 2,6 and 8,4,*
*(b) Students using quadrats positioned at coordinates on a grid, placed on a school field, to sample herbaceous plants*

*(b)*

The two coordinates used to locate the position of the quadrat can be generated from random number tables (Figure 12.27) or from the random number key of a scientific calculator.

## Sampling along transects

If a habitat is not uniform, but shows some sort of structure, then random sampling of the kind described above is not suitable, and would produce a sample which was unrepresentative of the structure. A good example is the rocky shore, where the location of organisms is determined by an environmental gradient. In this case it is necessary to take a series of samples along a predetermined line, at right angles to the gradient. This is called a **transect**.

A **belt transect** consists of a strip across the area to be studied, typically using $0.5 \times 0.5$ m quadrats. If these are placed continuously along the line, this is called a **continuous belt transect**. If they are placed at intervals, perhaps every 2 metres, the result is an **interrupted belt transect**. As with random sampling, the choice of quadrat size and the interval between quadrats depends on the type of organism being sampled, the size of the area under study, and the time available (Figure 12.28).

A **line transect** is a quick, but less informative method of sampling along an environmental gradient. Here a tape is stretched out along the transect, and any plant species which touch the tape are simply recorded, in order, along the line, either in a continuous or interrupted fashion as with the belt transect. This method is unlikely to be representative of the habitat unless it is repeated several times to check for consistency.

An element of random sampling can be introduced into a transect by using a **stratified** sampling technique. Here the habitat is divided into zones (strata) along the transect, and sampling is carried out at random (using coordinates chosen as above) within each stratum.

| 09 | 03 | 54 | 20 | 02 | 55 | 49 | 48 | 46 | 75 | 42 |
| 41 | 48 | 46 | 17 | 24 | 82 | 51 | 86 | 86 | 53 | 66 |
| 10 | 21 | 02 | 71 | 89 | 14 | 80 | 64 | 32 | 58 | 17 |
| 55 | 94 | 44 | 77 | 90 | 01 | 99 | 79 | 48 | 28 | 61 |
| 81 | 42 | 45 | 69 | 28 | 23 | 90 | 46 | 24 | 32 | 97 |
| 22 | 05 | 84 | 39 | 89 | 57 | 73 | 84 | 86 | 57 | 76 |
| 14 | 97 | 61 | 57 | 30 | 93 | 88 | 12 | 88 | 58 | 15 |
| 75 | 58 | 14 | 05 | 05 | 16 | 72 | 57 | 34 | 20 | 46 |
| 71 | 77 | 50 | 51 | 00 | 61 | 02 | 60 | 51 | 13 | 61 |
| 12 | 15 | 66 | 40 | 56 | 39 | 77 | 75 | 32 | 80 | 30 |
| 70 | 46 | 36 | 67 | 19 | 12 | 59 | 39 | 42 | 35 | 24 |
| 84 | 20 | 27 | 35 | 05 | 54 | 21 | 39 | 04 | 77 | 69 |
| 52 | 36 | 07 | 18 | 99 | 79 | 27 | 36 | 30 | 97 | 14 |
| 36 | 90 | 79 | 26 | 63 | 50 | 41 | 87 | 76 | 31 | 13 |
| 91 | 66 | 85 | 68 | 10 | 40 | 47 | 44 | 71 | 56 | 81 |
| 40 | 26 | 25 | 37 | 27 | 02 | 15 | 26 | 27 | 51 | 87 |
| 57 | 59 | 72 | 02 | 27 | 96 | 10 | 62 | 63 | 07 | 30 |
| 23 | 49 | 01 | 02 | 17 | 28 | 23 | 72 | 71 | 46 | 39 |
| 80 | 25 | 47 | 36 | 69 | 73 | 39 | 21 | 23 | 93 | 10 |

**Figure 12.27** *Part of a random number table*

## Remember

Students (and some teachers!) sometimes talk about 'throwing a quadrat', where in order to place a quadrat the investigator stands in the habitat, such as a field, closes his eyes and throws the quadrat over his shoulder. This is *not* a random method. It is biased by the choice of the place in which to stand, the direction of throw and so on, as well as being potentially dangerous to other students!

**Figure 12.28** *Students sampling with quadrats along an interrupted belt transect on a sheltered rocky shore*

**9** The green alga *Pleurococcus* lives on the bark of trees. It is thought to grow at higher densities on the north-east side of the trunk.

(a) Suggest a suitable sampling method to test this hypothesis. *(2 marks)*

(b) Suggest how differences in abiotic factors around the trunk could explain the distribution of the alga. *(2 marks)*

### Types of abundance measurements

Apart from biomass measurements (which are destructive and very time-consuming) there are four main ways that the abundance of organisms in a quadrat can be recorded. These are as density counts, frequencies, percentage cover and relative abundance scales. These each have their advantages and disadvantages.

The **density** of a species is found by counting the number of individuals in a quadrat and calculating the mean number of individuals per unit area. This is only possible if the species grows as clearly separate individuals, such as limpets or dandelion plants. Many species, particularly plants, do not grow in this way, and it may be impossible to decide upon what represents a single plant. For example, where does one grass plant start and another finish? Separate shoots may be connected by underground roots or stems. The method is also very time-consuming if there are large numbers per quadrat.

For species which are difficult to count, **frequency of occurrence** can be used. This is the probability of finding a particular species in a quadrat. For instance if a species is found in five out of 25 quadrat samples, its percentage frequency is 20%. A quadrat divided into a 10 × 10 grid can provide a percentage frequency for each sample. It needs to be decided before sampling begins whether the **root frequency** or **shoot frequency** (plants rooted, or leaves within the quadrat respectively) is to be measured. This measure of abundance is also greatly influenced by quadrat size, and does not take into account the density of the plants or their distribution.

Another method which can be used when densities are high or plants are difficult to count is **percentage cover**. This is a measure of the proportion of the ground covered by each species. It is fairly straightforward, but is inevitably an estimate, even when a gridded quadrat is used, and problems are encountered when plants occupy several layers which overlap. It is best used with a relatively 'two-dimensional' community, such as the barnacles on a rock surface (see Figure 12.13, page 386).

There are several **relative abundance scales**, tailored to fit a particular habitat, such as the ACFOR scale shown in Table 12.3.

| 1 Algae | 2 Lichens and *Lithothamnion* |
|---|---|
| A More than 30% cover | A More than 20% cover |
| C 5–29% cover | C 1–19% cover |
| F Less than 5% cover but zone still apparent | F Large scattered patches |
| O Scattered plants, zone indistinct | O Widely scattered patches, all small |
| R Only one or two plants present | R Only one or two small patches present |
| **3 Small barnacles and small winkles** | **4 Limpets and large winkles** |
| A 100 or more 0.01 $m^{-2}$ | A 5 or more 0.1 $m^{-2}$ |
| C 10–99 0.01 $m^{-2}$ | C 1–4 0.1 $m^{-2}$ |
| F 1–90.01 $m^{-2}$ | F 5–9 $m^{-2}$ |
| O 1–99 $m^{-2}$ | O 1–4 $m^{-2}$ |
| R Less than 1 $m^{-2}$ | R Less than 1 $m^{-2}$ |
| **5 Large barnacles *Balanus perforatus*** | **6 Dogwhelks, topshells and anemones** |
| A 10 or more 0.01 $m^{-2}$ | A 1 or more 0.1 $m^{-2}$ |
| C 1–9 0.01 $m^{-2}$ | C 5–9 $m^{-2}$, locally sometimes more |
| F 1–9 0.1 $m^{-2}$ | F 1–4 $m^{-2}$, locally sometimes more |
| O 1–9 $m^{-2}$ | O Less than 1 $m^{-2}$, locally sometimes more |
| R Less than 1 $m^{-2}$ | R Always less than 1 $m^{-2}$ |
| **7 Mussels and piddocks (score holes)** | **8 Tube worms such as *Pomatoceros*** |
| A More than 20% cover | A 50 or more tubes 0.01 $m^{-2}$ |
| C 5–19% cover | C 1–49 tubes 0.01 $m^{-2}$ |
| F Small patches, covering less than 5% of the rock surface | F 1–9 tubes 0.1 $m^{-2}$ |
| O 1–9 individuals $m^{-2}$. No patches | O 1–9 tubes $m^{-2}$ |
| R Less than 1 individual $m^{-2}$ | R Less than 1 tube $m^{-2}$ |

Such scales utilise estimates of the abundance of different organisms to classify each species into one of several categories, such as:

A = Abundant

C = Common

F = Frequent

O = Occasional

R = Rare

Notice that different groups in the table are classified using different criteria to define the scales – algae on a scale that combines numbers of individuals and percentage cover, barnacles as numbers per 0.01 $m^2$, and so on. Hence the term 'relative' is given to the scale. ACFOR scales are used for quantifying the

*Table 12.3 ACFOR scale for sampling rocky shore organisms (after Dr Keith Hiscock, Field Studies Council's Oil Pollution Research Unit)*

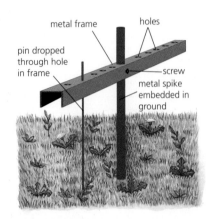

metal frame    holes

pin dropped
through hole
in frame
→ screw
metal spike
embedded in
ground

*Figure 12.29  A point quadrat*

*Figure 12.30  Student using a point quadrat*

abundance of other organisms such as terrestrial vegetation. The method is straightforward and relatively simple, but is very subjective. Two investigators will frequently decide upon different degrees of abundance for the same habitat.

### Point quadrats

The quadrats described so far are **area** quadrats. The **point quadrat** uses a sampling device consisting of pins mounted in a metal frame (Figures 12.29 and 12.30). The pins are lowered down from the frame until they touch the ground. Any organism (usually plants, but they can be sessile animals) that they touch is recorded as a 'hit' on a tally chart. The sample is a point, in other words the area of the quadrat is zero. The frame is then moved to the next sampling station, and the process repeated. As with area quadrats, the point quadrat can be used for both random sampling from an area, or sampling across a transect. When the method is used at intervals along a transect, it is common to repeat samples at the same interval along several parallel transects, so that the results at each interval can be combined.

There are usually 10 pins in the frame, and the presence or absence of a species at a number of points can be conveniently converted to a percentage value, by dividing the number of 'hits' by the total number of sampling points and multiplying by 100. This value can be proportional to the plant cover or relative numbers of plants, depending upon how the 'hits' are recorded (Figure 12.31).

Although the samples obtained with a point quadrat are a more objective measurement of plant cover than those from an area quadrat, the point quadrat has a number of limitations. It is only easy to use with species that are shorter than the height of the frame (although in theory you can sample the canopy of tall plants by lying under the frame and observing the plants through the holes!). It also overestimates some plants and underestimates others, depending on their leaf shape (e.g. long narrow leaves such as grasses may be overestimated). Most importantly, the final value obtained for each species is not an absolute measure of species density or actual percentage cover, but a relative value.

*Figure 12.31 Some ways in which a point quadrat can be used to record relative numbers of plants or plant cover*
*(a) Relative numbers of each species. Record the first 'hit' on each plant. If pin touches two plants of the same species record this as two 'hits' for that species,*
*(b) Total cover by all species. Record all 'hits' on each plant,*
*(c) Top canopy cover. Record only the first 'hit'*

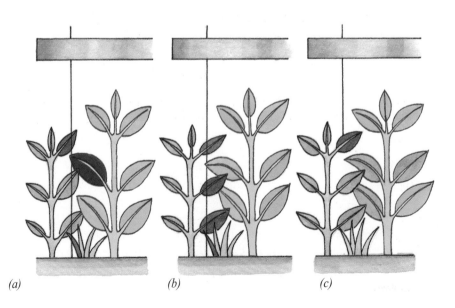

*(a)*    *(b)*    *(c)*

## 12.10 The chi-squared test of association

You have seen in Section 3.11, page 94 how the chi-squared ($\chi^2$) test can be used to compare observed frequencies with expected frequencies, for instance in the results of a genetic cross. There, the expected frequencies of a cross is known beforehand – the **expected** ratio of the offspring might be 9:3:3:1 for example. If we obtain **observed** frequencies which are 87:36:29:14, we can calculate the expected frequencies from the 9:3:3:1 ratio, and use the $\chi^2$ test to find out the probability of getting these observed frequencies by chance. In other words, test the '**goodness of fit**' to the 9:3:3:1 ratio.

There is another use of the $\chi^2$ test, where it is used to test whether there is an **association** between two attributes. This is commonly employed in certain types of ecological investigations where we suspect that certain organisms may be associated with a particular habitat. It is easiest to understand by considering the following example.

The terrestrial snail *Cepaea nemoralis* shows stable polymorphism (Section 4.8, page 130). It exists in three shell colours, yellow, brown and pink. Shells are also banded or unbanded (Figure 12.32). An investigation was carried out to compare the numbers of snails of different colours in two habitats. Habitat A was the floor of a deciduous woodland, while habitat B was a hedgerow. Snails were collected for three hours at each habitat, and put into two categories of colour, yellow and non-yellow. The results are shown in Table 12.4, which is called a **contingency table**.

From Table 12.4, it seems that there are disproportionate numbers of yellow snails in habitat B. In other words, there seems to be an association between snail colour and habitat. To test for this association we need to know the numbers of snails we would expect to find if there were no association. Unlike in the genetic cross, we do not have prior knowledge of predicted frequencies to start with, but we can estimate them from the observed totals for the four categories.

**Figure 12.32** *Variation in the snail* Cepaea nemoralis

The expected number of yellow snails from habitat A is:

$$\frac{83 \times 49}{122} \quad \text{i.e} \quad \frac{(\text{row total} \times \text{column total})}{\text{grand total}}$$

This is because overall 83 out of 122 snails were yellow, and so, as there are a total of 49 snails in habitat A, we would expect to find $(83/122) \times 49$ yellow snails in habitat A.

**Table 12.4** *Contingency table for* $\chi^2$ *test*

| Colour of snail | Observed number in habitat A | Observed number in habitat B | (row total) |
|---|---|---|---|
| yellow | 22 | 61 | 83 |
| non-yellow | 27 | 12 | 39 |
| (column total) | 49 | 73 | grand total = 122 |

| Colour of snail | Numbers in habitat A | | Numbers in habitat B | | (row total) |
|---|---|---|---|---|---|
| | Observed | *Expected* | Observed | *Expected* | |
| yellow | 22 | *33.3* | 61 | *49.7* | 83 |
| not yellow | 27 | *15.7* | 12 | *23.3* | 39 |
| (column total) | 49 | | 73 | | grand total = 122 |

*Table 12.5 Observed and expected numbers of snails in two habitats*

This gives an expected frequency of 33.3 snails. The same formula is then used to calculate the expected numbers of snails in each category in Table 12.4. These are shown in italics in Table 12.5.

We now need to state the **null hypothesis**. This is: 'there is no significant difference between the observed and expected numbers of snails', or alternatively, 'no association between snail colour and habitat'.

The value of $\chi^2$ is now calculated from:

$$\chi^2 = \sum \frac{(O - E)^2}{E}$$

where   O = observed frequencies
E = expected frequencies
$\Sigma$ = 'sum of'.

From Table 12.5:

$$\chi^2 = \frac{(22 - 33.3)^2}{33.3} + \frac{(61 - 49.7)^2}{49.7} + \frac{(27 - 15.7)^2}{15.7} + \frac{(12 - 23.3)^2}{23.3}$$

$$= 3.834 + 2.569 + 8.133 + 5.480$$

$$= 20.02$$

We can now find the significance of this value of $\chi^2$ from statistical tables. First we find the **degrees of freedom** (df) from the formula

df = (number of row categories – 1) × (number of column categories – 1)

= (2 – 1) × (2 – 1)

= 1

We can find the **critical value** of $\chi^2$ from a statistical table. Part of a table of critical values of $\chi^2$ is shown in Table 3.6, page 95. At a probability level of 0.05, with 1 degree of freedom, the calculated value of $\chi^2$ needs to be greater than 3.84 for us to be unable to accept the null hypothesis. In other words, on the basis of chance alone, a $\chi^2$ value of 3.84 will be found in 5 out of 100 cases. Our calculated value is 20.02, the probability is less than 5%, and so we cannot accept the null hypothesis, that there is no association between shell colour and habitat. Instead we can accept the alternative hypothesis, that colour and habitat *are* associated.

From Table 12.5, we see that there are higher than expected numbers of yellow snails, and lower than expected numbers of non-yellow snails in habitat B. The converse is true for habitat A.

### Box 12.1  Contingency tables with more than two columns or rows

$\chi^2$ tests of association can be used with more than two categories per row or column. For example, instead of yellow and non-yellow snails, we could have yellow, pink and brown. We would now have a $2 \times 3$ contingency table:

|  | Habitat A | Habitat B |
|---|---|---|
| Yellow | 22 | 61 |
| Pink | 12 | 7 |
| Brown | 15 | 5 |

The expected values are calculated for each of the six cells as for a $2 \times 2$ table, from:

(row total × column total) ÷ grand total.

$\chi^2$ is calculated as usual, from $\Sigma[(O - E)^2 \div E]$. The critical value is found from a statistical table, using $(2 - 1) \times (3 - 1) = 2$ degrees of freedom.

The problem with using the $\chi^2$ test with more than two rows or columns is that it is difficult to interpret the results if a significant association is found, since we cannot easily tell which cell(s) in the table are contributing most to the overall differences between observed and expected numbers.

## 12.11  Conservation

**Conservation** is the management of the Earth's resources so as to provide for the needs of humans, both now and in the future, at a **sustainable level**. It aims to ensure that resources are neither over-exploited nor destroyed.

Conservation is often confused with preservation. Although conservation often involves preservation of species or habitats, it is usually much more than this. Preservation is often a passive process, whereas conservation is much more proactive, involving preventative actions taken to limit damage to the biosphere. A simple example will illustrate the difference between preservation and conservation. **Culling** is sometimes used as a conservation measure. When a species is becoming over-abundant it can be reduced in numbers by deliberate killing of some of the population. This is carried out with some deer populations, to avoid them increasing beyond the carrying capacity of the habitat (see Section 12.1, page 381). Older or sick individuals are usually selected for culling. Culling could hardly be described as an act of preservation, yet it may preserve the equilibrium of the environment in the long term.

There are both ethical and practical reasons why conservation is important. Humans are unique in their ability to exploit and modify their environment, and our activities are damaging it in many ways. Most people would agree that humans do not have the 'right' to do this, and that our dominant position in the global ecosystem, along with our intelligence, gives us an ethical duty to maintain ecosystems and the diversity of organisms. However, even if an individual has no personal views on this, few people, from a purely selfish point of view, would want to see a biosphere completely dominated by agricultural land, industry and housing. Most people appreciate the aesthetic beauty of a tropical rain forest or a coral reef (Figure 12.33).

It is also important to remember that humans are only one part of the **biosphere**, the global ecosystem which is maintained by the involvement of all organisms in energy flow, food chains and the recycling of elements such as carbon and nitrogen (see *AS Biology*, Section 14.1, page 414). Damage at a local level to the smaller components of the biosphere may have long-term consequences for its overall stability. For example, continued destruction of the tropical rain forests may be contributing to global warming (see *AS Biology*, Section 15.24, page 468).

There are many practical reasons for conservation. Conservation of rain forests provides a supply of hardwood trees, and access to as yet undiscovered food plants and medicines (see *AS Biology*, Section 15.15, page 455). Reduction in the pollution of the oceans will help to maintain fish stocks as a food supply.

The key phrase which crops up over and over again in conservation is the need to manage ecosystems and habitats to maintain (or even *increase*) **biodiversity**, in the face of human activities which are having the opposite effect. These are global problems, and ultimately will require international actions to attempt to find solutions. However, we can explore some examples of conservation measures if we consider just one human activity, agriculture, in a temperate country such as the United Kingdom, and the adverse effects that it has on wildlife.

***Figure 12.33*** *Corals and other animals of the Great Barrier Reef, off the east coast of Australia*

Some agricultural practices which have led to environmental problems, including loss of habitats and a decrease in biodiversity are described below.

## The establishment of monocultures

A monoculture refers to the practice of growing the same crop on the same land for year after year. Often the crop occupies extensive areas of land. From the farmer's point of view this is highly advantageous, because it allows him to specialise in many ways which improve efficiency and increase profit. For instance, the types of specialised machinery needed can be reduced (Figure 12.34), precise inorganic fertilisers can be selected to supply the needs of the one crop, and pesticides selected which target the specific pests of the crop. However, monocultures deplete the soil of particular nutrients, which then need to be replaced by the application of more chemical fertiliser. They reduce the humus (organic matter) content of the soil, and may result in soil erosion. They also encourage the spread of particular insect or fungal pests of the one crop, or of particular weeds. This then results in increased use of pesticides.

*Figure 12.34* *Monocultures allow the use of specialised harvesting machinery*

## The removal of hedgerows

Hedgerows get in the way of heavy farm machinery, and are easily replaced by wooden or electric fences. Over the last 50 years there has been a massive reduction in the number of hedgerows on farmland in the UK, from over 800 000 kilometres in 1945 to about 345 000 kilometres today. Despite conservation measures, around 17 000 kilometres are still being lost each year. Much that has been removed, or lost through neglect, consisted of ancient hedgerow, hundreds of years old, containing diverse communities of plants and animals (Figure 12.35). Hedges have many environmental benefits. They form natural corridors for movements of small animals like birds, rodents and insects. They preserve the continuity of many food chains, and reduce the likelihood of soil erosion by rain or wind. Some species of plants and animals, such as the cirl bunting (a small bird) are totally dependent on hedges as a habitat.

### ! Did You Know

The approximate age of an ancient hedgerow can be estimated by the number of native species of woody plant present. Most hedgerows, when planted, consisted of only one species (often hawthorn). In a 30 metre length of hedgerow, one new species of shrub or tree becomes established approximately every 100 years, so a 30 metre length containing five species would be about 400 years old. From the middle of the 19th century hedgerows containing more than one species were planted, so this formula doesn't apply. If you want to make an estimate of the age of an ancient hedgerow, it is best to sample several 30 metre lengths of the hedgerow for a more reliable estimate of the numbers of woody plant species growing there. A hedgerow with five or more woody species is considered to be 'species rich'.

*Figure 12.35* *This ancient hedgerow is a habitat for hundreds of plant and animal species*

### Ploughing of arable land

This is carried out mainly in the winter months to prepare the soil for sowing of the next crop in the spring. Ploughing makes the soil susceptible to erosion by rain and wind, especially if monocultures are grown and hedgerows have been removed. Soil erosion is more likely if ploughing is carried out up and down the slopes of the land (compare this with 'contour ploughing' below).

### Drainage of wetlands

Wetlands include **salt marshes** which are areas periodically flooded by sea water or brackish water (a mix of fresh and sea water) at high tide, and **wet meadows**, which are low-lying areas of land, usually only flooded by fresh water during the winter months. Both types of land provide habitats for unique communities of organisms. Wet meadows are traditional over-wintering locations for many species of migratory and non-migratory waterfowl and other birds, and sites where distinct species of flowering plants and invertebrates are found (Figure 12.36). Both types of area have been 'improved' for agriculture by drainage, mowing and addition of inorganic fertilisers, resulting in a loss of these habitats and the communities they support. In addition, ponds on farms are frequently filled in or allowed to dry up by natural succession, losing these important small freshwater habitats.

### Use of inorganic fertiliser

Application of inorganic fertilisers such as nitrates and phosphates greatly increases crop yields, but their excessive use can have harmful environmental effects such as eutrophication, resulting in algal blooms and reduced diversity of species in rivers and lakes (Figure 12.2, page 380).

### Use of pesticides

Herbicides kill weeds and so reduce the effects of interspecific competition between crop plants and the weeds. Insecticides and fungicides kill pest species which would otherwise reduce crop yield or quality (see *AS Biology*, Chapter 15). However, pesticides may be non-specific, killing non-target organisms and

*Figure 12.36 The Somerset Levels extend inland from the coast of Somerset for some 29 000 hectares, making this region one of the largest and most important wetland areas of Britain. It has been drained for agriculture and peat extraction for hundreds of years, but this 'improvement' of the land has accelerated since 1945, with subsequent loss of unique habitats*

disrupting food chains. Some pesticides increase in concentration through food chains (bioaccumulation) and many are persistent, remaining active in the environment for long periods. They may be carried away by winds or run off into streams and rivers, so that they are carried far away from the place of their intended use.

### Some conservation measures which can be adopted to reduce the impact of farming on wildlife

The aims of conservation measures are to manage farming to allow sustainable yields, but to reduce the impact on the environment and maintain species diversity. They include the following examples, many of which appear minor by themselves, but when taken together can effect a major improvement.

*Figure 12.37* A beetle bank in cultivated crop

- Strips of uncultivated land can be established across fields of crops. These **green banks** or **beetle banks** (Figure 12.37) allow for the growth of wild flowers and attract insects, including species which act as natural predators of pests like aphids. Similarly **field margins** or **headlands** (strips over 2 metres wide) can be left uncultivated around the perimeter of a field, for the same purpose. Headlands also protect hedgerows from herbicide and insecticide sprays drifting from the cultivated land.

- Hedgerows can be maintained correctly by correct cutting and replanting to maintain a tall hedge with a thick base, providing food and shelter for wild flowers, invertebrates, birds and small mammals. Hedge trimming is best carried out in February, when the berries have been eaten from the hedge, but birds have not yet started to nest. It does not need to be repeated every year. The use of mechanical flails to cut the hedge can be very damaging, producing thin, low hedges which make poor habitats, and their use should be restricted. Since 1989, the planting of new hedgerows has been encouraged by subsidies from the government, and in 1998 a law was introduced which requires a farmer to seek permission to remove a 'designated' ancient hedgerow of over 20 metres in length. New hedgerows can be planted along contours, rather than across them, to help prevent soil erosion. Since subsidies were introduced, farmers now plant more hedgerows than they dig up, but there is still a net loss in the number of kilometres of hedgerows in the country.

 **?**

**10** If more hedgerows are planted than are dug up, why is there still a net loss of 17 000 kilometres per year? *(1 mark)*

- **Contour ploughing**, which involves ploughing along the contour lines of a sloping field rather than up and down hills across the contours, helps prevent erosion by surface run-off.

- Alternatives to the use of pesticides can be adopted. **Intercropping** involves the planting of one crop containing predators of a species which is a pest of an adjacent crop. For example cabbage plants can be intercropped with clover. The clover increases the populations of ground beetles, which reduce the numbers of pests of the cabbage plants. Deliberately introduced **biological control** agents may be employed, where a specific predator of the pest is released onto the crops (see *AS Biology*, Section 15.13, page 452). These methods reduce the need to use toxic pesticides. However, in practice more than one method may be required to achieve an economically practicable solution to pest problems. Combining these methods is called **Integrative Pest Management** or **IPM**.

● The risk of eutrophication can be reduced by adding the minimum quantities of inorganic fertiliser needed, and at times of year when it is most likely to be absorbed by plants. For instance, application of nitrate fertiliser in autumn will probably result in much nitrate ending up in water courses, since there are no crops growing at that time to take up the nitrate, and the seasonal rain will result in much being leached from the soil. Use of organic fertiliser, such as manure, produces a much slower release of inorganic ions, although it can produce other problems (see *AS Biology*, Section 15.5, page 443).

● **Permanent pasture** provides an important habitat for diverse communities of flowering plants, insects and other invertebrates (Figure 12.38). The modern practice of replacing permanent pasture with temporary grassland (**ley**) which is ploughed, enriched with artificial fertilisers and re-seeded, should be avoided. Meadows are sometimes cut several times in the spring by mowers, to produce **silage**. This produces a short, even 'sward' which does not promote species diversity. Greater diversity is achieved if cutting takes place later in the summer, for **hay**. Varying the length of the sward and the frequency of cutting also adds to the diversity of plant and animal species in the meadow. The grazing of different animals such as sheep and cattle also encourages plant diversity, as a result of their different feeding methods and habits, as well as the fertilisation of the ground by their dung. Permanent meadows should be protected form the invasion of small trees and bushes (scrub) which can be controlled by mowing or even burning.

> ### Definition
>
> Hay is grass which is mown and dried for use as an animal feed in the winter. Silage consists of grass (and other plants) which has been fermented anaerobically in a container called a silo. Silage is a compact, light-brown material with a higher carbohydrate and protein content than hay. It can be stored for several years and is also used as an animal feed.

*Figure 12.38* Permanent pasture – a traditional hay meadow in North Yorkshire containing diverse species of flowering plants and invertebrates

● Where meadows form wet grasslands during the winter months, the water levels should be controlled to ensure flooding occurs, allowing native and migrant over-wintering birds to continue to inhabit the area.

● **Coppicing** is a practice which increases species diversity in deciduous woodlands. It refers to the practice of cutting certain tree species down to ground level, forming stumps or **stools**, which grow new shoots known as **poles**. (Figure 12.39). The poles are harvested every few years, and different areas of a wood are maintained at different stages of the coppicing cycle. Coppicing has been carried out for hundreds of years, originally to produce fencing materials. The species most commonly coppiced are alder, ash, birch, hazel, poplar and willow. Coppicing has been revived as a way of producing biofuel from fast-growing trees (see *AS Biology*, Section 15.25, page 472), but it is also a conservation technique. By having areas of the wood at different stages in the coppicing cycle, it increases the diversity of **microhabitats** available in the wood. In addition, more light reaches the ground in a coppiced area, so that the diversity of ground flora is increased. Coppicing is also used to repair gaps in hedgerows.

*Figure 12.39  Coppiced woodland*

The real problem with establishing conservation measures is that there is often a conflict of interest between the aims of the conservationists and the farmers as food producers. With the global market for food, and the competition that this entails, the conflict is only really going to be resolved through international controls and **legislation** (enacting laws). Legislation can be a powerful means of protecting the environment. For example, the **Habitat Directive** of the European Commission was adopted in 1992 to protect habitats under threat. Governments of the member countries are required to set up and preserve designated **Special Areas of Conservation**, or **SACs**, which are supported by EU funding. The purpose of the SACs is to protect vulnerable natural habitats and species of plants and animals (excluding birds). Habitats which support significant numbers of wild birds are classified as **Special Protection Areas**, or **SPAs**. Throughout Europe, 168 natural types of habitat have been identified which need SAC designation, including 22 priority habitats in Britain, such as:

> **Remember**
> A **microhabitat** is a subdivision of a larger habitat. For example, a deciduous woodland habitat contains numerous microhabitats such as the leaf canopy, tree trunks, ground flora and leaf litter, (see *AS Biology*, Section 14.1, page 415).

● active raised bogs
● blanket bog
● Alpine and Northern British plant communities associated with slow-flowing water
● hard water springs depositing lime
● mountain grassland with matt-grass and many other species
● limestone pavements
● Caledonian Scot's Pine forest
● mixed woodland on alkaline soils associated with rocky slopes
● alder woodland on flood plains
● coastal dune heathland
● fixed dunes with grassland and herbs
● lime-deficient dune heathland dominated by crowberry

Photographs of some of these habitats are shown in Figure 12.40.

Coastal dune heathland

Limestone pavement

Raised bog

Fixed dunes

Blanket bog

Caledonian Scots pine forest

**Figure 12.40** *A variety of habitats from the Natura 2000 list*

The carefully specified descriptions of these habitats reflect their diversity. The Habitats Directive also lists 632 species of animals and plants in Europe which are in need of conservation, including 40 species in Britain, such as the lady's slipper orchid, yellow marsh saxifrage, marsh fritillary butterfly and bottle-nosed dolphin.

The network of SAC and SPA sites, aimed at conservation of habitats and endangered species for the twenty first century, has been given the title **Natura 2000**.

## Summary – ⑫ Populations and communities

● In ideal conditions, populations have the potential to increase in numbers exponentially. However, real populations rarely achieve this potential. Their size is limited by various factors.

● The average maximum population size is called the carrying capacity. The factors which limit growth of a population are known as environmental resistance.

● Mortality may change with the density of a population (density dependent) or may not be affected by it (density independent). Density-dependent factors are always biotic.

● Competition between organisms can occur between members of the same species (intraspecific) or members of different species (interspecific). Two species occupying one habitat cannot have the same ecological niche.

● Predator–prey graphs show fluctuations in numbers of a predator and its prey over time.

● The size of a population can be estimated by the mark-release-recapture method, which produces a number called the Lincoln Index.

● Succession is the progressive change in the composition of a community over a period of time. Succession is brought about by changes caused by the vegetation colonising the habitat. The first organisms to colonise a habitat are pioneers. Each distinct community in a succession is called a seral stage, and the final stable community is called the climax. There are different types of climax community, for example those dependent on climate (climatic climax) or those deflected by human interference (plagioclimax).

● Zonation of organisms occurs along an environmental gradient, such as a rocky shore or a shingle ridge.

● Species diversity is a measure of both species richness and abundance of each species present. It can be quantified using the Simpson's index.

● There is a need to avoid bias in sampling from an ecosystem. Random sampling with quadrats can be carried out using coordinates on a grid selected using random numbers. Sampling along a transect is applicable when the habitat contains an environmental gradient.

● The $\chi^2$ test can be used to test for an association between two attributes.

● Conservation is the management of the Earth's resources to provide for the needs of humans at a sustainable level, ensuring that resources are neither over-exploited nor destroyed.

● Agricultural practices employed by intensive agriculture, such as establishment of monocultures, destruction of hedgerows, ploughing, drainage of wetlands and the use of artificial fertilisers and pesticides have adversely affected natural habitats and communities.

● A number of conservation measures can be integrated into farming practices to encourage maintenance of habitats and biodiversity.

**? Answers**

1 Increasing concentration of the yeast's waste product, alcohol *(1)*. (Wine never exceeds an alcohol content of more than about 15%.)

2 The most likely factors are competition for food *(1)* between the sheep and with other herbivores, as well as predation *(1)*.

3 Great tits are very dependent on one food source, which can fail completely in some years. This is unlikely to happen with the sheep *(1)*. Abiotic factors such as winter cold are also more likely to result in deaths of the great tits *(1)*.

4 Intraspecific competition for resources occurs between members of the same species, while interspecific competition occurs between members of different species *(1)*.

5 (a) $N = (112 \times 140) \div 15 = 1045$ *(1)*.

   (b) Marking them on their backs would make them more conspicuous to predators *(1)*.

   (c) 15 is greater than 10% of the number marked in the first sample *(1)* so it is enough for an accurate estimate *(1)*.

6 Mutualism is a relationship between two species of organisms where both benefit *(1)*.

7 The barnacles *Chthalamus* and *Semibalanus (1)*. (Page 386)

8

| Species | n (community 3) | n − 1 | n(n − 1) |
|---|---|---|---|
| A | 20 | 19 | 380 |
| B | 20 | 19 | 380 |
| C | 20 | 19 | 380 |
| D | 0 | 0 | 0 |
| E | 0 | 0 | 0 |
| F | 0 | 0 | 0 |
| Total | N = 60 | Σn(n − 1) | = 1140 |

So $d = (60 \times 59) \div 1140 = 3.11$

*(1 mark for table, 1 mark for calculation, 1 mark for correct answer)*

9 (a) Small quadrats, e.g. 10 cm × 10 cm *(1)* placed in a continuous belt transect around the trunk *(1)*.

   (b) Cooler on north-east; less exposure to prevailing wind; less direct sunlight; less drying out *(max. 2)*.

10 These data for hedgerows planted and dug up do not include the kilometres lost as a result of their becoming derelict through neglect *(1)*.

## End of Chapter Questions

1 The graph shows the changes in the size of a bacteria population with time when grown in a flask containing a sterile solution of glucose.

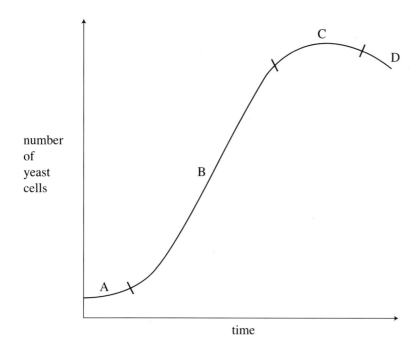

(a) Name the phase of growth and explain the shape of the curve at each of the periods A to C in the graph. *(6 marks)*

(b) With reference to named examples, explain how limiting factors determine the final size of the populations in nature. *(9 marks)*

C3.3

*(In this question, 1 mark is available for the quality of written communication.)*

*OCR 2000* *(Total 15 marks)*

2 (a) 'Set-aside' is the common name given to a European Policy under which farmers receive a subsidy for land taken out of cultivation. A study was carried out to investigate how the amount of time a set-aside field was left uncultivated would affect the species of birds feeding there.

N3.1
N3.2

**Table 1** shows the number of birds of different species feeding in a field which had been left uncultivated for one year.

| Species | Number of birds of that species feeding in the field |
|---|---|
| Greenfinch | 12 |
| Goldfinch | 8 |
| Wood pigeon | 3 |
| Chaffinch | 1 |

**Table 1**

(i) Use the formula $d = \dfrac{N(N-1)}{\Sigma n(n-1)}$

where   d = index of diversity

N = total number of organisms of all species

and   n = total number of organisms of a particular species

to calculate the index of diversity for the birds feeding in the field. Show your working.                  *(2 marks)*

(ii) Explain why it is more useful in a study of this sort to record diversity rather than the number of species present.                  *(2 marks)*

**Figure 1** is a graph showing the relationship between bird species diversity and plant species diversity in this study. **Figure 2** is a graph showing the relationship between bird species diversity and plant structural diversity for the same study. Structural diversity refers to the different form of plants such as herbs, shrubs and trees.

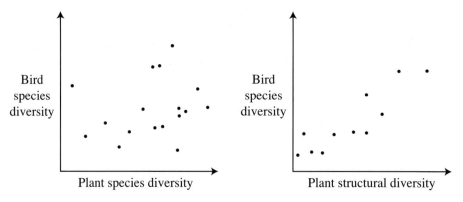

| **Figure 1** | **Figure 2** |

(b) Describe briefly how you could obtain the data that would enable you to calculate the diversity index for the species of plant growing on a set-aside field.                  *(3 marks)*

(c) (i)  Describe the relationships shown in **Figure 1** and **Figure 2**. *(2 marks)*

(ii) Explain why the plant structural diversity would increase with the amount of time the field was left uncultivated.                  *(2 marks)*

(iii) Suggest an explanation for the relationship between bird species · diversity and plant structural diversity shown in Figure 2.     *(2 marks)*

(d)  In another study of fields taken out of cultivation, the figures shown in **Table 2** were obtained.

| Time in years since cultivation stopped | Value of index of diversity for bird species |
|---|---|
| 5 | 2.1 |
| 15 | 3.2 |
| 20 | 5.6 |
| 25 | 4.1 |
| 40 | 4.8 |
| 60 | 9.4 |

**Table 2**

Predict what might happen to the bird species diversity in the study summarised in **Table 2** over the next 100 years. Explain how you arrived at your answer. *(2 marks)*

*AQA(A) 2000* *(Total 15 marks)*

**3** The wren is a small, insect-eating bird. The percentage change in size of the wren population from one year to the next was estimated over a number of years. The number of days with snow lying in the previous winter was also recorded. This information is shown on the graph below.

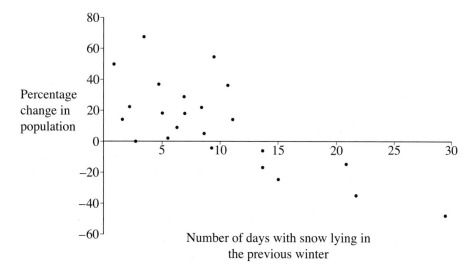

Number of days with snow lying in the previous winter

**(a) (i)** Describe the relationship between the number of days with snow lying and the change in population size. *(2 marks)*   N2.1

**(ii)** Suggest and explain a reason for this relationship. *(2 marks)*

N2.1

**(b)** A comparison was made between the number of breeding pairs of wrens each year and their breeding success.

| Number of breeding pairs of wrens / millions | Percentage increase in in population size |
|---|---|
| 1.2 | 55 |
| 1.9 | 48 |
| 2.5 | 35 |
| 2.9 | 25 |
| 3.9 | 2 |

Suggest an explanation for the relationship between the size of the breeding population and breeding success. *(2 marks)*

*AQA(B) 2000*

*(Total 6 marks)*

**4** A field was planted with grass and clover. After planting, half the field was used for cattle grazing and the other half was left ungrazed. Five years later a biologist collected samples of clover seeds from both parts of the field and grew them in an experimental plot. The diagram shows clover plants grown from seeds gathered from grazed and ungrazed areas. Both sets of seeds were germinated and grown in identical conditions.

Typical clover plant
from grazed part
of field

Typical clover plant
from ungrazed part
of field

**(a)** The clover plants grown from the seeds gathered in the ungrazed part of the field were noticeably taller than those grown from seeds collected in the grazed area. Suggest how this difference arose. *(4 marks)*

**(b)** It was suggested that, although these two kinds of clover differed in height, they were the same species. Describe how you could test this hypothesis. *(3 marks)*

**(c)** If the ungrazed portion of the field were left completely ungrazed for a much longer period than five years, what ecological changes would be observed, and why? *(5 marks)*

*AQA(B) 2000*                                                    *(Total 12 marks)*

**5** A survey was carried out on a rocky sea shore to determine the distribution of two species of marine molluscs, *Littorina saxatilis* (the rough periwinkle), and *L. littorea* (the common periwinkle). Both species are primary consumers. A profile of the rocky shore is shown on the diagram below. At low water mark, the shore is covered by sea water most of the time. The sea reaches high water mark twice a day.

N3.1
C2.2

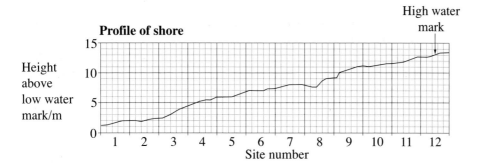

P.T.O

The sites of sampling were 10 metres apart, starting at the low water mark. The distributions were assessed by means of an abundance scale with 5 representing the greatest abundance. The results are shown as bar charts in the diagrams below.

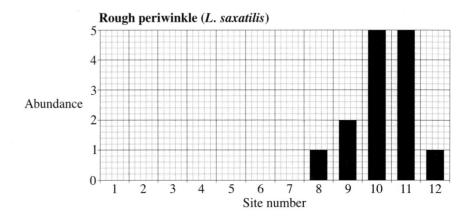

Rough periwinkle (*L. saxatilis*)

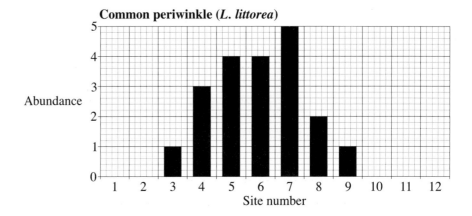

Common periwinkle (*L. littorea*)

**(a)** Compare the distribution and abundance of these two species on this rocky shore. *(3 marks)*

**(b)** Suggest which of the two species is likely to be more tolerant of desiccation. Explain your answer. *(2 marks)*

**(c)** Suggest *two* factors, other than desiccation, which might account for the difference in distribution of the two species. *(2 marks)*

*Edexcel 1998*

*(Total 7 marks)*

**6** The table shows data collected from a coastal area. A belt transect was used ☞○ **N3.1**
from the high tide line, inland across sand dunes, to a woodland behind the
dunes. Quadrats were taken at five metre intervals. The dominant species in
each quadrat was recorded.

| Quadrat number | sand grass couch | marram grass | lichens and mosses | heather | rye grass | birch | oak |
|---|---|---|---|---|---|---|---|
| 1 | + | | | | | | |
| 2 | +++ | + | | | | | |
| 3 | | | | | | | |
| 4 | ++ | ++ | | | | | |
| 5 | | + | | | | | |
| 6 | | +++ | | | | | |
| 7 | | +++ | | | | | |
| 8 | | ++ | + | | | | |
| 9 | | | ++ | + | | | |
| 10 | | | +++ | + | | | |
| 11 | | | +++ | +++ | | | |
| 12 | | | +++ | +++ | | | |
| 13 | | | +++ | +++ | | | |
| 14 | | | | | +++ | | |
| 15 | | | | | +++ | | |
| 16 | | | | | +++ | | |
| 17 | | | | | +++ | | |
| 18 | | | | | | + | |
| 19 | | | | | | + | |
| 20 | | | | | | ++ | + |
| 21 | | | | | | + | + |
| 22 | | | | | | | ++ |

*Key:*

|   | + | |
|---|---|---|
| ↓ | ++ | Increasing abundance of plants concerned. |
|   | +++ | |

**(a)** The plants recorded at this site illustrate a succession. With reference to
the table, explain what is meant by the term *succession*. *(4 marks)*

The quadrats containing rye grass are in an area used for agriculture. This
section of the transect illustrates a deflected succession.

**(b)** (i) Explain what is meant by a *deflected* succession. *(2 marks)*

(ii) Suggest how the deflected succession at this site may be maintained. *(2 marks)*

The numbers of a certain species of ground beetle in the woodland were estimated using a capture-recapture method. This involves catching a sample, counting the number caught, marking them in some way and releasing them (sample 1). After a period of time, a second sample (sample 2) is then captured and the number again counted, making a note of the number which are marked from the first sample. At one time of year, in the habitat under investigation, the figures recorded for samples of beetles were as follows:

| | |
|---|---|
| number in sample 1 | : 284 |
| number in sample 2 | : 267 |
| number found marked in sample 2 | : 63 |

**(c)** Use these figures to estimate the population size of beetles at this time. Show your working. *(2 marks)*

**(d)** Suggest how the beetles in sample 1 might have been marked. *(1 mark)*

**(e)** State two assumptions that are made when interpreting data from the capture-recapture method. *(2 marks)*

OCR 2000                                                          *(Total 13 marks)*

N2.2
N2.3
N3.2
N3.3

**7** The dog whelk, *Nucella lapillus*, is a marine snail found on rocky shores. It feeds on barnacles and molluscs by drilling holes into their shells or prising them open.

An investigation was carried out into the choice of prey species by the dog whelk. Sixty dog whelks were collected from a rocky shore where they had been feeding on barnacles. The dog whelks were given a mixture of two species of barnacle, *Semibalanus balanoides* and *Elminius modestus*. After 210 days, the number of each species of barnacle which had been eaten and the number remaining uneaten were recorded.

The observed results are shown in the table below.

| Species | Number eaten | Number remaining | Total |
|---|---|---|---|
| *S. balanoides* | 137 | 146 | 283 |
| *E. modestus* | 166 | 249 | 415 |
| Total | 303 | 395 | 698 |

*Data from Barnett BE (1979)*

A chi-squared ($\chi^2$) test was then carried out on these results to test the null hypothesis that dog whelks eat equal percentages of each of the two barnacle species.

The expected results for this test are shown in the table below.

| Species | Number eaten | Number remaining | Total |
|---|---|---|---|
| S. balanoides | 122.9 | 160.1 | 283 |
| E. modestus | 180.1 | 234.9 | 415 |
| Total | 303.0 | 395.0 | 698 |

The formula for calculating chi-squared ($\chi^2$) is given below.

$$\chi^2 = \sum \frac{(O - E)^2}{E}$$

where  O = observed values

E = expected values

**(a) (i)** Use the formula to calculate the value of $\chi^2$. Show your working.

*(2 marks)*

(ii) The table below gives some values for $\chi^2$ with one degree of freedom.

| P (%) | 20 | 10 | 5 | 2 | 1 |
|---|---|---|---|---|---|
| $\chi^2$ | 1.64 | 2.70 | 3.84 | 5.41 | 6.63 |

Does your calculated value of $\chi^2$ enable you to accept or to reject the null hypothesis? Give a reason for your answer. *(2 marks)*

**(b)** The dog whelks ate 48% of the *S. balanoides* and 40% of the *E. modestus*. Suggest long-term effects this could have on the populations of the two species. *(2 marks)*

**(c)** The investigation was repeated using dog whelks which had not been fed for a period of time. The results for this second experiment are summarised in the table below.

| Species | Number available | Number eaten | Percentage eaten |
|---|---|---|---|
| S. balanoides | 296 | 161 | 54 |
| E. modestus | 365 | 191 | |

(i) Calculate the percentage of the available *E. modestus* which was eaten.

*(1 mark)*

(ii) Describe the effects of starvation on the feeding behaviour of dog whelks. *(2 marks)*

*Edexcel 1999*

*(Total 9 marks)*

427

8 Explain briefly how the following techniques may be used as conservation measures:

(a) coppicing *(4 marks)*

(b) maintaining hedgerows *(3 marks)*

(c) Integrative Pest Management (IPM). *(2 marks)*

*(Total 9 marks)*

# Synoptic questions

1 Read the passage below and then answer the questions that follow.    N2.2

Evolutionary change develops when a mutation occurs and survives the selective process (that is, when it is found to be either neutral or advantageous). For example, a GCT codon might mutate to GAT and we would obtain leucine instead of arginine in the protein.

In about 20% of all mutations, because of the redundancy (degeneracy) of the code, a mutation might have no effect on protein structure. Thus a mutation from GCT to GCA would affect only the DNA and might well have no functional effects – no matter what the third base in the GC codon, we always obtain arginine in the protein.

The evolutionary process, then, involves a change (mutation) in the DNA which is incorporated into the ongoing gene pool of the evolving species and which can be reflected by a corresponding change in the amino acid sequence of the particular protein coded for by that gene.

We might state as a basic rule that such a process will have to produce divergence when any two populations become isolated from one another, as the relative rarity of mutations and the finite size of populations make it statistically improbable that identical changes will be available for natural selection to incorporate into the gene pools.

*Adapted from V Sarich: A molecular approach to the problem of human origins (1971)*

(a) Explain what is meant by each of the following terms.

    (i)  A GCT codon (paragraph 1)    *(1 mark)*

    (ii)  Redundancy (degeneracy) (paragraph 2)    *(1 mark)*

    (iii) Gene pool (paragraph 3)    *(1 mark)*

    (iv) Natural selection (paragraph 4)    *(1 mark)*

(b) Suggest *two* ways in which 'mutation from GCT to GCA' (paragraph 2) might arise.    *(2 marks)*

(c) Explain why such a mutation 'might well have no functional effects' (paragraph 2).    *(3 marks)*

(d) State *two* ways in which 'populations become isolated from one another' (paragraph 4).    *(2 marks)*

(e) Outline the possible consequences of the 'divergence' (paragraph 4) that may result from such isolation.    *(3 marks)*

*Edexcel 2000*    *(Total 14 marks)*

N2.1 **2** Some soyabean seedlings were grown in an atmosphere enriched with carbon dioxide.

An experiment was carried out to compare the uptake of nitrogen by these seedlings with that of seedlings grown in a normal atmosphere (control plants). Soyabeans belong to the Papilionaceae (legumes) and all the experimental plants had root nodules containing *Rhizobium*.

At the beginning of the experiment, the seedlings were 25 days old. The total amount of nitrogen incorporated into compounds in the plants was then measured at intervals until the plants were 100 days old.

The results of the experiment were shown in the graph below.

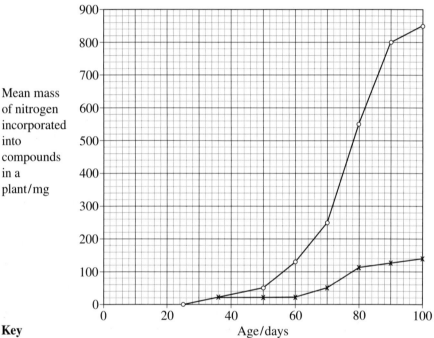

**Key**
o——o Enriched with carbon dioxide
×——× Normal atmosphere

**(a)** Of the nitrogen incorporated into compounds in the control plants, 75% was taken up from the soil. State the form in which this nitrogen was taken up by the plants. *(1 mark)*

**(b)** Explain how the control plants obtained the remaining 25% of their nitrogen. *(2 marks)*

**(c)** Compare the effect of the atmosphere enriched with carbon dioxide with that of the normal atmosphere on the mass of nitrogen incorporated into the seedlings. *(3 marks)*

**(d)** Suggest *one* reason for any differences you observe. *(1 mark)*

**(e)** A possible application of gene technology would be to incorporate genes for nitrogen fixation into cereal plants.

Suggest possible benefits of such an application. *(2 marks)*

*Edexcel 2000*

*(Total 9 marks)*

**3** North American populations of catfish, *Catostomus clarki*, produce two different forms of a particular enzyme, A and B. The graph below shows how the activity of each form of the enzyme varies with temperature. Enzyme activity is plotted as a proportion of its maximum activity, which is given the arbitrary value of 1.0.

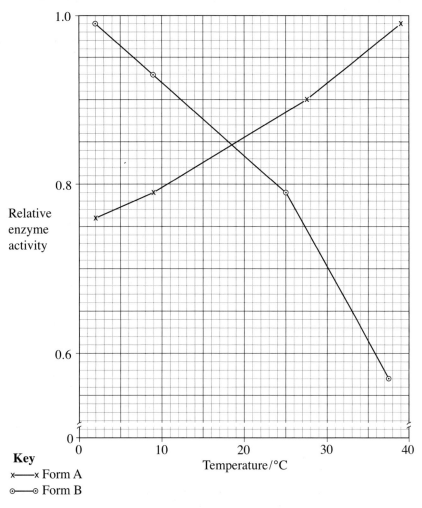

**Key**
x——x Form A
⊙——⊙ Form B

*Adapted from Edwards, Evolution in modern biology (1977)*

**(a)** Describe the differences in activity between the two forms of the enzyme

*(3 marks)*

**(b)** Production of the enzyme is controlled by a single gene, E. An allele $E^A$ codes for form A of the enzyme, and a difference allele $E^B$ codes for form B. The graph below shows the frequency of each allele in catfish populations between the southern and northern ends of the catfish's range. The southern end has a warm climate and the northern end a cold climate.

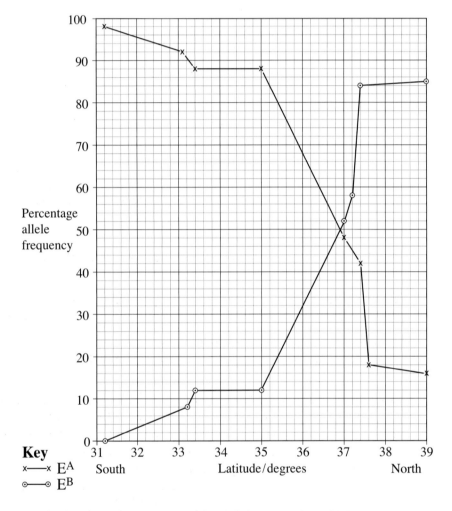

**Key**
x———x $E^A$
⊙———⊙ $E^B$

Suggest the process by which gene E came to have two different alleles. *(1 mark)*

**(c)** Describe and suggest an explanation for the distribution of the alleles between the southern and northern ends of the catfish's range. *(4 marks)*

*Edexcel 2000*

*(Total 8 marks)*

4 **(a)** Microorganisms present in a rabbit's gut are able to digest carbohydrates  N2.1
in the plant material that they eat. **Figure 1** shows the biochemical
pathways by which cellulose and starch are digested in the gut of a rabbit.

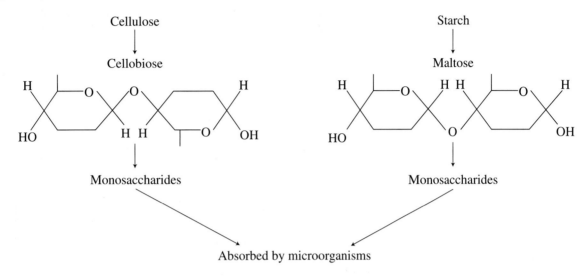

**Figure 1**

(i) Describe how a molecule of cellulose differs from a molecule of
starch. *(1 mark)*

(ii) Draw a diagram to show the molecules produced by digestion of
cellobiose. *(2 marks)*

(iii) Cellobiose and maltose are both disaccharides. Explain why amylase
enzymes produced by the rabbit are unable to digest cellobiose.
*(3 marks)*

**(b)** One way in which rabbits cause considerable damage to agricultural land
is by competing for plant material that would normally be eaten by
domestic animals. **Table 1** shows some features of the energy budgets of
rabbits and cattle living under the same environmental conditions. All
figures are kilojoules per day per kilogram of body mass.

| | Rabbits | Cattle |
|---|---|---|
| Energy consumed in food | 1272 | 424 |
| Energy lost as heat | 567 | 311 |
| Energy gained in body mass | 68 | 17 |

**Table 1**

(i) What is the purpose of giving these figures per kilogram of body
mass? *(1 mark)*

(ii) Explain the difference in the figures for the amount of energy lost as
heat. *(2 marks)*

(iii) Use the information in **Figure 1** to explain why all the energy
consumed in food cannot be converted to body mass or is lost as heat.
*(2 marks)*

**(c)** Rabbits were introduced to Australia in the middle of the last century. Their population grew rapidly and they are now major agricultural pests.

**Table 2** compares some features concerned with heat loss in cattle and rabbits at a temperature of 30°C.

| | Cattle | Rabbits |
|---|---|---|
| Percentage of body heat which is lost by evaporation | 81.0 | 17.0 |
| Core temperature of body | 38.2 | 39.3 |

**Table 2**

Use the information given in parts (b) and (c) of this question to explain:

(i)   how evaporation helps cattle to maintain a constant body temperature;                                                  *(2 marks)*

(ii)  the main way in which a rabbit would lose heat at an environmental temperature of 30°C;                                  *(2 marks)*

(iii) why rabbits are major agricultural pests in Australia;       *(2 marks)*

(iv)  why rabbits are better able to survive than cattle in the hot, dry conditions found in many parts of Australia.          *(3 marks)*

*AQA/A 2000*
*(Total 20 marks)*

5   The diagram below illustrates energy flow diagrams for **A** a deciduous forest and **B** a marine community. The units on the flow charts are kJ m$^{-2}$ day$^{-1}$. The major plants in the marine community are phytoplankton.

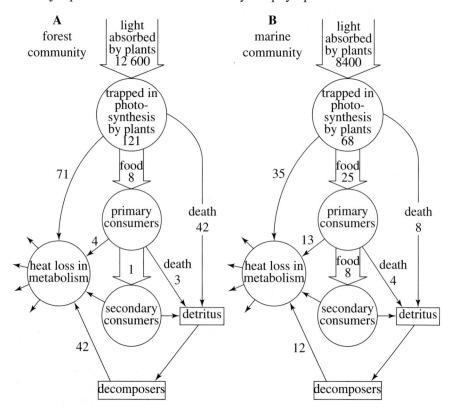

(a) Calculate the efficiency with which solar energy is trapped by the forest plants. *(1 mark)*

N2.2a

(b) With reference to the diagram, describe and explain the major differences between the energy flow in these two communities. *(8 marks)*

C2.3

(c) Suggest *two* ways in which the deciduous forest may be managed for timber production. *(2 marks)*

*OCR 2000*                                                                 *(Total 11 marks)*

6  The table below shows the rate of phosphate absorption by barley roots in a solution aerated with different mixtures of nitrogen and oxygen.

| Percentage of oxygen in aeration mixture | Phosphate absorption / $\mu$mol $g^{-1}$ $h^{-1}$ |
|---|---|
| 0.1 | 0.07 |
| 0.3 | 0.15 |
| 0.9 | 0.27 |
| 2.1 | 0.32 |
| 21.0 | 0.33 |

(a) State the conclusions that can be drawn about phosphate absorption from the data in the table. *(3 marks)*

N2.1

The graph below shows the rate of phosphate absorption by barley roots placed in solutions containing different concentrations of DNP (2,4-dinitrophenol). DNP is an uncoupler of the electron transport chain. Each solution was aerated with 21% oxygen.

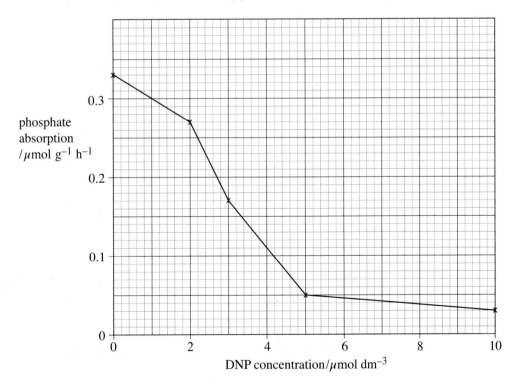

**(b)** With reference to the graph, describe and explain the effect on phosphate absorption of adding DNP to barley roots. *(4 marks)*

Malonate is an inhibitor of the Krebs cycle.

**(c)** (i) Predict the effect of adding malonate instead of DNP, on the uptake of phosphate by the barley roots. *(1 mark)*

(ii) Explain your answer to (i). *(3 marks)*

**(d)** Discuss the significance of phosphate for living organisms. *(6 marks)*

OCR 2000 *(Total 17 marks)*

**7** **(a)** Give **one** way in which the nutrition of:

(i) a heterotroph differs from that of an autotroph; *(1 mark)*

(ii) a chemoautotroph differs from that of a photoautotroph. *(1 mark)*

**(b)** The chemical structure of maltose is shown below.

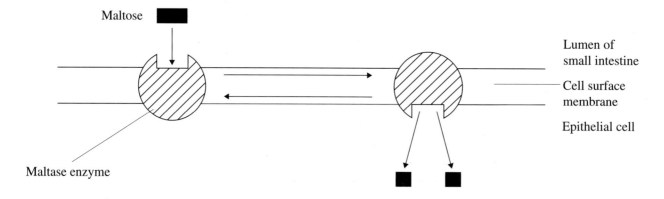

By means of a similar diagram, show the molecules produced when maltose is hydrolysed by the enzyme maltase. *(2 marks)*

The diagram shows how maltose is hydrolysed and the products transported into epithelial cells in the small intestine of a mammal.

Maltose

Lumen of
small intestine

Cell surface
membrane

Epithelial cell

Maltase enzyme

**(c)** The concentration of maltose in the gut affects its rate of uptake by the epithelial cells. Explain what causes:

(i) the rate of maltose uptake to be proportional to its concentration at low concentrations; *(1 mark)*

(ii) the rate of maltose uptake to remain constant at higher concentrations. *(2 marks)*

**(d)** The contents of the small intestine are constantly being mixed as a result of muscle action. What effect would you expect mixing to have on the rate of uptake of maltose? Give an explanation for your answer. *(2 marks)*

The digestive enzymes produced by an insect at difference stages in its life cycle are shown in the table.

| Stage in Life Cycle | Food | Enzyme | | | | |
|---|---|---|---|---|---|---|
| | | Exopeptidase | Lipase | Amylase | Sucrase | Maltase |
| **Caterpillar** (larva) | Leaves | ✔ | ✔ | ✔ | ✔ | ✔ |
| **Butterfly** (adult) | Nectar | | | | ✔ | |

**(e)** Describe the part played by exopeptidases in the digestion of proteins.
*(2 marks)*

**(f)** Explain how the differences in digestive enzymes shown in the table can be related to the fact that:

(i) growth takes place only in the larval stage of an insect's life;
*(2 marks)*

(ii) the adult stage of an insect's life is associated with reproduction and dispersal. *(2 marks)*

The graph shows some of the changes which occur in an insect pupa during metamorphosis.

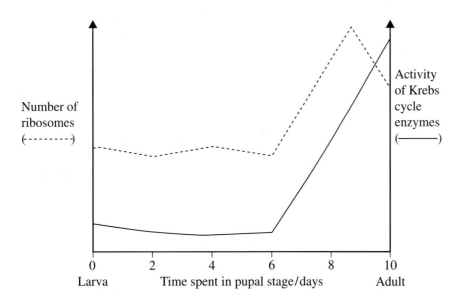

**(g)** During the pupal stage of an insect's life, the larval organs are broken down and the adult organs are formed. Explain how the change in the number of ribosomes shown on the graph can be related to these events.
*(2 marks)*

**(h)** Explain how the information about enzymes in the graph supports the hypothesis that the adult insect's tracheal system develops towards the end of the pupal stage. *(3 marks)*

*AEB 2000* *(Total 20 marks)*

**8** An indicator which turns from red to yellow in the presence of carbon dioxide was used in a class experiment designed to compare the rates of respiration of a range of different living organisms. The time it took to change colour with the same mass of each organism was recorded.

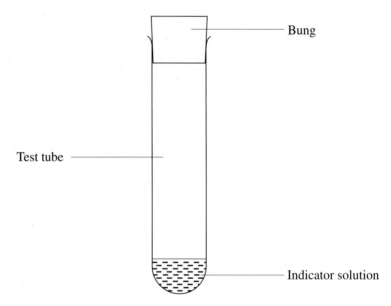

**(a) (i)** Show on the diagram the modification you would make to the apparatus to obtain results with blowfly larvae (maggots).

**(ii)** Before this apparatus was used to measure the rate of respiration of carrot root, the root was cut into small pieces. How might this have made the results more comparable with those obtained using maggots? *(3 marks)*

**(b)** Before the students introduced organisms into any tube, they were told to take a separate tube containing indicator and blow into it using a straw. This tube was bunged and kept on one side for the duration of the experiment. One student incorrectly described this tube as a 'control'. What was its actual purpose? *(1 mark)*

**(c)** The pooled class results are shown in the table below.  ⊸O N2.2a

| Tube | Organisms | Mean time taken for indicator to turn yellow/minutes | Relative rate of respiration |
|------|-----------|------------------------------------------------------|------------------------------|
| A | Blowfly larvae | 15.0 | |
| B | Carrot root | 50.0 | |
| C | Mushroom | 35.0 | |
| D | Yeast | 10.0 | |
| E | Cabbage leaves | no change | |

Using the information in the table, calculate values for the relative rates of respiration of organisms A to D, taking the rate of the organism with the slowest rate to be 1.0. Enter these rates in the appropriate spaces in the table above. *(2 marks)*

**(d)** Students in the class differed in their interpretation of the results obtained with cabbage leaves. Some thought they were caused because the leaves were photosynthesising (Hypothesis 1) but others thought that the rate of respiration was too slow to be detected by this experiment (Hypothesis 2).

Suggest a way in which you might modify the experiment to test each hypothesis. In **each** case use a different modification, and explain your reasoning in full. *(4 marks)*

*AQA/B 2000* *(Total 10 marks)*

# Answers to end of chapter and synoptic questions

Edexcel accepts no responsibility whatsoever for the accuracy or method of working in the answers given.

## Chapter 1

1  (a)  C *(1)*.

   (b)  A *(1)*.

   (c)  D *(1)*.

   (d)  E *(1)*.

   (e)  A *(1)*.

2  The inner membrane is the site of ATP synthesis in the mitochondrion *(1)*. The folds result in an increased surface area of the membrane *(1)* allowing more ATP synthesis.

3  Substrate level phosphorylation is the formation of ATP from ADP *(1)* using phosphate derived from substrates of the glycolysis pathway *(1)*. Oxidative phosphorylation is the formation of ATP from ADP and inorganic phosphate *(1)*, coupled with the electron flow (oxidations) in the electron transport chain *(1)*.

4  (a)  Potassium (or sodium) hydroxide *(1)*.

   (b)  Up *(1)*.

   (c)  Respiration produces heat *(1)*, which may cause the temperature to rise.

   (d)  Place both tubes in a (thermostatic) water bath *(1)*.

   (e)  The small syringe is used to re-set the level of the oil in the U-tube between each experiment (with the screw clip open) *(1)*.

5  (a)  A coenzyme is a small organic molecule which works in conjunction with an enzyme *(1)* as a carrier, transferring chemical groups or atoms from one substrate to another *(1)*.

   (b)  Alcoholic fermentation is a type of anaerobic respiration *(1)* where pyruvate from glycolysis is reduced to ethanol *(1)*. It is carried out by organisms such as yeast.

   (c)  A redox indicator is a chemical which undergoes a colour change on being oxidised or reduced *(1)*. It can therefore be used to monitor a redox (combined reduction/oxidation) reaction *(1)*.

6  (a)  (i)  Glucose – 6, pyruvate – 3 *(both for 1 mark)*; compound A – 6, compound B – 5, compound C – 4 *(all 3 for 1 mark)*.

       (ii)

| | Name of stage | Part of cell in which it occurs |
|---|---|---|
| Stage X | glycolysis | cytoplasm |
| Stage Y | Krebs cycle | mitochondrion/matrix |

*(2 marks – mark horizontally)*

   (b)  Passed to coenzymes/NAD or FAD *(1)* passed along series of carriers *(1)* on cristae/inner mitochondrial membrane *(1)* yield ATP *(1)* at decreasing energy levels *(1)* oxygen is final acceptor/water is produced *(1) (max. 3)*.

7  (a)  (i)  Both combine with oxygen *(1)* producing water *(1)*.

       (ii)  Inner membrane/cristae *(1)*.

   (b)  Glycolysis/Krebs cycle/link reaction *(1)*.

   (c)  Phosphorylation of substrate/named example/to start glycolysis *(1)*; active transport/ion pumps/named example *(1)*; muscle contraction *(1)*; light-independent reaction of photosynthesis (see Chapter 2) *(1)*; named anabolic reaction, e.g. protein synthesis, replication of DNA *(1)*; cell division *(1) (max. 2)*.

8  (a)  Carbon dioxide produced *(1)* which increases pressure in syringe/pushes down mixture *(1)*.

   (b)  Measure distance moved by meniscus over a set time *(1)*; repeat to find mean/plot graph and find slope *(1)*; need to eliminate air explained/boil solutions (except yeast) *(1)*; use buffer/keep pH constant *(1)*; have solutions at same temperature before taking into syringe *(1)*; use same concentrations/volumes of sucrose as glucose *(1)*; use same concentration/volume/mass of yeast *(1) (max. 4)*.

## Chapter 2

1  (a)  A = spongy mesophyll *(1)*, B = (upper) epidermis *(1)*.

   (b)  cells are vertically elongated *(1)* so trap light more efficiently *(1)*; many chloroplasts *(1)* so photosynthesise/trap light more efficiently *(1)*; fit closely together/no spaces *(1)* so trap light more efficiently *(1)*; thin cell walls *(1)* for easier gas exchange/light penetration *(1)*; cells close to upper surface *(1)* so trap light more efficiently *(1) (any two pairs of points)*.

2  (a)  Correctly drawn curve *(1)*, blue peak higher *(1)*.

   (b)  More photosynthesis *(1)* in certain wavelengths *(1)* corresponding to red and blue light *(1)* results in more oxygen *(1)* which attracts more bacteria *(1)* no photosynthesis in green/yellow light *(1) (max. 4)*.

   (c)  1 mark for any of the following points:

        RuBP is $CO_2$ acceptor; RuBP is 5C compound; reaction catalysed by carboxylase/Rubisco; temporary 6C compound formed; (splits to) form glycerate 3-phosphate/GP; use of NADPH/reduced NADP; NADPH from light-dependent stage; use of ATP; ATP from light-dependent stage; triose phosphate/3C sugar formed; converted to 6C sugars;

# Answers to end of chapter and synoptic questions

converted to sucrose/polysaccharides/starch/cellulose; regeneration of RuBP; uses more ATP *(max. 7)*.

*(plus 1 mark for quality of written communication: clear, well organised, using specialist terms)*

3  **(a)** Position of solvent front/distance moved by solvent *(1)*.

   **(b)** (i)  Either: provides energy *(1)* for reduction of GP *(1)* to convert GP to triose phosphate/triose phosphate to RuBP *(1)*

   or: provides phosphate *(1)* for formation of RuBP *(1)* *(max. 2)*.

   (ii)  Light absorbed by/excites chlorophyll; heat released when electrons return to chlorophyll; energy not passed to other molecules as in chloroplasts; energy conversion not efficient; some energy always lost as heat *(max. 2)*.

4  **(a)** ATP *(1)*.

   **(b)** (i)  Electrons raised to higher energy level; passed through chain of hydrogen acceptors; hydrogen ions; from photolysis/water *(max. 2)*.

   (ii)  Reduces; glycerate phosphate/GP *(2)*.

5  **(a)** 5, 3, 3 *(1)*.

   **(b)** (i)  Transfers energy *(1)*.

   (ii)  Supplies phosphate (and transfers energy) *(1)*.

   **(c)** NADPH and ATP/products of light-dependent reaction needed to convert *(1)* glycerate 3-phosphate to triose phosphate *(1)*.

6  **(a)** The compensation point *(1)*.

   **(b)** It is the point (for B) when $CO_2$ uptake by photosynthesis equals $CO_2$ loss by respiration (i.e. net exchange is zero) *(1)*.

   **(c)** Another factor is limiting; such as $CO_2$ concentration or temperature *(2)*.

   **(d)** Species A *(1)*.

   **(e)** Species A photosynthesises more than species B at low light intensities *(1)* (or converse).

## Chapter 3

1  **(a)** The position of a gene/allele on a chromosome *(1)*.

   **(b)** Genotype is the genetic make-up of an organism, i.e. what alleles are present. Phenotype is the appearance of the genotype *(1)*.

   **(c)** Genotypes – BbTt, Bbtt, bbTt, bbtt *(1)*. Phenotypes – brown taster, brown non-taster, blue taster, blue non-taster *(1)*.

   **(d)** iii and iv. The father must be brown eyed (BB or Bb) *(2)*.

   **(e)** Look at family pedigree over several generations; if the genes are linked, the alleles will be inherited together/DNA analysis if possible *(2)*.

2  **(a)** Different forms of a gene *(1)*.

   **(b)** (i)  $I^A I^O$, $I^B I^O$, $I^A I^B$, $I^O I^O$ *(2)*.

   (ii)  Parental genotypes: $I^A I^O$ and $I^B I^O$, gametes: $I^A$, $I^O$, $I^B$, $I^O$ *(1)*.

   Genotypes of children: $I^A I^B$, $I^A I^O$, $I^B I^O$, $I^O I^O$ *(1)*.

   Phenotypes of children: AB, A, B, O *(1)*.

3  **(a)** (i)  One copy, as only one chromosome is present from each pair as the chromosome number is halved at meiosis, i.e. haploid number *(1)*.

   (ii)  Two copies, as in a leaf cell the chromosomes are in pairs, the diploid number *(1)*.

   **(b)** (i)  RrTt *(1)*.

   (ii)  Yellow fruit, tall *(1)*.

   (iii)  RrTt × rrTt

|      | rT   | rt   |
|------|------|------|
| RT   | RrTT | RrTt |
| Rt   | RrTt | Rrtt |
| rT   | rrTT | rrTt |
| rt   | rrTt | rrtt |

*(2)*

50%/half *(1)*

4  **(a)** (i)  1 out of 8 *(1)*.

   (ii)  3 out of 4/75% *(1)*.

   **(b)** 0.0004 *(1)*.

   **(c)** In the heterozygote the allele is passed on; the heterozygote has an advantage; mutations can produce the recessive allele at a rate equal to its loss from the population *(any 2)*.

5  **(a)** They lay many eggs and have large numbers of offspring; short life cycle; results obtained quickly; clear features; obvious differences between male and females; small size; little space required *(any 3)*.

   **(b)** (i)  GgNn *(1)*.

   (ii)  Gn, gn *(1)*.

   **(c)** 9:3:3:1 *(1)*.

**6** **(a)** (i) $W^2$ is dominant, as the rats are resistant to warfarin in the heterozygous state *(1)*.

(ii) The difference is significant and not just due to chance *(1)*. 5.73 gives a probability of between 1 and 2% which is less than the critical value of 5% meaning that observed numbers are significantly different to those expected *(1)*.

**(b)** (i) $W^1W^1$ $\dfrac{28}{32}$ $\times 100 = 87.5\%$

$W^2W^2$ $\dfrac{4}{9}$ $\times 100 = 44.4\%$ *(3)*.

(ii) Homozygous dominant, $W^2W^2$ *(1)*.

(iii) $W^1W^1$ are killed by warfarin; $W^2W^2$ need more vitamin K in their diet *(2)*.

**(c)** If the use of warfarin has been discontinued then the homozygous recessive rats are no longer at a disadvantage/resistant rats are no longer at an advantage; therefore the numbers of non-resistant rats will increase; the number of heterozygotes will be unchanged, the homozygous dominant rats will now be at a disadvantage as they require large quantities of vitamin K and their numbers will fall. The selection pressure has changed from warfarin to vitamin K *(3)*.

**(d)** The resistant allele will still be present in the heterozygous rabbits, and as they do not require large quantities of vitamin K they are not selected against; the allele could also arise as a mutation *(2)*.

# Chapter 4

**1** **(a)** There are two peaks *(1)*.

**(b)** (i) Continuous variation; due to the complete range of different heights from one extreme to the other *(2)*.

(ii) The phenotype is affected both by genes and the environment/ by polygenic and environmental influences *(1)*.

**2** **(a)** $(1 \times 148) + (3 \times 153) + (9 \times 158) + (11 \times 163) + (4 \times 168) + (2 \times 173) + (2 \times 178) \div 32$ *(1)* $= 162.38$ cm *(1)*.

**(b)** A bar graph/ histogram with title, both axes labelled with units, suitable scale and accurate points *(5)*.

**(c)** Continuous variation *(1)*.

**(d)** The sample is small; different sexes are included; different diet; different racial group *(any 2)*.

**3** **(a)** An extra chromosome no. 21/47 instead of 46 *(1)*.

**(b)** Non-disjunction/failure of the chromosomes to separate at meiosis *(1)*.

**(c)** (i) 23 or 24 *(1)*.

(ii) 47 *(1)*.

**(d)** Amniocentesis/chorionic villus sampling; karyotype analysis to identify extra chromosome *(2)*.

**(e)** The sex of the child/other chromosome abnormalities *(1)*.

**(f)** Deletion; inversion; translocation; polyploidy *(any 2)*.

**4** **(a)**

| Animal | Number of chromosomes in one of the nuclei formed at the end of | | |
|---|---|---|---|
| | mitosis | first division of meiosis | second division of meiosis |
| Horse | 64 | 32 | 32 |
| Donkey | 62 | 31 | 31 |

*(2)*

**(b)** (i) 63 (32 from horse gamete + 31 from donkey gamete) *(1)*.

(ii) At meiosis; an odd number of chromosomes in the mule prevents pairing *(2)*.

(iii) Mule is infertile and cannot produce offspring *(1)*.

**5** **(a)** (i) Continuous *(1)*.

(ii) Many genes are involved/it is polygenic *(1)*.

**(b)** (i) Red males are more successful than black ones, i.e. mate more frequently; red males mate more often with red females and black males mate more often with black females *(2)*.

(ii) An increase in the allele for red wing cases; as red males mate more frequently than black ones *(2)*.

**6** **(a)** All evolved from a common ancestor *(1)*.

**(b)** Reproductive isolation enabled different mutations and selection to take place separately in each group; i.e. they adapted to the local conditions *(2)*.

**(c)** More easily isolated if they don't fly as they can only travel short distances; reproductive isolation occurs even if they are only a short distance apart *(2)*.

# Answers to end of chapter and synoptic questions

## Chapter 5

1 (a) Prokaryotae, Protoctista, Fungi, Animalia;

(b) Prokaryotae, Protoctista, Fungi, Plantae;

(c) Prokaryotae, Protoctista, Fungi, Plantae;

(d) Plantae;

(e) Protoctista, Fungi, Animalia, Plantae;

(f) Prokaryotae, Protoctista, Fungi;

(g) Prokaryotae, Protoctista, Plantae;

(h) Fungi;

(i) Fungi, Animalia;

(j) Protoctista, Animalia, Plantae *(10)*.

2 (a) (i) Suidae *(1)*.

(ii) *Sus (1)*.

(b) Same species as – can interbreed;
   – hybrids fertile as all pigs involved over many years;
   – may be variation of body length/mass within a species;
   – especially across range of distribution
   *(max. 3)*.

3 (a) (i) *Lepus (1)*.

(ii) Lagomorpha *(1)*.

(b) (i) Their habitats overlap but no evidence of interbreeding *(1)*.

(ii) Geographically isolated; adapted to slightly different conditions/different selection pressures; mutation may have taken place in one population/different mutations in each population *(max. 2)*.

4 (a) Protoctista *(1)*.

(b) (i) Eukaryotic cells/nuclei/mitochondria/ribosomes/cell membrane *(1)*.

(ii) Cell membrane/ribosomes/single celled *(1)*.

(c) Fungi do not have chloroplasts/are heterotrophic/do not photosynthesise; do not have cellulose cell walls/have cell walls of chitin; do not have undulipodia; store glycogen/do not store starch; do not show alternation of generations; have hyphae *(max. 2)*.

5 (a) Both have cell walls/cell vacuole; both produce spores; neither have a nervous system *(max. 1)*.

(b) No chloroplasts/photosynthesis in animals and fungi; no cellulose; both heterotrophic; both store glycogen/do not store starch *(max. 2)*.

(c) No mitochondria/membrane bound organelles; no nucleus; circular DNA; DNA not associated with protein *(max. 2)*.

6 (a) (i) Fungi *(1)*.

(ii) Protoctista *(1)*.

(b) (i) C, D and E *(1)*.

(ii) A, B, C and D *(1)*.

7 Fungi *(1)*.

Animalia *(1)*.

Plantae *(1)*.

Protoctista *(1)*.

8 (a) Larger groups divided into smaller groups *(1)*.

(b) (i) *Cercopithecus (1)*.

(ii) Cercopithidae *(1)*.

(c) All show some similarities as in same family; blue and green monkeys more similar, therefore more closely related than red and black-and-white colobus; all show some differences as different species *(max. 2)*.

## Chapter 6

1 (a) (i) Motor/effector *(1)*.

(ii) Cell body at one end *(1)*.

(b) (i) Synaptic knob/end bulb *(1)*.

(ii) Has vesicles containing neurotransmitter; arrival of action potential causes entry of calcium ions; vesicles move and fuse with presynaptic membrane; neurotransmitter is released; and diffuses across synaptic cleft; neurotransmitter fits into receptors on postsynaptic membrane; causing $Na^+$ ions to diffuse in; depolarising the postsynaptic membrane *(max. 4)*.

2 (a) (i) Membrane more permeable to $Na^+$ ions; $Na^+$ ions enter the membrane *(2)*.

(ii) Membrane less permeable to $Na^+$ ions/permeability to $K^+$ ions increases; $K^+$ ions leave axon *(2)*.

(b) Axon B is myelinated since small diameter and rapid rate of conduction; myelinated/salutatory conduction is faster *(2)*.

3 Motor; sensory; sodium; pump/carrier; resting; sodium; potassium *(7)*.

4 (a) Membrane has high permeability to $K^+$ ions; many $K^+$ ions move out; few sodium or chloride ions move in; membrane impermeable to protein which remains inside *(max. 3)*.

(b) Membrane becomes more permeable to $N^+$ ions; $Na^+$ ions diffuse in rapidly *(2)*.

5 (a) (i) Circle around one of the places shown as +40; inside of membrane has a positive charge *(2)*.

(ii) Arrow shown from left to right; neurotransmitter is only released from the left side of synapse from the vesicles (Y) *(2)*.

(b) X = mitochondria; these provide ATP as a source of energy to reform the neurotransmitter; Y = synaptic vesicles; contain the neurotransmitter which will be released to carry the impulse across the cleft of the synapse *(4)*.

(c) (i) Myelin/lipid *(1)*.

(ii) Speeds up transmission; insulates axon from other axons *(2)*.

6 (a) Presynaptic membrane becomes more permeable to calcium ions; calcium ions diffuse in; synaptic vesicles move and fuse with membrane; neurotransmitter released; neurotransmitter diffuses across cleft; binds with receptors on postsynaptic membrane; causes an increase in permeability of membrane to $Na^+$ ions *(max. 4)*.

(b) (i) Gates of sodium channels open; membrane becomes more permeable to sodium ions; $Na^+$ ions diffuse rapidly into neurone *(3)*.

(ii) +28 mV *(1)*.

(iii) 0.6 or 3/5 milliseconds *(1)*.

(iv) Diffusion of neurotransmitter across the cleft *(1)*.

(c) Mimics the action of acetylcholine; fits into acetylcholine receptors on postsynaptic membrane; stimulates the postsynaptic membrane *(max. 2)*.

7 (a) (i) 3 1 6 4 2 5 *(1)*.

(ii) A = calcium; B = sodium *(2)*.

(iii) Diffusion *(1)*.

(iv) Acetylcholine/noradrenaline *(1)*.

(b) (i) One impulse does not release enough neurotransmitter to produce an action potential; several impulses release more neurotransmitter and reach threshold; this produces an action potential *(max. 2)*.

(ii) Summation *(1)*.

8 (a) Curare – receptor sites on postsynaptic membrane blocked; no stimulus to muscle, so muscle does not contract *(2)*.

(b) Organophosphate – acetylcholine not broken down so stays in receptor; muscle is continuously stimulated *(2)*.

(c) Botulin – acetylcholine not released from vesicles; muscle not stimulated, so muscle does not contract *(2)*.

9 Unidirectionality: The structure of synapses means that impulses can only pass in one direction; vesicles containing neurotransmitter are only present on the presynaptic side; receptors for the neurotransmitter are only found on the postsynaptic membrane; therefore impulses can only pass from neurones A, B and I to neurone C and not the other way *(max. 2)*.

Convergence: Synapses can act as junctions; a number of neurones may form synapses with the same postsynaptic neurone; neurones A, B and I all converge to synapse with neurone C *(2)*.

Summation: This is the adding together of the effects of the impulses from a number of different neurones; in table Y it can be seen that if impulses are passed from A and B together, this does produce an impulse in neurone C; the effects of the two are added together (summated) to reach threshold; if impulses are only passed from A on its own, the effects are not enough to reach threshold and an impulse is not produced in neurone C *(max. 2)*.

Inhibition: Some synapses are inhibitory and they inhibit the production of an impulse in the postsynaptic neurone; in table Y it can be seen that impulses from A and B together do produce an impulse in C; impulses from A, B and I together do not, so I must be an inhibitory synapse; when I conducts an impulse, it prevents the impulses from A and B causing an impulse in neurone C *(max. 2)*.

10 agonistic; antagonistic; antagonistic; agonistic; agonistic; antagonistic; agonistic; agonistic *(8)*.

# Chapter 7

1 (a) (i) Blind spot *(1)*, (ii) choroid *(1)*, (iii) rhodopsin *(1)*.

(b) (i) M = circular muscles, N = radial muscles *(1)*.

(ii) Light intensity is reduced; the radial muscles respond by contracting/diameter of iris widens and lets more light into eye *(2)*.

(c) Inversion of the image correctly shown/image shown inverted on retina; light rays cross over in vitreous humour *(2)*.

2 (a) A = outer segment; B = inner segment *(2)*.

(b) Towards the outside/periphery of retina, not in fovea *(1)*.

(c) (i) Rhodopsin *(1)*.

(ii) Line drawn to label the vesicles in the outer segment *(1)*.

3 Low; rhodopsin; opsin; dark *(4)*.

4 (a) Arrow from bottom to top *(1)*.

(b) (i) Convergence and summation; a number of rods connect with one neurone to the brain; the effects of the low light received by a number of rods are added together *(max. 2)*.

(ii) Cones have high visual acuity; cones do not have convergence; each cone connects with one neurone to the brain *(max. 2)*.

5 (a) The chemical change in rhodopsin/the breakdown of rhodopsin; alters the permeability of the membrane causing a movement of ions; this produces a generator potential *(max. 2)*.

(b) They produce ATP/energy; required for synthesis of pigment/rhodopsin *(2)*.

**(c)** **(i)** Only stimulates the cone cells which contain the blue-sensitive pigment *(1)*.

**(ii)** Stimulates all types of cone cell *(1)*.

**6 (a)** Cerebrospinal fluid *(1)*.

**(b)** 85 mm ÷ 25; = 3.4 mm *(2)*.

**(c)** Sensory neurone drawn and labelled with cell body in correct place; motor neurone drawn and labelled with cell body drawn in correct place *(2)*.

**7 (a)** There is a greater rate/more rapid, movement in the light *(1)*.

**(b)** **(i)** Kinesis/photokinesis *(1)*.

**(ii)** There is a change in the rate, rather than the direction, of movement; it is faster in light *(max. 1)*.

**(c)** The light conditions are unfavourable, e.g. presence of predators; moving faster means that it soon moves out of the light. OR: The dark conditions are more favourable, e.g. fewer predators; moves slower in dark, therefore tends to stay in the dark for longer *(2)*.

**8** Cerebrum/cerebral hemispheres; control of involuntary activities/control of heart rate, breathing, etc.; cerebellum; coordination of temperature regulation/coordination of osmoregulation/secretion of ADH *(4)*.

**9 (a)** Heart rate: increased, decreased; Stroke volume: increased, decreased *(2)*.

**(b)** Constriction of arterioles; prevents blood removing the anaesthetic *(2)*.

**(c)** **(i)** Prevents stimulation of saliva production *(1)*.

**(ii)** Pupil will be dilated *(1)*.

## Chapter 8

**1 (a)** **(i)** Contraction of B pulls the jawbone up and closes mouth; contraction of A pulls the back of jawbone and opens mouth; they have antagonistic action/when one contracts the other relaxes *(3)*.

**(ii)** Muscle B is longer which gives more leverage for biting *(1)*.

**(b)** **(i)** P = thinner actin filaments only so looks paler; R = myosin filaments only – thicker than actin filaments so looks darker; Q = both actin and myosin filaments together so looks darkest *(3)*.

**(ii)** Diagram with narrower band P and narrower or no band R *(1)*.

**(c)** ATP produced by respiration; hydrolysis of ATP releases energy; this energy used to break the link between actin and myosin; and to reposition the myosin heads; myosin head now changes position, sliding actin past myosin; actin slid towards centre of sarcomere *(4)*.

**2** Myofibrils; myosin; actin; ATP *(4)*.

**3 (a)** Initiate muscle contraction; by unblocking the actin binding sites; and allowing myosin heads to attach; to form actomyosin bridges *(max. 2)*.

**(b)** **(i)** The more cross-bridges, the greater the force generated; the cross-bridges change position, pulling the filaments over each other and generating force *(2)*.

**(ii)** Diagram showing full overlap of actin and myosin *(1)*.

**4 (a)** A = I band; B = A band *(2)*.

**(b)** Myosin *(1)*.

**(c)** Accuracy of myosin; accuracy of actin *(2)*.

**(d)** Increases size/cross-sectional area of muscle; increase in size of muscle fibres; increase in number/size of myofibrils; increase in amount of proteins/myosin *(max. 3)*.

**5 (a)** **(i)** Moves the switch protein/tropomyosin; uncovering the binding site on the actin; allowing cross bridges to form *(max. 2)*.

**(ii)** Myosin head changes position; sliding the actin filaments past the myosin filaments *(2)*.

**(b)** ATP provides the energy to release myosin head from actin binding site *(1)*.

**6 (a)** **(i)** One nerve fibre and the muscle fibres that it supplies *(1)*.

**(ii)** The muscle fibres receive inputs from different motor nerves. The motor nerves are of different lengths *(1)*.

**(b)** Neurotransmitter released from the end plate; it diffuses across the cleft; and fits into receptors; on the muscle membrane/sarcolemma; altering the permeability of the membrane to sodium ions *(max. 3)*.

**(c)** **(i)** Taken to the liver; oxidised; to produce ATP; some is converted to glucose/glycogen *(max. 3)*.

**(ii)** Acts as a reserve store of ATP; muscle contraction uses lots of ATP; when ATP levels are low it provides phosphate; to change ADP to ATP *(max. 2)*.

**(d)** Posture has to be maintained for a long time; therefore important not to get fatigued *(2)*.

**(e)** **(i)** Few mitochondria; this limits aerobic respiration; capillaries are fewer; therefore less oxygen supplied *(4)*.

**(ii)** Myosin ATPase activity is high; more ATP is used for muscle contraction; there is a well-developed sarcoplasmic reticulum; therefore an improved supply of calcium ions *(4)*.

# Chapter 9

1 (a) In both, a change from the norm is detected and triggers a response restoring the norm; separate mechanisms are present which can restore the norm from different directions (2).

(b) Communication is hormonal in glucose regulation and nervous in thermoregulation; a coordination centre is involved only in thermoregulation (2).

2 (a) 30 mmol dm$^{-3}$ (1).

(b) (i) Blood glucose concentration increases (with some reference to figures); as absorbed from gut into blood (2).

(ii) Blood glucose concentration decreases (with some reference to figures); as taken up by liver/muscle/cells (2).

(c) Both increase up to a maximum at 30 minutes; then decrease; the rise in glucose stimulates insulin secretion and then insulin falls as glucose level falls (3).

(d) Low level of glucose stimulates the secretion of glucagon; from α-cells in Islets of Langerhans in pancreas; glucagon promotes the conversion of glycogen to glucose in liver; glucose released into blood (4).

3 (a) A – no response as time taken for food to be digested; B – increase as glucose absorbed from gut; C – decrease as high level stimulates β-cells of Islets of Langerhans in pancreas; to secrete insulin; which stimulates removal of glucose by cells/conversion of glucose to glycogen; D – level stable as glucose released slowly from the liver matching use in respiration; E – decrease as respiration rate increased with exercise; F – increase as low level stimulates α-cells in Islets of Langerhans in pancreas; to secrete glucagon; which stimulates conversion of glycogen to glucose/gluconeogenesis (max. 8).

(b) If too cold, molecules move slowly; slows down metabolic reactions/enzyme action – too slow to maintain life; if ice produced, this damages cells; if too warm, denatures enzymes; enzymes not able to catalyse metabolic reactions; may damage cell membranes/denature membrane proteins (max. 4).

4 (a) Body temperature fluctuates less than that of the environment; this allows the mammal to survive in varied climates; if body temperature too low, metabolic reactions take place too slowly; if body temperature too high, enzymes and other globular proteins are denatured (4).

(b) Control involves receptor(s), hypothalamus/coordination centre and effector(s); receptors in hypothalamus detect blood/body core temperature; receptors in skin detect skin temperature; initiate nerve impulses which are sent to thermoregulatory centre in hypothalamus; which consists of a heat gain centre and a heat loss centre; describe a specific effect; explaining how this results in heat loss/gain (max. 5)

(c) Control by hormones; high concentration of glucose stimulates production of insulin; insulin promotes increased rate of glucose uptake by cells; and conversion of glucose to glycogen in liver cells; low concentration of glucose stimulates production of glucagon; glucagon promotes conversion of glycogen to glucose/gluconeogenesis (6).

5 (a) (i) Glucagon secretion increases in A and decreases in B (1).

(ii) Insulin secretion rises less in A than in B; rise in insulin occurs more slowly in A than in B (2).

(b) (i) Glucose concentration rises as absorbed into blood; glucose rise promoted by glucagon release; glucose level drops slowly as used for respiration in cells; also as lost in urine (max. 3).

(ii) Glucose rise stimulates insulin secretion by β-cells; insulin promotes increased uptake by cells/conversion to glycogen in liver cells; as level falls, low level stimulates glucagon secretion by α-cells; glucagon promotes conversion of glycogen to glucose/gluconeogenesis (max. 3).

(c) Level would fall below that in group B; as no glycogen stores to mobilise; as glucose used for respiration; leads to coma (max. 3).

6 (a) (i) Stimulates increased production of volume/water; stimulates increased production of hydrogencarbonate ions;

(ii) Diluted by increased production of water (3).

(b) Bile (1).

(c) (i) Prevents stimulation of production of gastric juice by nervous reflex (1).

(ii) Acetylcholine is essential neurotransmitter at synapse; prevents stimulation of production of gastric juice by nervous reflex (2).

# Chapter 10

1 (a) (i) C, D and F (1).

(ii) B, D and E; these cells carry out active transport which requires energy from ATP (2).

(iii) C, D, E and F (1).

(iv) W and Z (1).

(b) Although sodium has been removed, water has followed by osmosis (1).

(c) Sodium ions have been removed in D but water has not left as the walls of D are impermeable to water (1).

(d) Water has left by osmosis but sodium ions have not been removed (1).

(e) It is decreased (made more negative) *(1)*.

(f) It cannot usually be filtered out of the blood as it is contained in red blood cells; it indicates that some red blood cells are damaged and leaking haemoglobin *(2)*.

2 (a) Arrow goes from blood, through pore, between podocyte processes into capsule *(1)*.

(b) Pores in capillary wall; thin basement membrane; gaps between processes of podocyte *(max. 2)*.

(c) If ratio = 1, filtration is most effective; small molecules filtered more effectively than large molecules *(2)*.

3 (a) Posterior pituitary; decreases – collecting duct/distal convoluted tubule *(2)*.

(b) (i) Not all glucose reabsorbed and some remains in filtrate in collecting duct *(1)*.

(ii) More chloride ions remain in filtrate *(1)*.

(i) or (ii) Lowering water potential gradient between filtrate and medulla tissues; so less water reabsorbed *(max. 1)*.

4 (a) Reduction in volume of urine produced; effect only lasts for two days *(2)*.

(b) Antidiuretic hormone/ADH *(1)*.

(c) Increases the permeability of the distal convoluted tubule and collecting duct to water; allowing water to be reabsorbed from the filtrate by osmosis; down water potential gradient between filtrate and medulla tissues *(max. 2)*.

5 (a) (i) 99.17%; (ii) 28 g; (iii) 52.8%; (iv) 0.0 g *(4)*.

(b) Chloride ions pumped out of ascending limb into medulla by active transport; sodium ions follow passively; water does not leave limb as it is impermeable to water; therefore medulla/interstitial tissues become very concentrated with solutes; water passes out of filtrate in collecting duct by osmosis (if walls permeable) *(max. 4)*.

6 (a) Increased respiration and heat production in active muscles; blood temperature rises; detected by 'warm' thermoreceptors in hypothalamus/receptors in skin detect rise of tissue temperature; heat loss centre switched on; nerve impulses transmitted to sweat glands increasing sweat production *(max. 5)*.

(b) Body loses water and ions when sweating; drinking water does not replace the ions; water potential of blood/body fluids becomes higher/less negative; water potential of blood becomes higher than that of brain cells; water enters brain cells by osmosis/down water potential gradient *(5)*.

(c) Excess water usually lost in urine; ADH makes walls of distal convoluted tubule/collecting duct; permeable to water; allowing reabsorption of water into blood rather than being lost in urine; water swells brain cells crushing them against skull/possible lysis *(4)*.

7 (a) Line continues from glucose reabsorption, directly proportional to plasma glucose concentration *(1)*.

(b) Proximal convoluted tubule *(1)*.

(c) Glucose reabsorbed by active transport; as glucose concentration increases, more carrier molecules in use, increasing rate of reabsorption; eventually above certain concentration/about 300 mg 100 cm$^{-3}$, all carriers in use and rate of absorption increases no more *(3)*.

## Chapter 11

1 (a) Cell growth involves increase in organic matter and size. Differentiation does not, but involves changes in the structure of the cell *(1)*.

(b) A meristematic cell is capable of dividing by mitosis/no large cell vacuole/thin flexible cell wall. A parenchyma cell is not capable of cell division/has large cell vacuole/cell wall hardened *(1)*.

(c) An amyloplast is an organelle containing starch grains which may be found in any plant cells. A chloroplast is an organelle found only in cells of the shoot system and contains chlorophyll (and sometimes starch grains) *(1)*.

(d) In fruit ripening the nature of the fruit wall changes but it does not fall. In fruit abscission the fruit falls off the plant *(1)*.

(e) A dormant seed is not capable of germination even when given the conditions needed for germination (warmth, water and oxygen). A packeted seed cannot germinate because it does not have water *(1)*.

(f) Synergism occurs when two or more substances complement each other's effects. Antagonism occurs when two substances oppose each other in their effect *(1)*.

(g) A photoreceptor 'detects' the presence of light by undergoing a chemical change. This change may initiate a chemical process but it does not bring about the response itself. A growth regulator controls a plant response *(1)*.

(h) Etiolation occurs when a plant grows in the dark, producing elongated frail stems and yellow unexpanded leaves. Normal growth takes place in the light, producing shorter sturdier stems with expanded green leaves *(1)*.

2 (a) IAA produced in tips of coleoptiles; no significant difference in the amounts produced in the light or dark *(2)*.

(b) (i) IAA moves from block A downwards through coleoptiles; more reaches B in darkness; more accumulates in C in light; suggests IAA moves away from light and accumulated on dark side *(4)*.

(ii) Stimulates greater elongation on dark side; causing growth curvature towards light; photosynthetic tissues receive more light *(max. 2)*.

**(c)** Remove apical bud – produces bushier plants; place cut stems in auxin rooting powder – adventitious root growth stimulated; use as selective weedkiller spray – broad-leaved weeds killed in in lawns or cereal crop; use to spray unpollinated flowers – seedless fruit produced; use to spray developing fruit – reduces fruit drop; use in nutrient agar in micropropagation techniques – stimulates growth of plantlets *(any 2 at 2 marks each).*

**3 (a)** All concentrations of IAA caused an increase in length; concentrations 0.0–0.01 have small effect; concentrations 0.01–1.0 give more marked effect; above optimum of 1.0, the effect is less marked *(max. 3).*

**(b)** IAA transported to coleoptiles more quickly at 30°C; enzyme's activity faster at 30°C giving increased elongation *(2).*

**(c)** (i) Tips produce IAA; they may produce varying amounts so that valid comparisons could not be made *(2).*

   (ii) Light affects the distribution of IAA *(1).*

**(d)** Any one of answers to Question 2(c) *(2).*

**4**

| Auxins | Gibberellins | |
|--------|--------------|---|
| ✓ | ✓ | |
| ✓ | ✗ | |
| ✗ | ✗ | |
| ✓ | ✗ | |
| ✗ | ✓ | *(5).* |

**5 (a)** Increase in gibberellin concentration gives increase in stem length; smaller effect at higher concentrations; reference to figures from table *(3).*

**(b)** (i) Possess gene for dwarfism/no gibberellin produced *(1).*

   (ii) All genetically identical *(1).*

**(c)** No controls for each treatment with gibberellin; only small number of plants used *(2).*

**(d)** Secreted by embryo after absorption of water; diffuses to aleurone layer and stimulates production of enzymes/ α-amylase; which hydrolyse food reserves producing, e.g. glucose for respiration *(3).*

**(e)** Less growth of parts which are not of economic importance; sturdy plants, do not need staking/can produce larger blooms; reduced cutting, e.g. of hedges *(max. 2).*

## Chapter 12

**1 (a)** A = lag phase; time of production of enzymes to use substrate; acclimatising to new environment; taking up nutrients *(max. 2).* B = exponential/log phase; cells dividing at fastest rate; using plentiful food supplies *(max. 2).* C = stationary phase; nutrients depleted/limiting, so growth stops; waste products build up; reaches carrying capacity/cell division = cell death; ref. to environmental resistance *(max. 2).*

**(b)** Example of organisms; examples of limiting factors; competition for resource; named resources; refs. to predation; refs. to disease; leads to differential survival; ref. to population numbers being held in check; increase in death rate/slowing in reproduction rate *(max. 8). (plus one mark for legible text with accurate spelling, punctuation and grammar)*

**2 (a)** (i) Understanding of $\frac{N(N-1)}{\Sigma n(n-1)}$; correct answer of 2.85 *(2).*

   (ii) Diversity gives both number of individuals and number of species; avoids difficulties of large number of rare species *(2).*

**(b)** Sample with quadrats/other acceptable method; random placing; count number of individuals of each species *(3).*

**(c)** (i) Figure 1 shows no correlation/relationship between bird species diversity and plant species diversity; Figure 2 shows positive correlation/as plant structural diversity increases so does bird species diversity *(2).*

   (ii) Succession; the longer the field is left uncultivated, the greater the proportion of shrubs and trees *(2).*

   (iii) Provides more niches; different species of birds feed/nest at different heights/in different places *(2).*

**(d)** (Continues to increase then) levels out; as climax is reached; maximum plant structural diversity *(max. 2).*

**3 (a)** (i) The more days of snow the lower the increase in population; above 12 days of snow decrease in population *(2).*

   (ii) Fewer birds to breed; explanation using data, e.g. food hidden under snow *(2).*

**(b)** More breeding pairs produces more competition; less breeding success *(2).*

**4 (a)** Natural variation in height of plants; in ungrazed area, tall plants competed successfully with other plants for light; taller plants more likely to survive; and pass on genes to next generation; proportion of taller plants in population increases *(max. 4).*

**(b)** Interbreed the two types of clover; see if offspring produced; if offspring are fertile they are the same species *(3).*

**(c)** Succession would occur; competition; arrival of new species; mechanism, e.g. airborne seeds; increased species diversity; modification of habitat, e.g. increased soil humus; development of shrubs; development of trees; plants which are not shade tolerant may die out *(max. 5).*

**5 (a)** *L. littorea* has greater range/*L. saxatilis* has narrower range; *L. littorea* found lower on shore/*L. saxatilis* found higher; *L. littorea* more abundant/*L. saxatilis* less abundant; *L. littorea* most abundant at site 7, *L. saxatilis* at sites 1 and 2; both occur at sites 8 and 9/ref. to overlap; neither occur at sites 1 and 2/on lower shore *(max. 3).*

**(b)** *L. saxatilis*/rough periwinkle; will be out of water/exposed for longer/less time in water/converse for *L. littorea (2)*.

**(c)** Food availability/food distribution; predation; parasites/disease; conditions for reproduction; temperature; competition; salinity; difference in wave action/tidal power *(max. 2)*.

**6 (a)** Sequence/change of communities/plants; which replace one another; in a given area; over a period of time; one sere changes/prepares environment for the next sere; until a climax community is reached; which is in equilibrium with local conditions; ref. to data from table *(max. 4)*.

**(b) (i)** Succession which does not proceed to the expected climax community; caused by a particular factor/by human intervention; plagioclimax *(max. 2)*.

**(ii)** grasses planted/re-seeded; maintained by grazing/cutting/burning; shrubs/trees do not establish *(max. 2)*.

**(c)** $\dfrac{\text{number in sample 1} \times \text{number in sample 2}}{\text{number marked in sample 2}}$ ;

$= \dfrac{284 \times 267}{63}$ ;  $= 1203$ to 1204 *(max. 2)*.

**(d)** Small dot of paint/other suitable material, on thorax/abdomen/wing case (not wings) *(1)*.

**(e)** Minimal births/deaths during investigation; minimal immigration/emigration during investigation/members of the population remain within the area; marked individuals mix thoroughly with rest of population; marking does not affect individuals in any adverse manner *(max. 2)*.

**7 (a) (i)**

$$\frac{(137-122.9)^2}{122.9} + \frac{(146-160.1)^2}{160.1} + \frac{(166-180.1)^2}{180.1} + \frac{(249-234.9)^2}{234.9}$$

*(1 mark for above or for correctly calculated simplified values)*

Answer 4.7 to 4.81 (depending on rounding errors) *(1)*.

**(ii)** Reject the null hypothesis; calculated value is greater than 3.84/less than critical value at 5% probability/less than 5% probability that difference is due to chance *(2)*.

**(b)** Numbers of both species would decrease; greater reduction in *S. balanoides*/ref. to extinction of *S balanoides*; in long term, *E. modestus* will increase as *S. balanoides* decreases/ *E. modestus* could replace *S. balanoides* *(max. 2)*.

**(c) (i)** 52.3/52 *(1)*.

**(ii)** More/greater percentage of prey eaten; less selection against *S. balanoides*/little/no preference/almost same % eaten/only 2% difference *(2)*.

**8 (a)** Coppicing involves cutting (certain species of) trees to ground level; resulting in growth of shoots/poles; this allows more light to reach the ground; increasing diversity of ground flora; different areas of wood coppiced at different times; provides a variety of habitats *(max. 4)*.

**(b)** Hedgerows help prevent soil erosion; and are habitats for plant and animal communities; maintaining food chains; they provide 'corridors' for movement of animals *(max. 3)*.

**(c)** Integrative Pest Management involves practices such as intercropping (or description); and biological pest control; which reduces the need to use harmful pesticides *(max. 2)*.

## Answers to synoptic questions

**1 (a) (i)** Sequence of bases guanine, cytosine and thymine on DNA, coding for one amino acid *(1)*.

**(ii)** There are more codons than amino acids *(1)*.

**(iii)** All the genes/alleles in a population *(1)*.

**(iv)** The process by which organisms most suited to the environmental conditions survive *(1)*.

**(b)** Substitution of base A for base T; insertion of A; deletion of T when next base is A; inversion if next base is A *(max. 2)*.

**(c)** All codons starting with GC code for same amino acid; often only first two bases are necessary to code for an amino acid/degeneracy; then no change in amino acid sequence in polypeptide; no change in shape of protein and so no change in function; the mutation may be in part of the DNA which does not code for a protein *(max. 3)*.

**(d)** Geographical isolation/description of barrier; behavioural isolation/description of e.g.; seasonal isolation/description; ecological isolation/description; difference in genitalia/pollination mechanisms *(max. 2)*.

**(e)** Isolated populations become progressively different from one another; separate mutations may occur in each group; populations eventually unable to interbreed; even if barriers removed; new species formed *(max. 3)*.

**2 (a)** Nitrates/ammonium ions *(1)*.

**(b)** *Rhizobium*/bacteria in root nodules fix atmospheric nitrogen; this converted to ammonia/ammonium ions; this combined with organic acids to make amino acids; ref. to nitrogenase *(max. 2)*.

**(c)** No difference until day 36; then more nitrogen fixed in $CO_2$ enriched atmosphere; the difference in the amount fixed increases with time; amount of nitrogen fixed in normal air levels off after 80 days, still increasing in enriched atmosphere; refs to numbers or calculations *(max. 3)*.

(d) More carbon dioxide so more photosynthesis *(1)*.

(e) Increase in yield; cereals could grow on low nitrate soil; less/no inorganic fertiliser needed; less nitrate leaching/less eutrophication *(max. 2)*.

3 (a) Activity of A increases with temperature between 0°C and 40°C; activity of B decreases over this rise in temperature; decrease in activity of B greater than increase of activity in A; activity about the same at 18°C *(max. 3)*.

(b) Mutation *(1)*.

(c) $E^A$ higher frequency than $E^B$ in south; frequency of $E^A$ decreases northwards/$E^B$ increases northwards; enzyme A more common in higher temperatures/enzyme B more common in lower temperatures; explain in terms of natural selection *(max. 4)*.

4 (a) (i) Cellulose monomers are β-glucose, starch monomers are α-glucose *(1)*.

(ii) Two recognisable monosaccharides; correct groups produced by hydrolysis *(2)*.

(iii) Cellobiose and maltose have different shapes; cellobiose will not fit/form complex with; active site of rabbit enzyme *(3)*.

(b) (i) To take difference in size of animals into account *(1)*.

(ii) Heat lost from surface; rabbit smaller and so has larger surface area to volume ratio *(2)*.

(iii) Some energy used by microorganisms; for respiration and growth *(2)*.

(c) (i) Evaporation uses latent heat to convert water to vapour; this heat is drawn from the body *(2)*.

(ii) Lose it directly to a cooler environment; by radiation/convection *(2)*.

(iii) No natural predators of rabbits in Australia to keep numbers in check; consume more food per kilogram than cattle; competition for limited resources *(max. 2)*.

(iv) Rabbits have higher body temperature than cattle so do not have to lose as much heat; can lose more heat by radiation as smaller; do not rely on sweating as much as cattle; so less dependent on drinking water *(max. 3)*.

5 (a) 0.96% *(1)*.

(b) Ref. to differences in absorption of light between leaves and phytoplankton; ref. to limiting factors of photosynthesis and the effects on energy flow in the two communities; higher energy flow through consumer food chain in B; higher energy flow through detritus food chain in A; larger population of decomposers in forest; inedible wood in forest; smaller biomass of secondary consumers in A; ref. to efficiency of transfer of energy between trophic levels; any other valid point *(max. 8)*.

(c) Replanting of large trees which have been felled; description of coppicing *(2)*.

6 (a) Absorption increases with percentage of oxygen in mixture; oxygen needed for aerobic respiration/ref. to oxidative phosphorylation; active transport involved; energy from ATP needed for absorption; only 2% needed as little increase above this *(max. 3)*.

(b) Absorption reduced; ref. to supporting figures; ref. to small change between 5 and 10 units; DNP allows electrons down ETC without ATP production; so less ATP to release energy for active transport; so less active transport *(max. 4)*.

(c) (i) Less absorption *(1)*.

(ii) Krebs cycle produces NADH; NADH supplies $H^+/e^-$ for oxidative phosphorylation; less ATP/energy for active transport *(3)*.

(d) Component of DNA/RNA; ref. to role of DNA/RNA; component of ATP/ADP; ref. to role of these in energy transfer; component of sugar phosphates; ref. to role of these in respiration/photosynthesis; component of cyclic AMP; ref. to role of this as secondary messenger; component of phospholipids; ref. to role of these in membranes; bones; teeth *(max. 6)*.

7 (a) (i) Organic molecules are taken in by heterotrophic organisms and broken down. This releases energy from the molecules taken in. (Autotrophs build up organic molecules from inorganic using an external energy source.) *(1)*.

(ii) Chemautotrophs use energy released from chemical reactions whereas photoautotrophs use light energy *(1)*.

(b) Two alpha glucose molecules correctly drawn due to addition of water; H on the top and OH on the bottom on both sides of both molecules *(2)*.

(c) (i) At low concentrations maltose uptake limited by maltose concentration in the gut *(1)*.

(ii) At high maltose concentrations uptake is limited by number of enzymes; once all maltase enzymes are occupied they become the limiting factor *(2)*.

(d) Mixing will increase uptake as increases likelihood of collisions between maltose and maltase; peristalsis maintains a concentration gradient bringing in more maltose *(2)*.

(e) Exopeptidases hydrolyse peptide bonds between amino acids; found at the ends of polypeptide chains; this produces dipeptides and amino acids *(max. 2)*.

(f) (i) Growth requires full range of food materials especially proteins; only the larval stage produces exopeptidase and lipase to make use of this range of food *(max. 2)*.

(ii)  Reproduction and dispersal require energy released during respiration; sucrase hydrolyses sucrose into sugars which can be respired and provide this energy *(2)*.

**(g)**  Ribosomes required for protein synthesis; rise in ribosome number indicates increase in protein synthesis after day 6 as adult organs are formed *(2)*.

**(h)**  Oxygen required for Krebs cycle; tracheal system provides oxygen; once tracheal system is developed, oxygen can reach cells and aerobic respiration including Krebs cycle is possible; hence the activity of Krebs cycle enzymes increases after 6 days once oxygen is available *(max. 3)*.

**8 (a)** (i)  A metal gauze or cage containing maggots above the liquid *(1)*.

(ii)  Cutting the root increases the surface area; so there is a similar surface area for gas exchange in both maggots and roots *(2)*.

**(b)**  The colour provides a standard to establish a positive reaction *(1)*.

**(c)**  3.3, 1, 1.4, 5 *(2 marks for all correct, 1 mark for 3 correct)*.

**(d)**  1 Tube must be covered to block the light, e.g. metal foil/black paper; to prevent the leaves from photosynthesising *(2)*. 2 Placing tube in a water bath at approx. 30°C, will increase the rate of respiration as the enzymes are nearer their optimum; this will show if respiration is detectable *(2)*.

# Index

RuPB *see* ribulose 1,5 bisphosphate

SACs *see* Special Areas of
Conservation

saliva 310

salmon 130

salt marshes 412

saltatory conduction 192, 202–3

sampling methods 402–6

sarcolemma 274

Schwann cells 189, 192

scrub and wood community 399

sea anemones 178

secretin 311

seeds 361, 368

segmented worms 179

selection pressure 121, 125

selective advantage 124

selective breeding *see* artificial
selection

selective reabsorption 325, 330

selective weedkillers 358

senescence 363

sensitivity 228, 241, 242

sensory cortex 256

sensory neurones 189, 245

sequoia 166

sere 391

sex chromosomes 76–84

sex determination 85

sex-linked characteristics 77–8, 83,
89

sexual dimorphismm 158

sexual selection 123

shade plants 50

shivering 295

shoots 355

shrubs 392

sickle cell anaemia 68–9, 128–9

silage 414

Simpson's index 401

skeletal muscle 269, 273–4, 282

sliding filament hypothesis 277

slow twitch muscle fibres 282

smooth muscle 269

snails 130

snapdragons 67–8

sodium channel proteins 209

sodium/potassium pumps 195, 196

somatic cell gene therapy 64

somatic cells 106

somatic nervous system 259

sound energy 1

SPAs *see* Special Protection Areas

spatial summation 213, 242

Special Areas of Conservation
(SACs) 415

Special Protection Areas (SPAs) 415

speciation 105, 136–44

species 105, 136, 155–9

species diversity 400–2

speech 258

spiders 139, 179

spinal cord 244–6

spinal reflexes 247

spongy mesophyll 34

stabilising selection 126–9, 132

standard deviation 126

statoliths 348, 356

stems 353, 361

stimuli 211, 227, 347

stomata 33

stratified sampling 403

striated muscle *see* skeletal muscle

stroma 35

strychnine 217

subcutaneous fat 304

substrate level phosphorylation 11

succession 390–4, 400

summation 212–13

sun plants 50

sustainability 409

sweating 295

sympathetic nerves 234, 260

sympatric speciation 136, 138

symplast pathways 49

synapses 189, 206–17

synaptic bulbs 191, 208

synergism 366

synovial joint 271

synovial membrane 271

systematics 160

T tubules 274

target organs 294

taxes 250

taxonomy 160

TCA (tricarboxylic acid) cycle
*see* Krebs cycle

tears 260

temporal isolation 138

temporal summation 213

tendons 270

test cross 61, 72

thalassaemia 129

thermoreceptors 228, 307

thermoregulation 301–9

thin-layer chromatography 40

thresholds 203–4

thylakoid lumen 35

thylakoids 35

*Tilapia* 143–4

toadstools 174

transduction 31, 228

transects 403

translocation mutations 107

trees 166, 392

tricarboxylic acid (TCA) cycle *see*
Krebs cycle

trichromatic theory 243–4

triose phosphate 47

trisomy 107–9

tropisms 350–1

tropomyosin 276, 279

troponin 276

TTC (triphenyl tetrazolium chloride) 23

ultrafiltration 325, 326, 328, 329
unlinked genes 88
urea 322
uric acid 323, 324
urinary system 324–5
urination 261, 324
urine 325

variation 59, 88
   environmental causes 110
   genetic causes 106
ventricles 252
vertebrates 180
vesicles 208
vision 230-4, 243-4, 260
visual acuity 241, 242
visual cortex 243
voluntary muscle *see* skeletal muscle
voluntary nervous system 259

Wallace, Alfred 110
water balance 321–46
Wernicke's area 258
wet meadows 412
wetlands 412
white adipose tissue 304
withdrawal reflex 247, 248–9
woody plants 392

xanthophylls 37
xerophytic adaptations 397
xerosere 391
xylem 33

Y-linked alleles 82
yeast 18–19, 175

Z lines 274
Z-scheme 42
zebronkey 138
zonation 395–400
zone of cell differentiation 350
zone of cell elongation 350
zoos 147
Zyban 216
Zygomycota 174

# Acknowledgements

We are grateful to the following for permission to reproduce the photographic material:

Front cover:

*top:*    Stone (Frans Lanting)

*centre:*  Stone (Art Wolfe)

*bottom:*  Science Photo Library (Andrew Syred)

1.9, Hulton Archive; 1.13a, 5.14d1, 5.15a, 5.15c1, 11.15, Dr Jeremy Burgess/SPL; 1.13b, 1.15, 3.7, 5.15c3, 6.33, Andrew Lambert; 1.16, 8.15a,b, 9.7b, Action Plus; 2.4, Dr Kari Lounatmaa/SPL; 2.14, Glen Baggott, Birkbeck College; 2.21, Claude Nuridsany & Marie Perennou/SPL; page 53, John Adds; 3.1, 5.15d3, Science Photo Library; 3.11a, Christian d'Orgeix, www.branchcreekpoultry.com; 3.11b,c, Barry R. Koffler/www.feathersite.com; 3.13, Adrian Thomas/A-Z Botanical Collection; 3.15a, Dr Tony Brain/SPL; 3.15b, 4.24, 5.13e, 5.14c2, 10.8c, Eye of Science/SPL; 3.25a,b,c, 6.4, 11.3b, CNRI/SPL; 3.30, 4.13, 7.33, Mary Evans Picture Library; 3.31, 7.21,Topham Picturepoint; 3.32, Ralph Eagle/SPL; 3.33a, Peter O'Toole/OSF; 3.33b, Foto Natura Stock/FLPA; 3.33c, Laurie Campbell/NHPA; 3.36, M Watson/Ardea; 3.37a, Werner Layer/Bruce Coleman; 3.37b, Colin Milkins/OSF; 3.37c, Roger Jackman/OSF; 3.38a, S Maslowski/FLPA; 3.38b, John Watkins/FLPA; 3.39, Courtesy of the USDA-NRCS-Wetland Science Institute, Western Wetland Flora: Field Office Guide to Plant Species, 1992, USDA, Natural Resources Conservation Service, unpaginated; 4.3, John Walmsley; 4.7a,Konrad Wothe/OSF; 4.7b, Doug Allan/OSF; 4.8a, David M Dennis/OSF; 4.14a, Martin B Withers/FLPA; 4.14b, G Hyde/FLPA; 4.15, 5.4a, 12.5, 12.20h, Mark Hamblin/OSF; 4.16, 5.17e3, Richard Packwood/OSF; 4.23, LLT Rhodes/Animals Animals/OSF; 4.28a, Paul Franklin/OSF; 4.28b, KG Vock/OSF; 4.30a, 5.17a1,c1,c2,c3, 9.1a, Heather Angel/Natural Visions; 4.30b, 12.13b, NA Callow/NHPA; 4.32, Derek Bromhall/OSF; 4.35, 5.13c, 5.14c1, Sinclair Stammers/SPL; 4.38, 5.15d1,d2, 11.22, 11.23a,b, 11.25, 11.26a, 11.37a, 12.35, Holt Studios International/Nigel Cattlin; 4.41, Holt Studios International/Inga Spence; 5.2a, 5.16a3, 5.16b1,b2,c3, Garden & Wildlife Matters; 5.2b, MJK/Photos Horticultural; 5.3a, Alan G Nelson/OSF; 5.3b1,b2,b3, David Thompson/OSF; 5.4b, EA Janes/NHPA; 5.4c, Henry Ausloos/NHPA; 5.5a, 12.40e, William S Paton/OSF; 5.5b, © Hanne & Jens Eriksen/VIREO; 5.5c, Staffan Widstrand; 5.5d, WM Griffin/OSF; 5.5e, David Tipling/OSF; 5.10, GJ van Rooij/University Museum, Amsterdam; 5.11, Steffie Shields/Garden Matters; 5.13b, Courtesy of Dr Norman Pace, Boulder, Colorado; 5.14a1, Astrid & Hanns-Frieder Michler/SPL; 5.14a2,b2, Eric Graves/SPL; 5.14b1, Michael Abbey/SPL; 5.14d2, Rob Walls/Ardea; 5.14e1, Andrew Syred/SPL; 5.14e2, Jean-Paul Ferrero/Ardea; 5.15b1, 5.16c2, 5.16d1,d2, 11.21b, 11.26b, Photos Horticultural; 5.15b2, Maurice K Walker/Garden Matters; 5.15b3, John & Irene Palmer/Garden Matters; 5.15b4, Biophoto Associates; 5.15c2, David Scharf/SPL; 5.16a1,c4, Colin Milkins/Garden Matters; 5.16a2, Peter Reynolds/FLPA; 5.16b3, Jeremy Hoare/Garden Matters; 5.16c1, Nancy Rothwell/Garden Matters; 5.16d3, Ian West/OSF; 5.17a2, 12.32b, GI Bernard/OSF; 5.17a3, Aldo Brando Leon/OSF; 5.17b1, John Clegg/Ardea; 5.17b2, ©London Scientific Films/OSF; 5.17d1, Julie Swales/FLPA; 5.17d2, JS&EJ Woolmer/OSF; 5.17d3, Marty Cordano/OSF; 5.17d4, 12.20b, Alastair/OSF; 5.17e1, 12.33, David B Fleetham/OSF; 5.17e2, Jack Dermid/OSF; 5.17e4, David W Breed/OSF; 5.17e5, Mike Hill/OSF; 6.6, 6.28, Quest/SPL; Associated Press; 6.8b, 6.8a, Press Association/Topham Picturepoint; 6.16, Ken Lucas/Ardea; 6.23, Pr S Cinti/CNRI/SPL; 6.26, Dr Martyn Gorman/SPL; 6.32, Lesley Howling/Collections; 7.2, P Davey/FLPA; 7.8a, James Marchington/Ardea; 7.8b, Michael Freeman/Bruce Coleman Collection; 7.8c, Nigel Luckhurst; 7.10, Frances Furlong/OSF; 7.22, Allsport; 7.26, Geoff Tompkinson/SPL; 7.30, Andrew Ward/Life File; 7.32a,b, The Natural History Museum, London; 7.35, St Mary's Hospital Medical School/SPL; 8.5, 9.3a, 10.8e, 10.11b, Biophoto Associates/SPL; 8.6, Biology Media/SPL; 9.1b, Claude Steelman/OSF; 9.7a, Mark Clarke/SPL; 9.8, Michael Fogden/OSF; 9.10a, Zig Leszczynski/Animals Animals/OSF; 9.10b, Andrew Henley/Natural Visions; 9.11a, Robert A Tyrrell/OSF; 9.11b, G Cappelli/Panda M159226D; 9.12a, ©Okapia/OSF; 9.12b, John Brown/OSF; 9.14b, Manfred Kage/SPL; 10.5a, SIU/SPL; 12.2, 12.20c, David Tipling/OSF; 12.9, Don Smith/FLPA; 12.10, Anthony Bannister/NHPA; 12.13a, DP Wilson/FLPA; 12.17, Sheila Apps/Garden & Wildlife Matters; 12.18, 12.23, 12.24, 12.26h, 12.26b, 12.30, Phil Bradfield; 12.19a, S Jonasson/FLPA; 1219b, Ragnar Sigurdsson; 12.20a, Lon E Lauber/OSF; 12.20d, David C Fritts/OSF; 12.20e, Tony Morrison/South American Pictures; 12.20f, Edward

Parker/OSF; 12.20g, Ben Osborne/OSF; 12.20i, Doug Allan/OSF; 12.21, Julian Cremona; 12.32a, Tim Shepherd/OSF; 12.32c, Archie Allnutt/OSF; 12.34, Massey Ferguson/Holt Studios International; 12.36, MB Withers/FLPA; 12.37, Courtesy of The Game Conservancy Trust; 12.38, 12.39, Bob Gibbons/Holt Studios International; 12.40a, Terry Heathcote/OSF; 12.40b, Barrie Watts/OSF; 12.40c, Tony Bomford/OSF; 12.40d, Michael Leach/OSF; 12.40f, Peter Wilson/Holt Studios International

Garden Matters = Garden and Wildlife Matters Photo Library
NHPA = Natural History Photographic Agency
FLPA = Frank Lane Picture Agency/Images of Nature
OSF = www.osf.uk.com (Oxford Scientific Films)
SPL = Science Photo Library

Other copyright material:
figure 7.9, copied from a diagram previously published in Penguin Science Survey B, 1963